DEFECTS IN FUNCTIONAL MATERIALS

DEFECTS IN FUNCTIONAL MATERIALS

Editors

Francis Chi-Chung Ling
University of Hong Kong, Hong Kong

Shengqiang Zhou
Institute of Ion Beam Physics and Materials Research, Germany

Andrej Kuznetsov
University of Oslo, Norway

NEW JERSEY • LONDON • SINGAPORE • BEIJING • SHANGHAI • HONG KONG • TAIPEI • CHENNAI • TOKYO

Published by

World Scientific Publishing Co. Pte. Ltd.
5 Toh Tuck Link, Singapore 596224
USA office: 27 Warren Street, Suite 401-402, Hackensack, NJ 07601
UK office: 57 Shelton Street, Covent Garden, London WC2H 9HE

British Library Cataloguing-in-Publication Data
A catalogue record for this book is available from the British Library.

DEFECTS IN FUNCTIONAL MATERIALS

ISBN 978-981-120-316-9 (hardcover)
ISBN 978-981-120-317-6 (ebook for institutions)
ISBN 978-981-120-318-3 (ebook for individuals)

For any available supplementary material, please visit
https://www.worldscientific.com/worldscibooks/10.1142/11352#t=suppl

Desk Editor: Rhaimie Wahap

Typeset by Stallion Press
Email: enquiries@stallionpress.com

Printed in Singapore

Preface

Functional materials cover a wide range of materials — such as metals, semiconductors, polymers, oxides, etc. — demonstrating suitable electrical, optical and magnetic properties to be used for different applications. Indeed, such materials exhibit a number of physical phenomena including ferroelectricity, piezoelectricity, superconductivity, magnetism, dielectric and optoelectronic properties, etc. As a result, functional materials find a variety of applications in sensors, lasers, displays, printable electronics, solid-state lighting, energy harvesting and storage, catalysis, etc. The research on functional materials has attracted much of attention in recent years, and its advancement nutrifies the development in such cross-disciplinary branches as life-science, energy, and information technologies.

Defects play a crucial role in determining the materials' electrical, optical and magnetic properties. Defects may be introduced intentionally by additional processing or be present in materials upon the fabrication. Defects could be in forms of the lattice imperfections or impurities, simple point defects or defect complexes. In any case, due to the periodicity breaking, the electronic states associated with such defects may occur in otherwise forbidden energy gap, with the corresponding levels lying shallow or deep relative to the conduction/valence band edges. For example, in a semiconductor, the carrier concentration can be elevated by introducing the shallow level defects or compensated by the deep level defects. Another charismatic example is the carrier lifetime — a critical parameter in a number of semiconductor technologies — which may be tuned in a wide range by incorporating or avoiding appropriate defects. Accordingly, defects affect materials' optical properties, as may be revealed from the optical absorption or luminescence spectra. Notably, some of the magnetic phenomena in wide band gap semiconductors may be associated with the presence of defects too. Overall, selecting materials for each specific function, requires an accurate control over the defects' distribution in the materials. Thus, defects characterization and tuning are of paramount importance for the applications of functional materials. However, in spite of the importance of these issues, the understanding of defects in many materials is far from being complete, not least accounting for that in new materials and for the new detect functionalization concepts.

This book aims to highlight recent progress in understanding how defects affect electrical, optical and magnetic properties in functional materials, already resulted (or potentially resulting) in novel applications in the fields of quantum computing, optoelectronics, photovoltaic, magnetism, etc. Specifically, the first two chapters review some of the conventional and novel methodologies to study defects in functional materials, e.g. in 2D-materials using transmission electron microscopy (TEM) and scanning tunneling microscopy (STM). This part is followed by seven topical review chapters on: defects in perovskites, specifically emphasizing the impacts on the transport and optical properties critical for solar cell and light emitting diode applications; color center defects in wide band gap semiconductors and their applications in quantum technology; density functional theory study of native defects, impurities and defect complexes in InN — a material for high speed electronic applications; dopant- and impurity-induced defects and their impacts on the electrical properties of ZnO a promising material for applications in optoelectronics, spintronic etc.; ferromagnetism in ordered B2 alloys induced by antisite defects; magnetism in SiC induced by defects; defect associated ferromagnetism in ZnO based materials.

We trust this book can be of use for a wide and cross-disciplinary research community dealing with functional materials, as well as for the students in the field to use this text for studies, referencing, and inspirations.

Editors
Francis Chi-Chung Ling
The University of Hong Kong

Shengqiang Zhou
Helmholtz-Zentrum Dresden-Rossendorf

Andrej Kuznetsov
University of Oslo

September 2019

We trust this book can be of use for a wide and cross-disciplinary research community dealing with functional materials as well as for the students in the field to use this text for studies, references, and inspirations.

[illegible]

Contents

CHAPTER 1

Studying Properties of Defects

FRANCIS CHI-CHUNG LING[*,§], SHENGQIANG ZHOU[†,¶] and ANDREJ KUZNETSOV[‡,||]

[*]Department of Physics, The University of Hong Kong,
Pokfulam Road, Hong Kong, China
[†]Helmholtz-Zentrum Dresden-Rossendorf,
Institute of Ion Beam Physics and Materials Research,
Bautzner Landstrasse 400, 01328 Dresden, Germany
[‡]Department of Physics, Center for Materials Science
and Nanotechnology, University of Oslo, Sem Saelands vei 24,
0316 Oslo, Norway
[§]ccling@hku.hk
[¶]s.zhou@hzdr.de
[||]andrej.kuznetsov@fys.uio.no

1. Introduction

Functional materials cover a wide range of materials — such as metals, semiconductors, polymers, oxides, etc. — demonstrating suitable electrical, optical and magnetic properties to be used for different applications. Indeed, such materials exhibit a number of physical phenomena including ferroelectricity, piezoelectricity, superconductivity, magnetism, dielectric and optoelectronic properties, etc. As a result, functional materials find a variety of applications in sensors, lasers, displays, printable electronics, solid-state lighting, energy harvesting and storage, catalysis, etc. The research on functional materials has attracted much of attention in recent years, and its advancement nutrifies the development in such cross-disciplinary branches as life-science, energy, and information technologies.

Making use of the materials for each specific function requires a full control over electrical, optical and/or magnetic properties, which are crucially determined by the presence of defects in the materials. Such defects may be introduced intentionally or unintentionally during the materials synthesis and/or processing steps. Notably, despite of the importance of defects, the understanding of defects in many materials and the relevant applications are far from being complete.

Admittedly, there is a big number of very good reviews and books devoted to defects studies keeping published along the decades, confirming strong and renewing interest to this subject. For example, within the last two decades, excellent reviews on defect identification and defect properties were published by Agullo-Lopez [1], Kuzmany [2], Stavola [3], Stavola [4], McCluskey [5], Schroder [6], Janotti and Van de Walle [7], Grundmann [8], Pajot [9], Pajot [10], Vines and Kuznetsov [11], Freysoldt *et al.* [12], Stavola and Fowler [13], McCluskey and Haller [14], and Tuomisto [15] — to name just a part of the literature.

Complementing the existing literature, this chapter introduces a selection of electrical, optical, structural, and magnetic spectroscopic methods for defects studies frequently used in functional materials, including those further discussed in the subsequent chapters of the present book. Particular attention is given to the topics of the CV and Hall measurements, deep level transient spectroscopy (DLTS), luminescence spectroscopy, positron annihilation spectroscopy (PAS), and electron spin resonance (ESR); also listing a number of relevant references for further reading.

2. Electrical Characterizations

The electrical characterizations of materials is of paramount importance for a wide range of applications as being comprehensively discussed for example by Look [16], Schroder [6], Grundmann [8], McCluskey and Haller [14]. The current-voltage (*I-V*) and the capacitance-voltage (*C-V*) measurements are basic methods to assess the materials electrical properties, to be further refined in the corresponding spectroscopies. All these methods in one or another variation, require a fabrication of the ohmic and rectifying contacts.

Specifically, in a non-degenerate semiconductor, the electron and hole concentrations are given by:

$$n = N_C \exp\left[\frac{E_F - E_C}{kT}\right] \tag{1}$$

$$p = N_V \exp\left[-\frac{E_F - E_V}{kT}\right] \tag{2}$$

where E_F is the Fermi level, N_C and N_V are the effective densities of states for the conduction band and valence band respectively. Notably, the np product is given by

$$np = N_C N_V \left(-\frac{E_g}{kT}\right) = n_i^2 \tag{3}$$

i.e., independent of E_F, with n_i labelling the intrinsic carrier concentration.

A donor D donates an electron to the conduction band, i.e. $D^0 \rightleftharpoons D^+ + e^-$. The neutral donor D^0 and ionized donor D^+ concentrations are respectively: $N_D^0 = N_D f_{\text{Fermi}}(E_D)$ and $N_D^+ = N_D(1 - f_{\text{Fermi}}(E_D))$, where $f_{\text{Fermi}}(E_D) = [1 + \exp(E_D - E_F)]^{-1}$ is the Fermi-Dirac distribution, N_D and E_D are respectively the concentration and ionization energy of the donor. The ratio between concentrations of D^0 and D^+ is given by:

$$\frac{N_D^0}{N_D^+} = g \exp\left(\frac{E_F - E_D}{kT}\right) \tag{4}$$

if the degeneracy (g) of the donor is taken into account.

Assuming reverse bias conditions in a Schottky diode fabricated on n-type semiconductor with the donor concentration much exceeding the acceptor concentration (i.e. $N_D \gg N_A$), the diode depletion width W is given by:

$$W = \sqrt{\frac{2\epsilon_0\epsilon_r(V_{bi} + V_R)}{eN_D}} \tag{5}$$

where V_{bi} is the built-in potential and V_R is the applied bias. The corresponding junction capacitance C is given by:

$$C = A\sqrt{\frac{e\epsilon_0\epsilon_r N_D}{2(V_{bi} + V_R)}} \tag{6}$$

where A is the area of the metal contact. Notably,

$$\frac{d\left(\frac{1}{C^2}\right)}{dV_R} = \frac{1}{A^2}\frac{2}{e\epsilon_0\epsilon_r N_D} \tag{7}$$

Thus, the N_D can be found from the slope of the $1/C^2$ against the V_R plot. Similar discussions hold for the acceptor doped p-type semiconductors.

Temperature dependent dc conductivity measurement can be used to determine the energy levels of the defect in the band gap. This method requires good ohmic contacts fabricated on the sample. The dc conductivity is given by:

$$\sigma(T) \sim \sum_i \exp\left(\frac{-E_{Di}}{kT}\right) \tag{8}$$

where k is the Boltzmann constant, T is the temperature, E_{Di} is the ionization energy of the ith defect. The ionization energies of the defects can be obtained via the Arrhenius plot $\log\sigma(T)$ against $1/T$.

Hall effect measurement can be used to distinguish type of carriers (i.e. electron or hole), and to obtain the materials' carrier concentration and carrier mobility. Ohmic contacts in the van-der-Pauw configuration are needed for Hall effect measurement. The Hall coefficient for n-type materials ($n \gg p$) is $R_H = -r/en$ and that for p-type material ($p \gg n$) is $R_H = r/ep$, with r being the scattering factor depending on the magnetic field and temperature; conventionally $r = 1$. The Hall coefficient can be found experimentally by: $R_H = tV_H/BI$, where t is the sample thickness, V_H is the Hall voltage, B is the magnetic field and I is the current. The Hall mobility is given by: $\mu_H = R_H\sigma_H$, where σ_H is the conductivity. The above description of Hall-effect measurement applies for homogeneous sample structures. The interpretation of the Hall-effect data for non-homogeneous and/or multi-layered samples is possible too as developed by Look [16].

Deep level transient spectroscopy (DLTS) is used to identify deep traps in the materials and devices. A rectifying contact (for example a Schottky junction or a p–n junction) is needed for the DLTS characterization. The concentration, energy level and capture cross section of the deep trap can be obtained from the DLTS study. A forward bias is applied to fill all the traps (for simplicity assuming one deep donor trap of density of N_{T0}). A reverse bias is then applied to create a depletion region at the rectifying contact. The filled traps in the depletion region will emit electrons to the conduction band depending on the temperature with the emission rate of: $e_n = \sigma_{\text{cap}} v N_C \exp[-(E_C - E_T)/kT]$, where σ_{cap} is the electron capture cross section of the trap, $v = \sqrt{3kT/m^*}$ is the electron velocity, m^* is the effective electron mass, E_T is the energy level of the deep trap, N_C is the effective density of states of the conduction band exhibiting a temperature dependence of $\sim T^{3/2}$. This implies that $e_n \sim T^2 \exp[-(E_C - E_T)/kT]$. The rate equation for the ionized deep donor N_T^+ is: $\frac{dN_T^+(t)}{dt} = e_n(N_{T0} - N_T^+(t))$ and the solution is: $N_T^+(t) = N_{T0}[1 - \exp(-e_n t)]$ if all the traps are filled initially as the reverse bias is applied. Since the capacitance follows $C \sim \sqrt{N_T^+}$ if the reverse bias is fixed, the $N_T^+(t)$ evolution can be revealed by the capacitance transient, i.e. $C(t) \sim 1 - \frac{1}{2}\exp(-e_n t)$. Figure 1 shows the schematic capacitance transients $C(t)$ with different emission rates e_n's of 0.01, 0.1 and 1 (in arbitrary reciprocal time unit).

The rate-window approach is usually adopted to extract the emission rate from the capacitance transient [17, 18]. The DLTS signal is defined as a difference of the capacitance taken at two consecutive

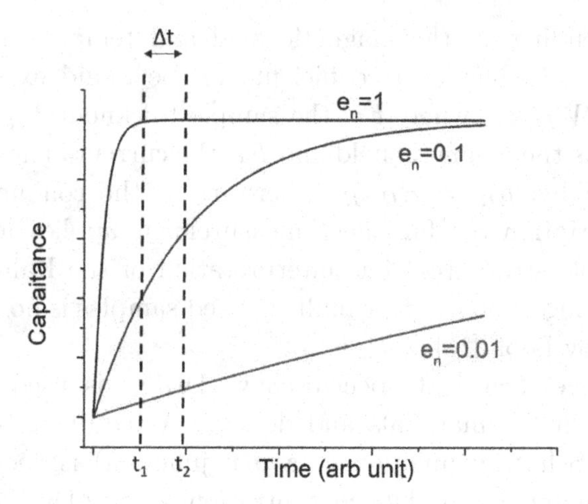

Figure 1. Schematics of the capacitance transient resulted from thermally excited deep electron traps in the depletion region while applying the reverse bias. Three cases with different emission rates namely $e_n = 0.01$, 0.1 and 1 (in arbitrary reciprocal time unit) are shown.

times t_1 and t_2 (with $t_2 > t_1$):

$$\begin{aligned} \Delta C = C(t_1) - C(t_2) &\sim \exp(-e_n t_2) - \exp(-e_n t_1) \\ &= \exp(-e_n t_2)[1 - \exp(-e_n \Delta t)] \end{aligned} \tag{9}$$

where $\Delta t = t_2 - t_1$ is called the rate window of the measurement (see Fig. 1). It can be seen from Fig. 1 that for the transient with the high emission rate (i.e. $e_n = 1$), ΔC is effectively zero. For the transient with the low emission rate (i.e. $e_n = 0.01$), ΔC is small as compared with the one with the moderate emission rate (i.e. $e_n = 0.1$). It can be shown that maximum DLTS signal ΔC maximum occurs at: $e_{n,\max} = (t_1 - t_2)[\ln(\frac{t_1}{t_2})]^{-1}$.

Accordingly, the DLTS spectrum is obtained by fixing the rate window Δt and measuring the ΔC as a function of temperature. A peak appearing in the DLTS spectrum at the temperature of T_{peak} associates with the maximum DLTS signal and thus the corresponding emission rate at the temperature of T_{peak} is given by:

$$e_n(T_{\text{peak}}) = (t_1 - t_2)\left[\ln\left(\frac{t_1}{t_2}\right)\right]^{-1}. \tag{10}$$

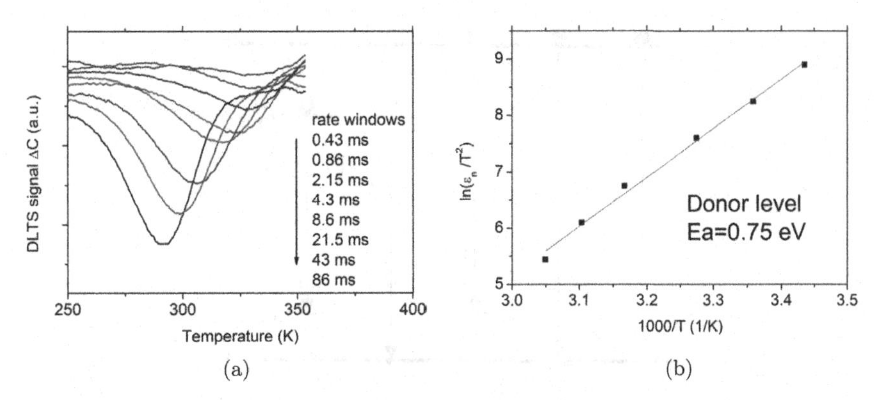

Figure 2. (a) Examples of the DLTS spectra taken from GaN Schottky diode using different rate windows; (b) The corresponding Arrhenius plot, i.e. $e_n T^{-2}$ against $1/T$.

Thus applying different rate windows Δt will result in peaks occurring at different temperatures (for example, Fig. 2(a) which is the DLTS spectra of a GaN film). E_T can be found by plotting Arrhenius plot of $e_n T^{-2}$ against $1/T$ (Fig. 2(b)), with the corresponding emission rate e_n given in Eq. (10))

Additionally, there are other variants of the junction spectroscopic methods like Laplace deep level transient spectroscopy (LDTLS) which is the high resolution version of DLTS, admittance spectroscopy, optical DLTS, minority carrier transient spectroscopy (MCTS) and etc. More details of the junction spectroscopies can be found in Peaker *et al.* [19] and references therein.

3. Optical Characterization

There are many techniques used to characterize the materials optical properties, as reviewed for example by Davies [20], Dragoman and Dragoman [21], Schroder [6], and Grundmann [8]. These methods include studying the luminescence, optical absorption, and inelastic Raman scattering, etc. Configuration coordinate diagram of defect can be used to understand the processes of luminescence, light absorption and nonradiative capture, as reviewed by Alkauskas *et al.* [22].

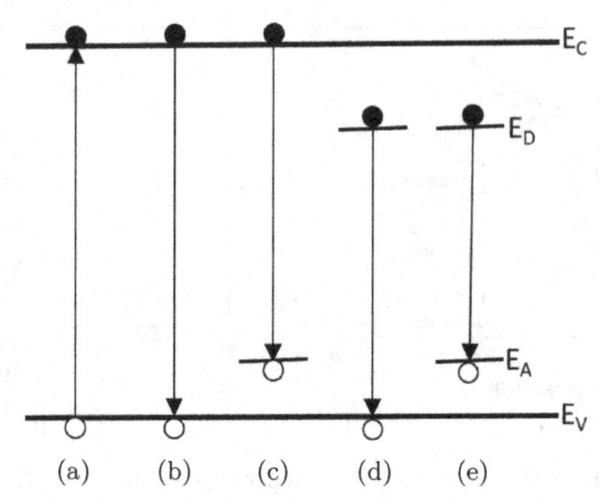

Figure 3. (a) The electron hole pair generation; and (b–e) transitions that could lead to photon emissions.

In particular, the luminescence spectroscopy in forms of either photoluminescence (PL) or cathodoluminescence (CL), is widely employed method to identify optical active defects. A laser with photon energy larger than the material band gap is usually employed to excite electrons from valance band to conduction band for PL, while the excitation in CL is achieved via the accelerated electrons (Fig. 3(a)). The depth resolved CL can also be carried out to obtain the defect depth profiles by varying the electron incident energy while keeping the excitation power constant [23, 24], while the depth information of electron energy loss can be obtained from the Monte Carlo simulations [25].

Characteristic "defect emission" occurs as a result of the electron transitions from the conduction band to the defect level, or from the defect level to the valance band (Figs. 3(c) and (d)), so that the photon energy emitted can be assigned to a specific defect [26]. Otherwise, the donor-acceptor pair (DAP) emission is originated from the recombination between an electron of a neutral donor and a hole of a neutral acceptor (Fig. 3(e)) and the emitted photon energy is:

$$hf = E_g - E_D - E_A + \frac{e^2}{4\pi\epsilon_0\epsilon_r R} \tag{11}$$

where E_g is the band gap, while R is the distance between the donor and acceptor, E_D and E_A are the donor and acceptor energy levels, respectively. Free recombination between free electrons and holes yields photons with energy equal to the band gap. However, due to the Columbic attraction, an electron at the conduction band and a hole at the valance band can also form a hydrogen like state called free exciton. The photon energy emitted from the free exciton recombination is:

$$hf = E_g - E_X \tag{12}$$

where E_X is the exciton binding energy. For materials with low exciton binding energy (like GaAs, $E_X \sim 4.2\,\text{meV}$), free exciton emission is only observed at low temperatures. For materials having large exciton binding energy (like ZnO, $E_X \sim 60\,\text{meV}$), free exciton emission is observed at the room temperature. Excitons can be bound to donors, acceptors or other potential fluctuations. For example, the transition energy for the donor bound exciton (D^0X) is given by:

$$hf = E_g - E_X - Q \tag{13}$$

where Q is the binding energy of the exciton to the donor.

4. Structural Characterization

There are many spectroscopic techniques to reveal the structural properties of defects in materials. High resolution transmission microscopy (HRTEM) and scanning probe microscopy (SPM) are capable of visualizing the defect structure in atomic scale (Schwander *et al.* [27]; Jäger and Weber [28] and references therein) and their applications in studying defects in 2D nanomaterials are discussed in Chapter 2 of the present book. Nuclear and ion-beam analysis methods can be used for obtaining depth impurity profiles as well as lattice defect distributions [29, 30]. For example, Rutherford backscattering spectroscopy (RBS), RBS with channeling, (RBS/C), nuclear reaction analysis (NRA), secondary ion mass spectrometry (SIMS) are widely employed techniques for studying defects in functional materials.

Positron annihilation spectroscopy (PAS) is a non-destructive probe for neutral or negatively charged vacancy type defects in semiconductors, reviewed by Schultz and Lynn [31], Krause-Rehberg and Leipner [32], Coleman [33], and Tuomisto and Makkonen [34]. Positron is the anti-particle of electron having the same mass but opposite charge. Positron-electron annihilation produces two gamma photons (511 keV because of mass-energy conservation), if the positronium process is not involved. Positronium (the hydrogen-like state of a positron and an electron) does not form in metals or semiconductors because the bound positron-electron pair in the degenerate electron gas would polarize the medium and then screen the positron-electron interaction. However, positronium could form in solid surfaces, amorphous materials, some molecules and ionic solids.

In more details, positrons injected into the sample from a monoenergetic positron beam or a β^+ radioactive source like ^{22}Na rapidly thermalized (~10 ps) and then undergo diffusion. The positron could annihilation in the delocalized state (or bulk state) during the diffusion. Alternatively, neutral or negatively charged vacancy type defect acts as trap for the positively charged positron. If the positron binding energy of the positron trap is large enough to prohibit thermal de-trapping, the positron will finally annihilate in the trapped state. This implies that the annihilation from the different positron states are in competition. The principal of PAS is that the outgoing gamma photons originated from the annihilation between the positron trapped in the defect and its surrounding electron, carry the information of electronic environment of the defect site, which is a fingerprint of the defect. Using the monoenergetic positron beam as the positron source, the positron energy can be varied up to ~30 keV, i.e. corresponding to the implantation depth of up to several hundred nm e.g. in Si. Thus, the defect depth profile can be obtained by doing sequential measurements varying positron incidence energies. PAS has been used to study the correlations between the materials properties and the vacancy type defects in different materials. For example, Krause *et al.* [35] studied the EL2 and its metastable state in GaAs; Lawther *et al.* [36] studied the compensating defect complexes in Group-V heavily doped Si;

Tuomisto *et al.* [37] identified zinc vacancy (V_{Zn}) as the dominant acceptor in undoped ZnO; Ling *et al.* [38] identified the shallow acceptors in undoped GaSb; Khalid *et al.* [39] correlated magnetic data in undoped ZnO with V_{Zn}-related defects. Kilpeläinen *et al.* [49] studied the thermal evolution of the defect complexes in P-doped SiGe.

Two techniques are typically used in PAS, namely the Doppler broadening spectroscopy (DBS) and the positron lifetime spectroscopy (PLS). DBS measures the Doppler broadening of the line shape of the annihilation photopeak of the annihilation gamma ray energy spectrum. Doppler broadening spectrum reveals the electronic momentum distribution seen by the positrons, i.e. $\rho(\overrightarrow{p}) \sim \sum_i |\int \overrightarrow{dr} \exp(-i\overrightarrow{p} \cdot \overrightarrow{r})\psi_+(\overrightarrow{r})\psi_i(\overrightarrow{r})|^2$, where the summation includes all the electronic states i. $\psi_+(\overrightarrow{r})$ is the positron wave function and $\psi_i(\overrightarrow{r})$ is the electron wave function with state i. The positron momentum is negligible as compared to that of its annihilating electron counterpart, and thus the total momentum of the positron-electron pair before the annihilation is effectively the electronic momentum (p in Fig. 4). Because of the linear momentum conservation, one of the annihilation gamma photons is Doppler blue shifted ($p_{\gamma,1}$ in Fig. 4) and the other one is red shifted ($p_{\gamma,2}$ in Fig. 4). The Doppler shifted energy is given by $\Delta E = p_z c/2$, where c is the speed of light and p_z is the longitudinal momentum component of p in the direction of the 511 keV annihilation photon emission (i.e. z-direction in Fig. 4). For the Doppler broadening spectrum obtained from a single detector, the high momentum information is usually hidden behind the background noise. Introduction of a second detector for coincidence gating improves the resolution and the signal-to-noise ratio. Coincidence Doppler broadening spectroscopy (CDBS)

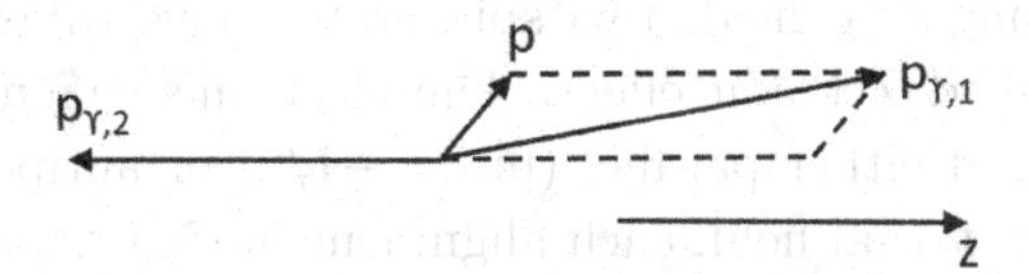

Figure 4. Schematics of the DBS momentum conservation, accounting for the momentum before the annihilation p (i.e. effectively the electron momentum), and the momenta of the two annihilation gamma photons $p_{\gamma,1}$ and $p_{\gamma,2}$.

can thus be used to explore the annihilation events originated from the high momentum electrons (i.e. the core electrons). As the core electron momentum distribution is the characteristic of a given element, CDBS are used to study the impurity decoration of vacancy type defects (for example Uedono *et al.* [40], Johansen *et al.* [41], and Rauch *et al.* [42]).

PLS measures the positron lifetime distribution in the sample. The positron lifetime is inversely proportional to the overlapping of the electron and positron density, i.e. $\frac{1}{\tau} \sim \int |\psi_+(\vec{r})|^2 n_-(\vec{r})\overrightarrow{dr}$, where $n_-(\vec{r})$ is the electron density. Since the electron density at the open volume defect is lower than the delocalized bulk state, the characteristic positron lifetime of the defect state is longer than that of the bulk state. The simple trapping model is normally used to analyse the positron lifetime spectrum; as such the positron lifetime spectrum modelled with the exponential components having different decaying time constants. Assuming for simplicity a system having only one positron defect trap, the spectrum is given by: $I_1 \exp(-\frac{t}{\tau_1}) + I_2 \exp(-\frac{t}{\tau_2})$, where τ_1 and τ_2 are the constants. The long lifetime component with the decay time constant τ_2 (i.e. $\tau_2 > \tau_1$) is the positron defect trap component. Thus, τ_2 is the characteristic positron lifetime of the defect state and is the fingerprint of the defect. Application of PAS to study the vacancy type defects in SiC is discussed in Chapter 8.

5. Magnetic Characterization

The resonant absorption of electromagnetic radiation by unpaired electrons is known as electron spin resonance (ESR) [43]. An electron has a spin S of 1/2 and an associated magnetic moment. In an external magnetic field, two spin states have different energies and this is called Zeeman effect. The electron's magnetic moment (m_S) aligns itself either parallel ($m_S = -1/2$) or antiparallel ($m_S = +1/2$) to the external field, each alignment having a specific energy: $E = m_S g_e \mu_B B_0$, where B_0 is the external field, g_e is the g-factor for the free electron, μ_B is the Bohr magneton. For unpaired free electrons, the separation between the lower and the upper state is

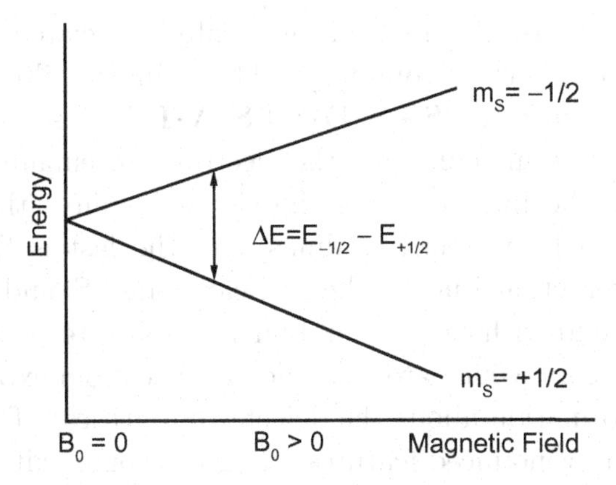

Figure 5. Schematics of the energy splitting of an unpaired electron under external magnetic field.

$\Delta E = g_e \mu_B B_0$. In classical theory, the g-factor is given by Landé formula in which the electron spin g_e factor equals to 2. Experimentally, it was found that the electron spin g_e factor for a free electron is ~2.0023, indeed close to its theoretical value. As such, both μ_B and g_e may be seen as constants. Therefore, the splitting of the energy levels is directly proportional to the magnetic field's strength (see Fig. 5). An unpaired electron can transfer between two energy levels by either absorbing or emitting a photon of energy $h\nu$ setting the resonance condition at $h\nu = \Delta E$. This result is the fundamental equation for the ESR spectroscopy technology: $h\nu = g_e \mu_B B_0$. In principle, this equation holds for a large combinations of frequency (ν) and magnetic field (B_0) values. Practically, most of the ESR measurements are performed with microwaves in the 9–10 GHz region. The ESR spectrum is usually taken by fixing the microwave frequency and varying the magnetic field. At the condition of the gap between two energy states matching the energy of the microwaves, the unpaired electrons can jump between their two spin states. Following Maxwell–Boltzmann distribution, there are typically more electrons in the lower state, leading to a net absorption of energy. This absorption is monitored and converted into a spectrum. For the microwave frequency of 9.388 GHz, the resonance should occur at the magnetic field of about $B_0 = h\nu / g_e \mu_B = 0.3350\,\mathrm{T}$.

In real systems, electrons are generally associated with one or more atoms of their surroundings. The spin Hamiltonian can be written as $H = \mu_B \mathbf{B} \cdot \mathbf{g}_e \cdot \mathbf{S} + \mathbf{S} \cdot \mathbf{D} \cdot \mathbf{S} + \mathbf{S} \cdot \mathbf{A} \cdot \mathbf{I}$ (Loubser and van Wyk [44]). The first term represents the electronic Zeemann interaction, the second is the interaction of the electron spin with the crystal fields produced by the surroundings, and the last is the magnetic hyperfine interaction due to the nuclear spins. **S** and **I** represent operators related with electron and nuclear spins, respectively. $\mathbf{g}_e$, **D** and **A** are tensors, which are to be determined from experiment and carry the information about the defects (impurities). These interactions will add pronounced features and great complexity in the ESR spectrum of a real system. The unpaired electron can gain or lose angular momentum and therefore change the value of the g-factor to be different from that for free electrons. For a spin system with $\mathbf{S} > 1/2$, the spin-spin interaction among the unpaired electrons arises, splitting the energy level in absence of the external magnetic field. The magnetic moment of nucleus with a non-zero nuclear spin ($\mathbf{I} \neq 0$) will affect any unpaired electrons which have wavefunctions overlapping with that atom. This leads to the hyperfine interaction, splitting the ESR resonance signal into doublets, triplets and so forth. Note that the split of the energy levels caused by nuclear spins is often smaller than that of the electronic Zeeman interaction and spin-spin interaction. Moreover, the g-factor and hyperfine interaction are generally not the same for all orientations of an unpaired electron in an external magnetic field. This anisotropy depends on the electronic structure of the environment of the unpaired electron, thus containing information about the atomic or molecular orbital with which the unpaired electron is associated.

For over six decades, ESR has played a key role in the study of point defects in semiconductors and many other materials even in liquids. Under optimized conditions with narrow resonance line, the sensitivity can be very high, i.e. as low as the defect concentration of $\sim 10^{10}\,\text{cm}^{-3}$ at room temperature. The sensitivity can be increased at lower temperatures. The ESR information, which can be obtained, about the defects and impurities in semiconductors was nicely summarized by Loubser and van Wyk [44]:

(1) In the same crystal, different types of paramagnetic centers result in well distinguishable ESR spectra and the intensity of a particular center is proportional to the concentration of this kind of centers.
(2) The shift of the g-factor is a measure of the orbital contribution to the magnetic moment: is generally negative for electrons and positive for holes.
(3) From the resolved fine-structure lines, one can obtain the number of unpaired electrons associated with the defect.
(4) The hyperfine interaction due to the nucleus of spin $\mathbf{I}$ associated with the defects leads each fine-structure line split into $2\mathbf{I}+1$ hyperfine lines. Therefore, impurities with non-zero nuclear spin can be identified.
(5) The anisotropy of the spectra, e.g., the angular dependence of $\mathbf{g}$ as the crystal is rotated in the external field, gives information about the symmetry of the defects.
(6) The widths of the ESR resonance lines contain information about the magnetic and exchange interaction between defects and about spin-lattice relaxation.

The above-mentioned information is essential for the defect identification and understanding its electronic structure. In many cases, the identification of point defects in semiconductors needs additional support from theoretical modelling. A careful correlation between ESR experiments and theoretical calculations allows an unambiguous identification of the defect. Along with the new development of optical and electrical detection methods, the sensitivity of ESR can be further increased (Kennedy and Glaser [45] and Spaeth [46]). For example, single isolated nitrogen-vacancy pair in diamond can be resolved and optically detected (Abe and Sasaki [47]) using the optical confocal microscopy.

Figure 6 shows another example, which is the ESR spectra of Se^+ in isotopically pure Si (Nardo [48]). The electronic g factor is $g = 2.0057$. All stable Se isotopes ${}^{X}Se^{+}(X = 74, 76, 78, 80)$ have zero nuclear spin and the ${}^{77}Se$ isotope has nuclear spin $\boldsymbol{I} = 1/2$ and the isotropic hyperfine coupling of $\boldsymbol{A} = 1.6604\,\text{GHz}$ with the donor electron spin. The central line (around $g = 2$) is corresponding to

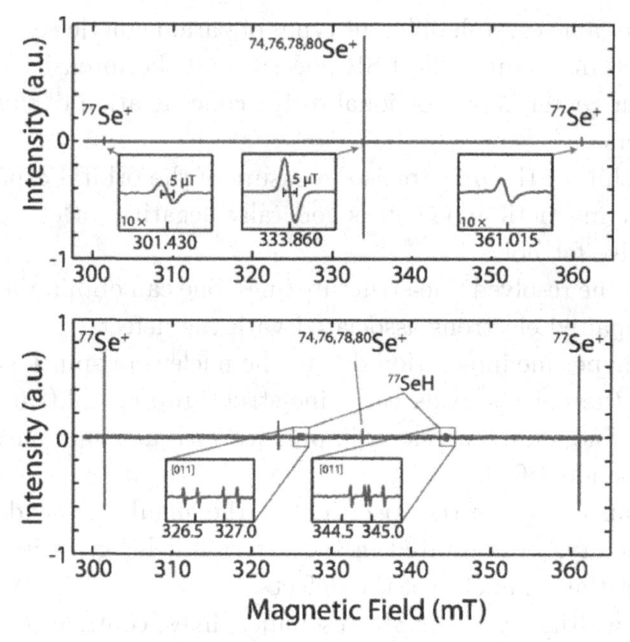

Figure 6. ESR spectra of Se^+ in ^{28}Si for ^{28}Si:Se (upper, nature Se) and ^{28}Si:^{77}Se (lower). The natural abundance of ^{77}Se is 7.5%, and the remaining isotopes (92.5% abundance) possess zero nuclear spin. In the lower panel, the ^{77}Se-doped sample shows predominantly the hyperfine-split lines arising from coupling to the $I = 1/2$ nuclear spin of ^{77}Se. The additional features in the spectrum correspond to the residual isotopes of Se and the presence of Se-H pairs. The measurement was performed at 23 K using the microwave frequency of 9.38 GHz (Nardo [48]).

Se isotopes with zero nuclear spin and the hyperfine-split lines are corresponding to ^{77}Se. The remaining features in the $^{77}Se^+$ ESR spectrum are due to Se-H pairs. Those resonance peaks have a very small linewidth ($<5\,\mu$T) due to Si isotopic purification.

With the development of sensitive magnetometry, such as superconducting quantum interference device (SQUID), the paramagnetism induced by defects or impurities can also be measured in an intergral method. By fitting the experimental data according to the Curie theory, one also can get the momentum number and get some information about the defects. For instance in ion or neutron irradiated SiC, the paramagnetism from defects was measured by SQUID magnetometry. The magnetization is proportional to the concentration of defects. In some cases, the paramagnetic centers can couple with each other ferromagnetically. Defect induced ferromagnetism has been observed in various materials, such as the results in Chapters 7, 8 and 9.

6. Conclusions

As a general conclusion, we wish to emphasize that the defect characterization involves (i) observing the defect related phenomena with specific electrical, optical or magnetic properties, (ii) revealing the nature of the responsible defects, and (iii) studying the creation and evolution of the defects in a material or device fabrication process. These include the experimental spectroscopic methods characterizing the electrical, optical, magnetic and structural properties of the defects and materials. Notably to obtain a comprehensive picture of the defect, correlative studies between different experimental spectroscopic methods are preferable to increase the chances for non-ambiguous identifications.

References

[1] Agullo-Lopez F., Catlow C. R. A. and Townsend P. D. (1988) *Point Defects in Materials.* Academic Press.
[2] Kuzmany H. (1998) *Solid-State Spectroscopy — An Introduction.* Springer Verlag.
[3] Stavola M. Ed. (1999) *Identifications of Defects in Semiconductors. Semiconductors and Semimetals **51A** and **51B**.* Academic Press.
[4] Stavola M. (1999(a)) Vibrational spectroscopy of light element impurities in semiconductors. *Identification of Defects in Semiconductors*, Stavola M. (Ed.), *Semiconductors and Semimetals Vol. 51B 153*, Academic Press.
[5] McCluskey M. D. (2000) Local vibrational modes of impurities in semiconductors. *J. Appl. Phys.* **87**, 3593.
[6] Schroder D. K. (2006) *Semiconductor Material and Device Characterization.* 3rd *Edn.* John Wiley & Sons.
[7] Janotti A. and Van de Walle C. G. (2009) Fundamentals of zinc oxide as a semiconductor. *Rep. Prog. Phys.* **72**, 26501.
[8] Grundmann M. (2010) *The Physics of Semiconductors, An Introduction including Nanophysics and Applications.* 2nd *Edn.* Springer-Verlag.
[9] Pajot B. (2010) *Optical Absorption of Impurities and Defects in Semiconducting Crystals, I. Hydrogen-like Centers.* Springer Verlag.

[10] Pajot B. and Clerjaud (2013) *Optical Absorption of Impurities and Defects in Semiconducting Crystals, II Electronic Absorption of Deep Centers and Vibrational Spectra.* Springer Verlag.

[11] Vines L. and Kuznetsov A. Yu. (2013) *Bulk Growth and Impurities, Semiconductors and Semimetals Volume 88*, pp. 67–104.

[12] Freysoldt C., Grabowski B., Hickel T., Neugebauer J., Kresse G., Janotti A. and Van de Walle G. (2014) First-principles calculations for point defects in solids. *Rev. Mod. Phys.* **86**, 253.

[13] Stavola M. and Beall Fowler W. (2018) Novel properties of defects in semiconductors revealed by vibrational spectra. *J. Appl. Phys.* **123**, 161561.

[14] McCluskey M. D. and Haller E. E. (2018) *Dopants and Defects in in Semiconductors.* 2nd *Edn.* Taylor and Francis.

[15] Tuomisto F. (Editor), *Characterisation and Control of Defects in Semiconductors, IET Institution of Engineering and Technology*, 2019.

[16] Look D. C. (1989) *Electrical Characterization of GaAs Materials and Devices.* John Wiley & Sons.

[17] Lang D. V. (1974) Deep-level transient spectroscopy: A new method to characterize traps in s emiconductors. *J. Appl. Phys.* **45**, 3023–3032.

[18] Mooney P. M. (1999) Defect identification using capacitance spectroscopy. *Identification of Defects in Semiconductors*, Stavola M. (Ed.), *Semiconductors and Semimetals Vol. 51B 93*, Academic Press.

[19] Peaker A. R., Markevich V. P. and Coutinho J. (2018). Junction spectroscopy techniques and deep-level defects in semiconductors. *J. Appl. Phys.* **123**, 161559.

[20] Davies G. (1999) Optical measurements of point defects. *Identification of Defects in Semiconductors*, Stavola M. (Ed.), *Semiconductors and Semimetals Vol. 51B 1*, Academic Press.

[21] Dragoman D. and Dragoman M. (2002) *Optical Characterization of Solids.* Springer Verlag.

[22] Alkauskas A., McCluskey M. D. and Van de Walle C. (2016) Defects in semiconductors — Combining experiment and theory. *J. Appl. Phys.* **119**, 181101.

[23] Dong Yufeng, Tuomisto F., Svensson B. G., Kuznetsov A. Yu. and Brillson L. J. (2010) Vacancy defect and defect cluster energetics in ion-implanted ZnO. *Phys. Rev. B* **81**, 201R.

[24] Wang Zilan *et al.* (2019) Vacancy cluster in ZnO films grown by pulsed laser deposition. *Sci. Rep.* **9**, 3534.

[25] Drouin D. *et al.* (2007) CASINO V2.42-a fast and easy-to-use modelling tool for scanning electron microscopy and microanalysis users. *Scanning* **29**, 92.

[26] Børseth T. M., Svensson B. G and Kuznetsov, A. Yu. (2006) Identification of oxygen and zinc vacancy optical signals in ZnO. *Appl. Phys. Lett.* **89**(26), 262112.

[27] Schwander P., Rau W.-D., Kisielowski C., Gribelyuk M. and Ourmazd A. (1999) Defect processes in semiconductors studied at the atomic level by

transmission electron microscopy. *Identification of Defects in Semiconductors*, Stavola M. (Ed.), *Semiconductors and Semimetals Vol. 51B, 226*, Academic Press.
[28] Jäger N. D. and Weber E. R. (1999) Scanning tunnelling microscopy of defect in semiconductors. *Identification of Defects in Semiconductors*, Stavola M. (Ed.), *Semiconductors and Semimetals Vol. 51B, 262*, Academic Press.
[29] Wang Y. Q. and Nastasi M. eds. (2010) *Handbook of Modern Ion Beam Analysis.* 2nd *Edn.* Materials Research Society.
[30] Jeynes C., Webb R. P. and Lohstroh A. (2011) Ion beam analysis: A century of exploring the electronic and nuclear structure of the atom for materials characterization. *Reviews of Accelerator Science and Technology* **4**, 41.
[31] Schultz P. J. and Lynn K. G. (1988) Interaction of positron beams with surface, thin films, and interfaces. *Rev. Mod. Phys.* **60**, 701.
[32] Krause-Rehberg R. and Leipner H. S. (1999) *Positron Annihilation in Semiconductors Defect Studies.* Springer Verlag.
[33] Coleman P. (Ed.) (2000) *Positron Beams and Their Applications.* World Scientific.
[34] Tuomisto F. and Makkonen I. (2013) Defect identification in semiconductors with positron annihilation: Experimental and Theory. *Rev. Mod. Phys.* **85**, 1583.
[35] Krause R., Saarinen K., Hautojärvi P., Polity A., Gärtner G. and Corbel C. (1990) Observation of a monovacancy in the metastable state of the EL2 defect in GaAs by positron annihilation. *Phys. Rev. Lett.* **65**, 3329.
[36] Lawther D. W., Myler U., Simpson P. J., Rousseau P. M., Griffin P. B. and Plummer J. D. (1995) Vacancy generation resulting from electrical deactivation of arsenic. *Appl. Phys. Lett.* **67**, 3575.
[37] Tuomisto F., Ranki V. and Saarinen K. (2003) Evidence of the Zn vacancy acting as the dominant acceptor in n-type ZnO. *Phys. Rev. Lett.* **91**, 205502.
[38] Ling C. C., Lui M. K., Ma S. K., Chen X. D., Fung S. and Beling C. D. (2004) Nature of the acceptor responsible for p-type conduction in liquid encapsulated Czochraiski-grown undoped gallium antimonide. *Appl. Phys. Lett.* **85**, 384.
[39] Khalid M., Ziese M., Setzer A., Esquinazi P., Lorenz M., Hochmuth H., Grundmann M., Spermann D., Butz T., Brauer G., Anwand W., Fischer G., Adeagbo A., Hergert W. and Ernst A. (2009) Defect-induced magnetic order in pure ZnO films. *Phys. Rev. B* **80**, 035331.
[40] Uedono A., Ishibashi S., Watanabe T., Wang X. Q., Liu, S. T., Chen G., Sang L. W., Sumlya M. and Shen B. (2012) Vacancy-type defects in $In_xGa_{1-x}N$ alloys probed using a monoenergetic positron beam. *J. Appl. Phys.* **112**, 014507.
[41] Johansen K. M., Zubiaga A., Tuomisto F., Monakhov E. V., Kuznetsov A. Yu and Svensson B. G. (2011). H passivation of Li on Zn-site in ZnO: Positron annihilation spectroscopy and secondary ion mass spectroscopy. *Phys. Rev. B* **84**, 115203.

[42] Rauch C., Makkonen I. and Tuomisto F. (2011) Identifying vacancy complexes in compound semiconductors with positron annihilation spectroscopy: A case study of InN. *Phys. Rev. B* **84**, 125201.
[43] Watkins G. D. (1999) EPR and ENDOR studies of defects in semiconductors. *Identification of Defects in Semiconductors*, Stavola M. (Ed.), *Semiconductors and Semimetals Vol. 51A 1*, Academic Press.
[44] Loubser J. H. N. and van Wyk J. A. (1978) Electron spin resonance in the study of diamond. *Rep. Prog. Phys.* **41**, 1201.
[45] Kennedy T. A. and Glaser E. R. (1999) Magnetic resonance of epitaxial layers detected by photoluminescence. *Identification of Defects in Semiconductors*, Stavola M. (Ed.), *Semiconductors and Semimetals Vol. 51A 93*, Academic Press.
[46] Spaeth J.-M. (1999) Magneto-optical and electrical detection of paramagnetic resonance in semiconductors. *Identification of Defects in Semiconductors*, Stavola M. (Ed.), *Semiconductors and Semimetals Vol. 51A 45*, Academic Press.
[47] Abe E. and Sasaki K. (2018) Magnetic resonance with nitrogen-vacancy centers in diamond — microwave engineering, materials science, and magnetometry. *J. Appl. Phys.* **123**, 161101.
[48] Nardo R. L., Wolfowicz G., Simmons S., Tyryshkin A. M., Riemann H., Abrosimov N. V., Becker P., Pohl H. J., Steger M., Lyon S. A., Thewalt M. L. W. and Morton J. J. L. (2015) Spin relaxation and donor-acceptor recombination of Se^+ in 28-silicon. *Phys. Rev. B* **92**, 165201.
[49] S. Kilpeläinen, K. Kuitunen, F. Tuomisto, J. Slotte, H. H. Radamson, and A. Yu. Kuznetsov, *Phys. Rev. B* **81**, 132103 (2000).

CHAPTER 2

Defect Physics in 2D Nanomaterials Explored by STEM/STM

JINHUA HONG*, MAOHAI XIE†,‡ and CHUANHONG JIN*,§

*State Key Laboratory of Silicon Materials, School of Materials Science and Engineering, Zhejiang University, Hangzhou 310027, China
†Physics Department, The University of Hong Kong, Pokfulam Road, Hong Kong
‡mhxie@hku.hk
§chhjin@zju.edu.cn

1. Introduction

Structural disorders such as point defects (vacancies, anti-sites, interstitials, and dopants), dislocations, and grain boundaries are commonly present in crystalline materials, which play a key role in dominating the mechanical, opto/electronic, and chemical properties of the crystal materials [1–3]. For instance, the success of modern microelectronics is based on the manipulation of the charge carriers in semiconductor channels of field effect transistors through defect engineering. This helps to realize p-to-n conversion and to control the electric and optical performance of semiconductors. In general, structural defects break the translational symmetry of crystals leading to unique electronic states modifying the intrinsic electronic structure, and thereby considerably tailor the electronic, optical, magnetic, as well as the chemical/catalytic properties of crystals, as reflected by optical spectroscopy — such as ultraviolet–visible spectroscopy (UV–Vis), photoluminescence (PL) spectroscopy, X-ray absorption spectroscopy (XAS), alternating current (AC) conductance, or magnetic properties measurements.

Defect exploration, control, and engineering have always been at the heart of modern materials science and industrial applications such as traditional semiconductor microelectronics, metal refining, and catalyst design. Besides the traditional 3D solids, such a structure–property philosophy of defects remains a common but challenging issue to be explored even in low-dimensional crystal materials, such as nanocrystals, quantum dots, nanotubes, or ribbons and graphene-like two-dimensional (2D) materials. Hence, direct imaging of the defect structure will be of crucial significance to detect the localized defect state and its electronic properties. The whole picture of structure–property correlation in defects may also lead to the discovery of novel nanophysics in nanomaterials such as magnetism/spin crossover and single-atom catalysis.

In this chapter, five sections will be included in defect characterization: (1) a brief introduction of scanning transmission electron microscopy (STEM) and scanning tunneling microscopy (STM) in direct probing; (2) STEM characterization of point defects (vacancy and antisite) in 2D transition metal dichalcogenides; (3) real-time experimental observation of defects' migration in monolayer MoS_2; (4) STEM and STM/scanning tunneling spectroscopy (STS) characterization of inversion domain boundaries in molecular beam epitaxy (MBE)-grown $MoSe_2$; (5) STEM and STS of the domains of bilayer $MoSe_2$ to reveal the stacking band structure diversity.

2. Instrumentation and Technique

Among the available microscopy techniques for direct atom probing in real space, STEM and scanning probe microscopy (SPM) are suitable to visualize the structure of defects at the atomic scale. Both microscopic techniques can enable us in direct atomically resolved imaging of the defect structure in nanomaterials, even with a proper temporal resolution to capture the defect dynamics. The difference lies in the principles: the former STEM utilizes the electron–atom scattering of the fast-incident electron beam (30–300 keV) penetrating the sample to image atomic structures including defects both within a solid and on the surface; the SPM like scanning tunneling microscopy (STM) is based on the quantum tunneling effect of the valence electrons from the surface atoms of a solid sample. In the quantum limit system such as graphene-like 2D materials with only surface and no volume states, all atoms and defects appear as the surface, forming an ideal platform for the defect exploration using both techniques. Tremendous research examples on 2D materials system have demonstrated the versatility of both routes to identify atomic defects and the electronic states induced.

2.1. *Principles of ADF-STEM and EELS in a TEM*

Modern electron microscopy has developed into an era of aberration correction in electron optics. Owing to recent decade's commercialization and improvement of probe aberration corrector, high-angle annular dark field scanning transmission electron microscopy (HAADF-STEM) has become a standard atomic resolution imaging mode to directly visualize the structure of the crystal samples. In HAADF imaging (Z-contrast imaging), the detector receives only electrons after a large-angle elastic scattering between incident electrons and sample atoms, as shown in Fig. 1(a). In this incoherent imaging, the intensity and contrast will not change drastically with the sample thickness and the defocus compared to high-resolution (HR) TEM and bright-field (BF) STEM, but obey an approximate elastic Rutherford scattering formalism $\propto Z^2$, where Z is the atomic number of the atomic column imaged. Heavy and light atoms will give very contrasting brightness in different columns in the imaging. Thus, HAADF-STEM imaging directly reflects the real atomic structure and resolves the species of the atoms in a compound material, even without any image simulation. A more accurate scaling

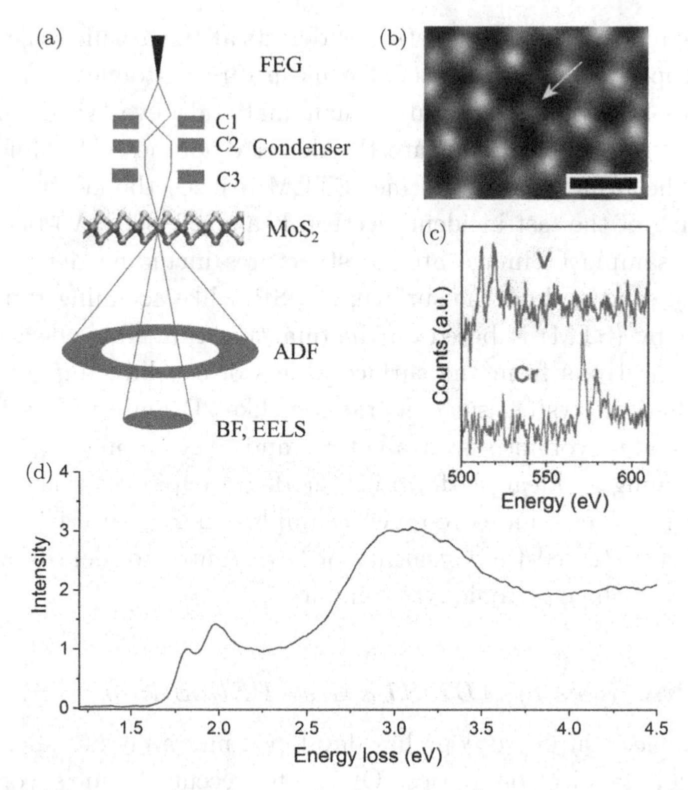

Figure 1. Electron scattering geometry, imaging and spectroscopy in a TEM. (a) Setup of STEM imaging and HAADF, BF detectors, and EELS spectrometer. (b), and (c) Typical ADF-STEM image and core-loss EELS of doped monolayer MoS_2. Reproduced from Robertson *et al.* (2016) with permission. (d) Monochromated valence EELS of monolayer MoS_2. Reproduced from Suenaga *et al.* (2015) with permission.

relation of HAADF-STEM imaging can be utilized for the atom-by-atom structural and chemical analysis of light-element 2D materials. Along with the high-angle scattered signal, low-angle inelastic scattered electrons and transmitted electrons can be collected for annular dark/bright-field (ADF/ABF) imaging simultaneously with the HAADF imaging and spectroscopic analysis (Figs. 1(b)–1(d)). In the ADF-STEM and ABF-STEM imaging, electrons are partially coherent and the yielded contrast in the imaging will change with the sample thickness and defocus. Although the contrast does not directly indicate the atomic structure, they can be interpretable after a quantitative image simulation.

While imaging by annular detectors, the post-column spectrometer receives the low-angle scattered and transmitted electrons after an energy transfer to the target atom in the electron–sample interaction, to yield the electron energy loss spectroscopy (EELS). Through an inelastic scattering, the energy transfer from the fast-incident electron to the target atom will excite the core electron in the K and L shells, or the valence electrons forming the band structure, into unoccupied electronic states above the Fermi level of the sample. According to the Fermi's golden rule, the scattering intensity, or cross section, mainly depends on the energy gap and the density of unoccupied states to accommodate the excited electron. This will give rise to the cross section of low loss being several orders of magnitude higher than that of core loss. Generally, low-loss signals with a large cross section are dependent on the electronic structure of the sample and behave as interband transition (Fig. 1(d)) and plasmon excitation related with the valence electrons of the specimen. In the high-loss regime of EEL spectrum, chemical species (Fig. 1(c)) and unoccupied electronic states of the target nanostructures can be quantitatively determined.

In recent years, instrumental performance has advanced to atom-by-atom spectroscopic analysis of the crystal specimen even with single-atom accuracy, higher sensitivity, and energy resolution. Owing to the special versatility in the light element analysis, the valence state of the target elements can be determined from a quantitative measurement of the chemical shift of the characteristic K, L and M edges in the core-loss EEL spectrum. Meanwhile, chemical environment and magnetic structure of the target atom can also be deduced from the energy loss near edge fine structure (ELNES) at the single-atom level. In defective crystals, ELNES can be employed to unveil the defect-induced nanophysics such as valence/spin states and coordination crystal fields of single transition metal dopants or other defects. Specifically, high-energy resolution EELS has been demonstrated as a versatile technique to measure bandgaps, plasmon excitations, or even the vibrational modes of various crystal materials.

2.2. *Principles of STM and STS*

STM/STS is a common atom probe used in electronic state imaging and analysis in surface science, based on the quantum tunneling effect between the sample surface and the STM tip. It is particularly suited for studying the defects in surfaces due to its high spatial resolution as well as the fact that electronic states introduced by defects will contribute to the tunneling current and thus be revealed by the difference in STM contrast between defects and the surrounding defect-free region. More importantly, by performing scanning tunneling spectroscopy measurements, one can derive the local density of states (LDOS) by taking the differential conductance spectra at the defects and its localization, thus providing direct evaluations of the electronic properties of the defects. Instead of repeating on the working principles of STM/S here, readers are referred to a few excellent monographs or book chapters on this celebrated technique ['Methods of Experimental Physics, Vol. 27, Scanning tunneling microscopy' Joseph A. Stroscio & William J. Kasiser (eds.), Academic Press 1993; 'Introduction to Scanning Tunneling Microscopy (2nd ed.)' C. Julian Chen, Oxford University Press 2008; 'Scanning Probe Microscopy, Atomic Force Microscopy and Scanning Tunneling Microscopy', Bert Voigtlaender, Springer 2015; 'Scanning Probe Microscopy, Analytical Methods' Roland Wiesendanger (ed.), Springer 1998; 'Scanning Tunneling Microscopy and Its Application' Chunli Bai, Springer 2000].

3. Atomic Defects in 2D Transition Metal Dichalcogenides

In the post-graphene era of 2D materials, the diverse layered transition metal dichalcogenides (TMDs) [4–7] are a large 2D family with unique structure, opto/electronic, and valleytronic properties, especially in valleytronics [8–11] and electronics [12–14] application. Among them, IV–VI group $MX_2 (M = \mathrm{Mo, W}; X = \mathrm{S, Se})$ has three types of crystal phases: 1T phase as a metal and 2H and 3R phases as semiconductors. The hexagonal $MoS_2/MoSe_2$ in the 2H phase has been extensively investigated in materials science and finds wide applications in industries [15] as lubricants and hydrodesulfurization catalysts.

3.1. *Point defects in monolayer MoS_2*

Two-dimensional layered MoS_2 is a typical semiconductor with a well-known cross-over from indirect to direct bandgap when the thickness decreases from bulk to monolayer, as a result of quantum confinement effect. Semiconducting MoS_2 (E.g., 1.3–1.8 eV) can be promising building blocks of photodetectors, gas sensors, and opto/electronic devices. To synthesize MoS_2 atomic layers in large scale, chemical vapor deposition (CVD) [16, 17] has been demonstrated as a feasible route to realize the scalable nanoelectronic applications based on large-size high-quality thin films. However, plenty of point defects and grain boundaries are still inevitably present in the atomic thin layers after the CVD growth.

3.1.1. *Vacancies and antisite defects*

Diverse intrinsic point defects in CVD-grown MoS_2 were first systematically characterized by Zhou *et al.* [1] by using atomic resolution HAADF imaging. Due to the large intensity difference of Mo and S_2 column in this Z-contrast mode, it is easy to directly assign the type of the point defects. Single-site sulfur vacancies were found to be the most common defects, including mono-vacancy (V_S) and double-vacancy (V_{S2}) with only one or two S atoms missing from the S sublattice. Other less common defects observed include extended Mo vacancies such as V_{MoS3} and V_{MoS6}, and antisite defects with Mo atom replacing S2 column (Mo_{S2}) or S2 occupying the Mo site ($S2_{Mo}$), but with a much lower frequency. Through HRTEM imaging, Komsa *et al.* [18] observed the structure of single vacancies V_S and V_{S2} in monolayer MoS_2 with atomic resolution, which could be readily created by electron beam irradiation at an acceleration voltage of 80 kV. Atomic S vacancies get generated and agglomerated in the monolayer under the electron beam irradiation. These vacancy sites could accommodate impurity atoms to form substitutional dopants, such as N, P, As, and Sb in V-A group behaving as acceptors and F,

Cl, Br, I in VII-A group as donors, respectively. This electron beam-mediated substitutional doping could serve as a route to engineer the local electronic structure of TMDs.

Using atomically resolved ADF imaging, Jin *et al.* found plenty of antisite defects emerging in physical vapor-deposited (PVD) MoS_2 monolayers. Figure 2 is an image gallery to demonstrate all types of antisite defects in monolayer MoS_2 including Mo replacing S sublattice (Mo_S, Mo_{S2}, $Mo2_{S2}$) and S substituting Mo sublattice ($S_{Mo}, S2_{Mo}$). In the ADF imaging mode, these two different categories of antisites can be easily distinguished and even quantitatively analyzed. The experimental atomically resolved ADF-STEM images of antisites agree well with the simulated images based on density functional theory (DFT) relaxed structures. The DFT relaxed atomic model of antisite Mo_S in Fig. 2k still retains the three-fold symmetry, while antisite Mo_{S2} (in Fig. 2l) has an obvious off-center characteristic because of the deviation of the antisite Mo atom from the center of the triangles formed by the three nearest-neighboring Mo atoms. The different structural symmetries between Mo_S and Mo_{S2} give rise to their contrasting magnetic properties.

3.1.2. *Defect species vs sample synthesis methods*

Intrinsic structural defects emerge inevitably in the sample growth within finite time according to the thermodynamic theory. Hong *et al.* [19] found the primary point defects in monolayer MoS_2 changed with the growth methods, physical vapor deposition (PVD) [11, 20–22], mechanical exfoliation (ME) [23] and chemical vapor deposition (CVD) [16, 17, 24, 25], as shown in Fig. 3. It is observed that antisite defects Mo_{S2} with Mo replacing the S2 sublattice are the dominant point defects in PVD MoS_2, while in ME and CVD monolayers, sulfur vacancies V_S are the most common defects. Different atomic growth mechanisms [19] have been outlined to account for the difference in the primary defect species in the different growth methods.

The predominant defects such as antisite Mo_{S2} and Mo_S in PVD samples are statistically analyzed at a density of $(2.8 \pm 0.3) \times 10^{13} cm^{-2}$ and $7.0 \times 10^{12} cm^{-2}$, corresponding to an atom percent

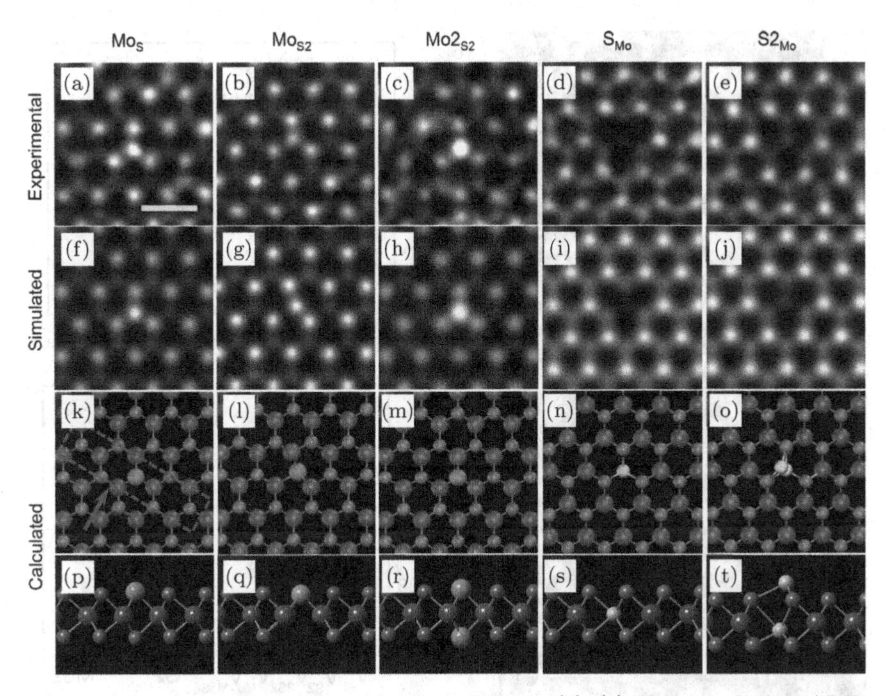

Figure 2. Atomic structures of antisite defects. (a)–(c) High-resolution ADF-STEM images of antisite Mo_S, Mo_{S2}, and $Mo2_{S2}$, respectively. The former two antisites are dominant in PVD-synthesized MoS_2 single layers. Scale bar: 0.5 nm. (d) and (e) Atomic structures of antisite defects S_{Mo} and $S2_{Mo}$, respectively. (f)–(j) Simulated STEM images based on the theoretically relaxed structures of the corresponding point defects in (a)–(e). (k)–(t) Top and side views of DFT-relaxed atomic model of all antisite defects. Light blue: Mo atom; gold: S2 atoms. Reproduced from Jin *et al.* (2015) with permission.

of 0.8% and 0.21%, respectively. The dominant point defect V_S vacancy in the CVD samples has a statistical concentration of $(1.2 \pm 0.4) \times 10^{13} cm^{-2}$. As both concentrations of primary defects (vacancies or antisites) are remarkably high (0.8–0.2%), it is naturally expected that they will considerably tailor the electronic structures [19].

3.1.3. *Local magnetism induced by antisite defects*

In the nonmagnetic MoS_2, sulfur vacancies V_S, V_{S2} will not induce any local magnetism [19] to the monolayer or multilayer. Through

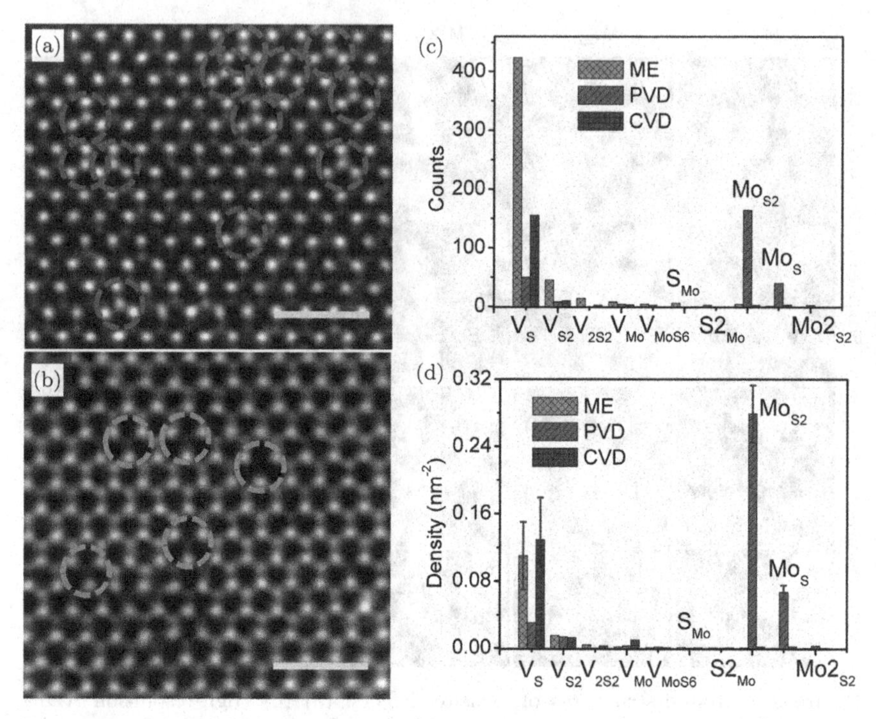

Figure 3. Atomically resolved ADF-STEM images to reveal the distribution of different point defects. (a) Antisite defects in PVD MoS_2 monolayers. Scale bar: 1 nm. (b) Vacancies including V_S and V_{S2} observed in ME monolayers, similar to those observed for the CVD sample. Scale bar: 1 nm. (c) and (d) Histograms of various point defects in PVD, CVD, and ME monolayers. ME samples are in green, PVD samples in red, and CVD in blue. Reproduced from Jin *et al.* (2015) with permission.

advanced first-principles calculation, Jin *et al.* found that only antisite Mo_S will give rise to local magnetism in monolayer MoS_2, while antisite Mo_{S2} is nonmagnetic. The calculated electronic structures of the most common antisites Mo_S and Mo_{S2} are shown in Fig. 4. The defect states behave as nearly flat band dispersion within the intrinsic bandgap, indicating the excessive involvement of Mo d electrons (four Mo atoms within the ansite defect). Further, the density distribution of the extended electron wave function around the antisite forms a "superatom" with a radius of roughly 6 Å, representing the hybridization of Mo d and S p orbitals (Figs. 4(e) and 4(f)).

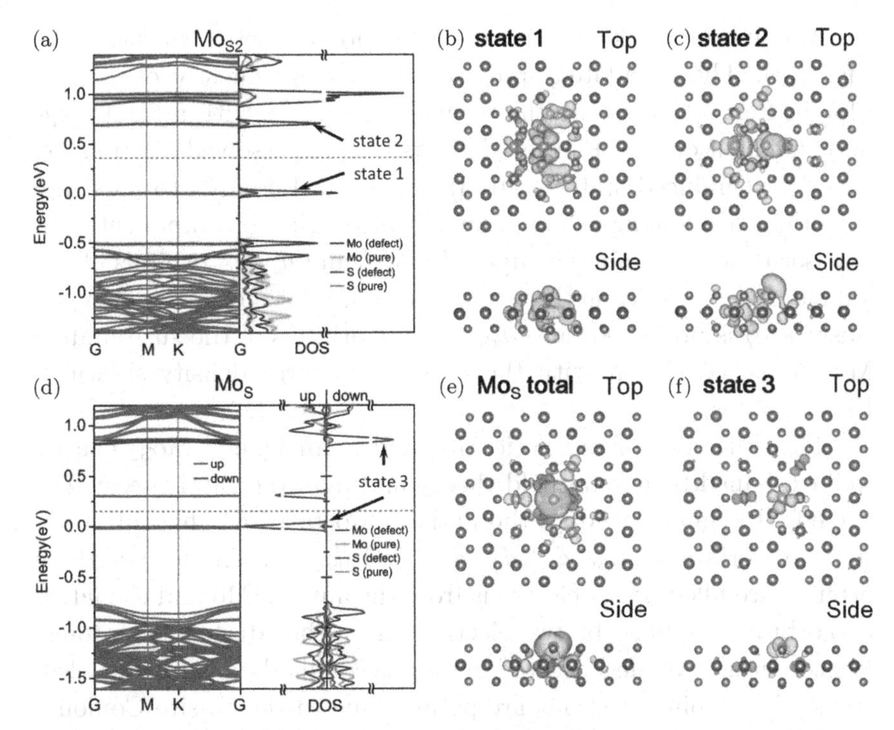

Figure 4. Electronic properties of the predominant antisite defects in the MoS_2 monolayer. (a) Band structure and the corresponding density of states (DOS) of the antisite defect Mo_{S2}. The gray bands are from normal lattice sites, similar to the conduction band and valence band of a perfect monolayer, while the discrete red bands show the localized defects states. The DOS is projected onto the atoms around the defect (defect) and those in the middle plane of two adjacent defects (pure), respectively. The gray dash line indicates the position of the Fermi level. (b) and (c) Real-space distribution of the wave functions of the two defect states below and above the Fermi energy. (d) The band structure and DOS of antisite Mo_S, with a color scheme similar to that in (a) but the two spin components are colored in red (spin-up) and blue (spin-down), respectively. (e) Spin density of antisite Mo_S, as defined $\rho_{\mathrm{up}} - \rho_{\mathrm{down}}$, charge densities ρ_{up} and ρ_{down} are spin-resolved for spin-up and down components, which are represented by yellow and blue isosurfaces, respectively. (f) Spin-resolved real-space distribution of the wave function of the two marked defect states (state 3) in (d). The isosurface value in (b), (c), (e) and (f) is $0.001\mathrm{e} \cdot Bhor^{-3}$. Reproduced from Jin *et al.* (2015) with permission.

Antisite Mo_{S2} is calculated to be nonmagnetic, as shown in Fig. 4(a). The calculated spin-polarized charge density of antisite Mo_S in Fig. 4(e) presents the magnetic structure with a total magnetic moment of $2\mu_B$. Figure 4(f) plots the spin-resolved distribution of a defect-induced state (state 3) marked in Fig. 4(d) to illustrate the origin of the magnetism. The occupied spin-up component (yellow isosurface) is mainly composed of d_{xy} and d_{x2-y2} orbitals of the antisite Mo atom, while the unoccupied spin-down component (cyan isosurface) is projected onto d_{xy} and d_{z2} orbitals of the surrounding Mo atoms, consistent with the total spin charge density shown in Fig.4(e).

The difference of magnetic Mo_S and nonmagnetic Mo_{S2} can be well explained by crystal field theory and hybrid orbital theory. For three-fold symmetric Mo_S, the antisite Mo takes d^4s hybridization with five orbitals, i.e.,$s, d_{xz}, d_{yz}, d_{xy}$, and d_{x2-y2}. The former three orbitals are filled by six electrons from the antisite Mo and the latter two orbitals are filled by two electrons from the adjacent Mo atoms. Hence, these two degenerate orbitals accommodate unpaired electrons whose spin directions are parallel due to the on-site Coulomb repulsion according to Hund's rules. Thus, these two unpaired electrons lead to the magnetic moment of $2\,\mu_B$. For antisite Mo_{S2}, the off-center characteristic in structure gives rise to the absence of orbital d_{xz} in the formation of hybridization. Hence, the antisite Mo takes d^3s hybridization forming four hybridized orbitals, originated from $s, d_{xy}, d_{x2-y2}, d_{yz}$, filled by eight electrons. As a result of the d^3s hybridization, antisite Mo_{S2} is non-magnetic. The structural symmetry breaking of antisite Mo_{S2} makes a big difference in the magnetic properties [19], in contrast with the antisite Mo_S.

3.2. *Capturing the dynamics of point defects in MoS_2*

Besides the high sub-atom spatial resolution for static imaging, TEM also has a moderate temporal resolution in the order of millisecond to second. Modern fast camera techniques have been developed to allow for ms-frame-rate recording of image slices together with atomic resolution. Recent advancements in both spatial and temporal dimension have brought the electron microscopy into the so-called 4D TEM era.

Atomic diffusion on surfaces and inside solids is the most elementary process in materials behaviors such as phase transition [26], nanomaterials growth [27–29], defect evolution, surface reconstruction, and heterogeneous catalysis [30]. Real-time TEM or STM would provide us a proper time window to directly observe the atomic migration or molecular dynamics which is of great significance in many of these material processes.

3.2.1. *Mo adatom*

In the monolayer MoS_2 system, Jin *et al.* [31] used time-sequential ADF-STEM imaging to track the defects' evolution and atomic migration. As shown in Fig. 5, this chemically sensitive ADF-STEM imaging visualizes an obvious time sequence of the hopping of Mo adatom on the monolayer substrate. Statistical analysis also indicates its random migration on the lattice without any directional preference. Three types of Mo adatom configurations were frequently observed: on top of Mo sites (T_{Mo}), above the center of the hexagon or the hollow site (H), and on top of S sites (T_S), shown in Figs. 6(a)–6(f), respectively. They correspond to DFT-derived ground state, metastable configurations of Mo adatom in the surface migration on the monolayer. The statistical counts of all these states (Figs. 6(g) and (h)) agree well with the DFT-calculated stability sequence that T_{Mo} is the most stable ground-state configuration, H is the first

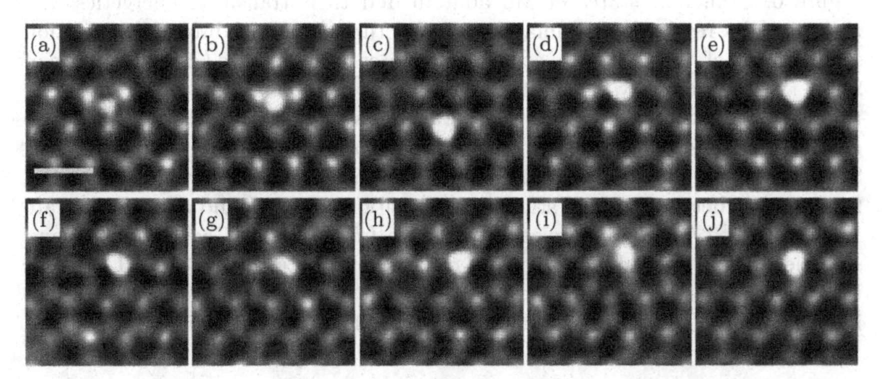

Figure 5. The migrating Mo adatom defects. (a)–(j) Experimental time sequential of ADF images of Mo adatom hopping as an example. Time interval: 3 s. Scale bar: 0.5 nm. Reproduced from Jin *et al.* (2017) with permission.

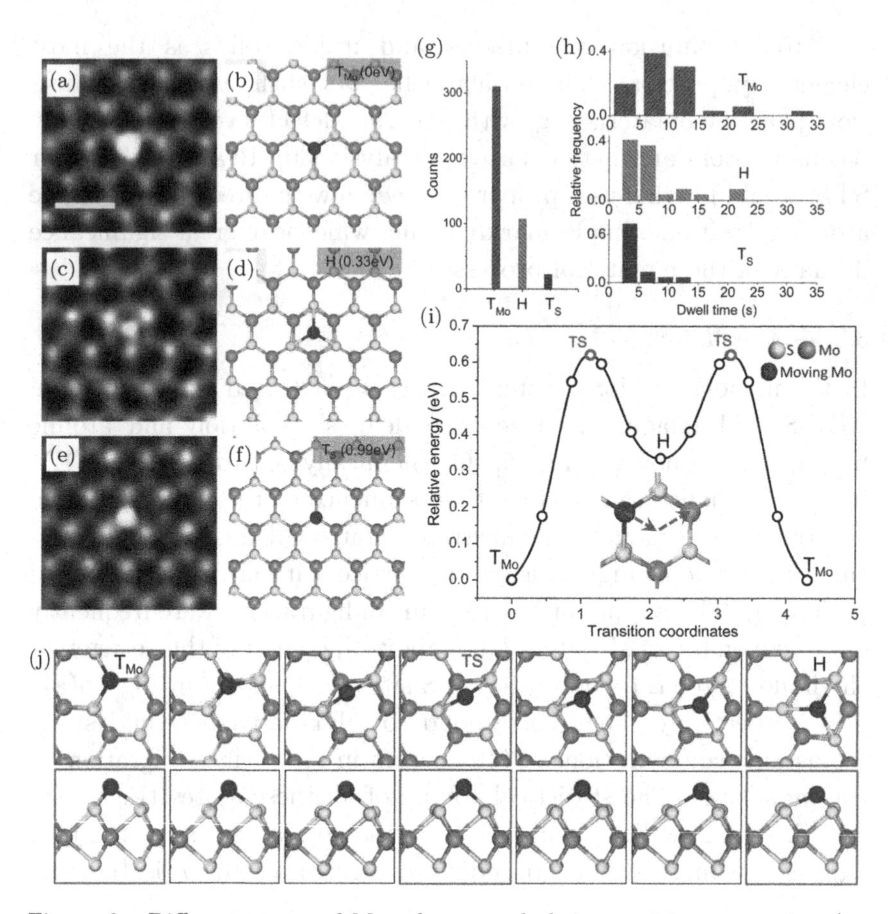

Figure 6. Different states of Mo adatom and their transition energetics. (a–f) Atomically resolved ADF images and structure models of different adsorption states on top of Mo site (T_{Mo}), at hexagon-center or hollow site (H), and on top of S site (T_{S}), respectively. Scale bar: 0.5 nm. False color is used to better illustrate the adatom configuration. The relative energies of different adatom states are given. (g) Statistical counts of different adsorption states in (a), (c), (e). (h) Statistical dwell time of different adatom structures. (i) DFT revealed energetics for evolution between ground state T_{Mo}, transition state TS, and metastable state H. (j) Detailed atomic dynamics of transition from T_{Mo} to H with top view and side view of the 3D atomic structure, respectively. Note the $T_{\mathrm{Mo1}} \rightarrow H \rightarrow T_{\mathrm{Mo2}}$ transition in (i) is symmetric, and hence, only the first half process (left red arrows in (i)) is drawn. Reproduced from Hong *et al.* (2017) with permission.

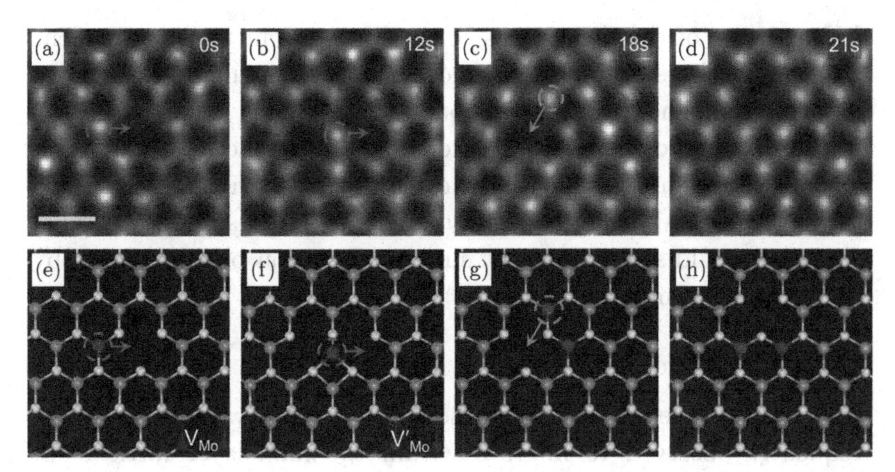

Figure 7. Atomic-scale migration of vacancy defects in monolayer MoS_2. (a)–(d) Time-lapse series of experimental ADF-STEM images of an Mo vacancy. Scale bar: 0.5 nm. (e)–(h) DFT-relaxed atomic structures corresponding to the evolution in **a**–**d** from the Mo vacancy (V_{Mo} in (a)) to its metastable state (V'_{Mo} in (b)) and then to another Mo vacancy V_{Mo} in (c). Note the vacancy migration is actually the motion of a neighboring Mo atom, shown in blue or purple. Reproduced from Jin *et al.* (2017) with permission.

metastable state, and TS is the second metastable state. All these adatom configurations have a three-fold symmetry structure with local magnetic moments $> 2\,\mu_{\mathrm{B}}$, according to the DFT calculation. They are all highly spin polarized and localized mainly on the Mo adatom, with a minor contribution from the neighboring S atoms.

The experimental ADF-STEM images of the ground-state T_{Mo}, first-metastable H, and second-metastable T_{S} also provide important configurations as the input for the DFT calculation, to reveal more details of the structure transition and energetics involved in the surface migration. Figures 6(i) and 6(j) show the energy profile and corresponding structural evolution in the primary kinetic pathway $T_{\mathrm{Mo1}} \rightarrow H \rightarrow T_{\mathrm{Mo2}}$ with a migration barrier of 0.62 eV. Another secondary kinetic pathway $T_{\mathrm{Mo1}} \rightarrow T_{\mathrm{S}} \rightarrow T_{\mathrm{Mo2}}$ is also found but with a much lower frequency in the experimental observations, with a barrier calculated to be 1.1 eV. The $T_{\mathrm{Mo1}} \rightarrow H \rightarrow T_{\mathrm{Mo2}}$ pathway on MoS_2 surface is preferred and acts as the dominant pathway, due to the tendency of the *d*-electrons of the Mo adatom to form covalent bonds with the surface S in a most-stable triangular–prismatic or metastable octahedral coordination [32].

This difference in energy barriers for different adatom migration pathways also agrees well with the contrasting experimental statistics of T_{Mo} and T_{S}. As both the energy barriers are not so large, thermal activation could still induce the surface migration but influenced by the electron beam irradiation.

3.2.2. *Mo vacancy*

Compared to the mobile Mo adatom defects, Mo vacancies are less frequently observed to migrate within the monolayer lattice. Jin *et al.* utilized high acceleration voltage to observe the evolution of the vacancies through time-elapsed ADF-STEM imaging series [31]. Figure 7 shows one time-sequential example of the migration of Mo vacancies with initial, metastable, and final states all imaged in one series of vacancy hopping. This Mo vacancy migration is actually the movement of neighboring Mo lattice atom. In the corresponding structure models in Figs. 7(e)–7(h), the migrating Mo lattice atom neighboring the Mo vacancy is highlighted in blue and purple, with arrows indicating the distance and direction of the next hopping. Again, in the vacancy hopping, the defect migration still obeys the random walk behavior without directional preference, typical of the Brownian motion of particles. Also, the Mo vacancy hopping only occurs within the sublattice of Mo and would never enter the S sublattice. Consistent with the DFT calculation, statistical analysis confirms the ground-state V_{Mo} and metastable V'_{Mo} (Figs. 8(a) and 8(b), which suggest us a most likely kinetic pathway for vacancy migration.

As shown in Figs. 8(c) and 8(d), the DFT-calculated energetics in the $V_{\mathrm{Mo}} \rightarrow V'_{\mathrm{Mo}} \rightarrow V_{\mathrm{Mo}}$ pathway give an initial migration barrier of 2.9 eV, which is much higher than the simple surface migration of Mo adatom. This contrasting energy barrier is easy to understand since the migration of Mo atom in the vacancy migration case is confined within the central Mo atomic plane of the sandwiched trilayers.

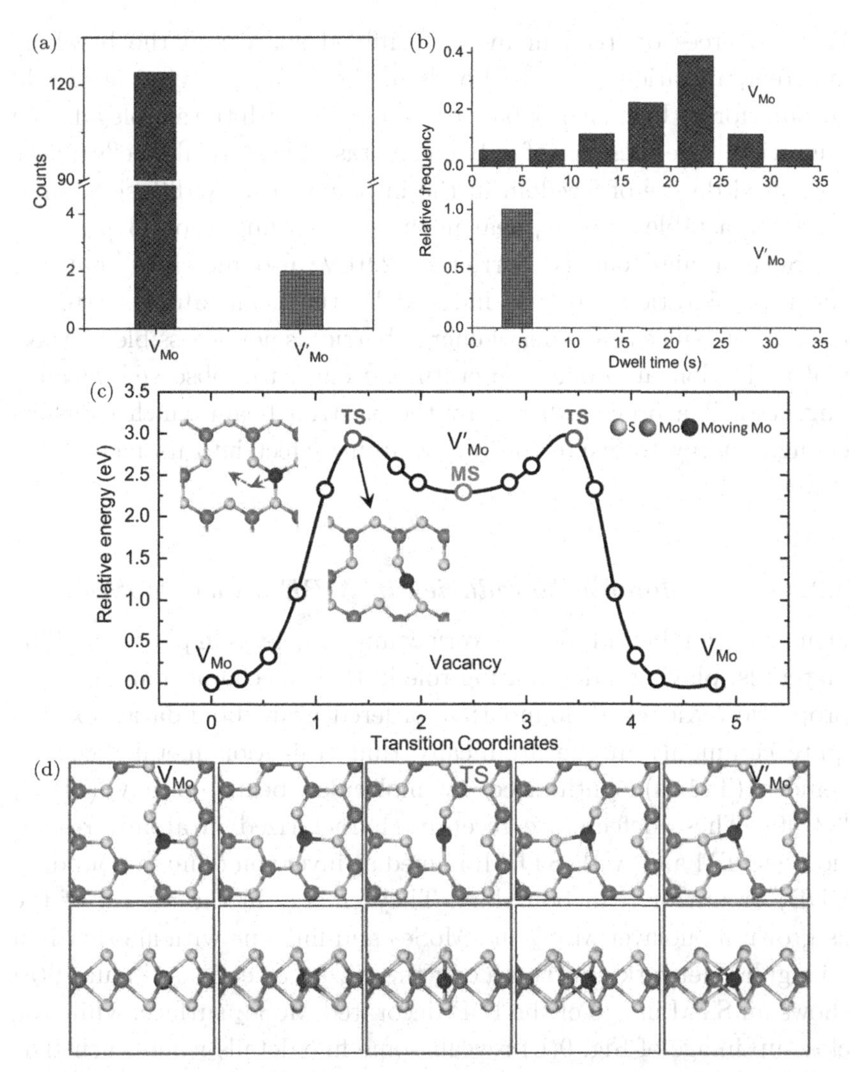

Figure 8. States of vacancy and their evolution. (a) and (b) Statistical counts and dwell time of vacancies and their metastable states during the migration of an Mo vacancy. (c) DFT-calculated migration pathway of Mo vacancy. The dynamic process is shown by the inset atomic models with arrows illustrating the migration pathway of the neighboring Mo atom. (d) Detailed atomic dynamics of Mo vacancy migration with top-view and side-view, respectively. Note the $V_{\text{Mo}} \rightarrow V'_{\text{Mo}} \rightarrow V_{\text{Mo}}$ transition in **c** is symmetric: hence, only the first half process of vacancy migration (right red arrow in (c)) is drawn in detail in (d). Reproduced from Jin *et al.* (2017) with permission.

Fewer degrees of freedom in the confined space and the breaking and reorganization of Mo–S bonds in the vacancy migration would account for its high energy barrier, compared with the simple adatom migration. The presence of only one metastable state also reflects the decreased degree of freedom in the in-plane vacancy diffusion, since more metastable states appear in the adatom migration [31].

Such a high energy barrier of 2.9 eV also indicates that the vacancy migration must be induced by the beam–atom scattering interaction, since this order of energy barrier is not accessible by thermal activation at room temperature. Hence, the observed vacancy migration is a process driven by the electron beam which transfers enough energy to excite the target atom/defect into its metastable states.

3.3. *Grain/domain boundaries in MBE-grown $MoSe_2$*

Domain/grain boundaries are very common defects in polycrystalline materials, playing a dominating role in their mechanical and electric properties. Xie *et al.* found that ordered grain boundaries existed quite commonly in the atomically thin transition metal dichalcogenides (TMDs) synthesized by molecular beam epitaxy (MBE) [33–40]. These defects are recently characterized in atomic resolution by STM and ADF-STEM, named as inversion domain boundary (IDB) or mirror twin boundary. They emerge in the matrix of the as-grown monolayer MoS_2 and $MoSe_2$ and link one with another in a triangular network and run along the zigzag directions. Figure 9(a) shows an STM image of the IDB-decorated $MoSe_2$ surface, while the close-up image of Fig. 9(b) reveals some fine details where each IDB defect manifests by two closely spaced mirror-symmetric lines with intensity modulations of period ~1 nm. From the STS measurements (see Fig. 10(f)), one notes that these IDBs introduce mid-gap states at the Fermi level; thus, they act as metallic channels, where charge density wave transition may occur at a low temperature [33, 42], and thus lead to the intensity modulations. Another cause of the modulations has been suggested to reflect the quantum confinement effect of the metallic states or the Tomonaga-Luttinger liquid [33–35].

Figure 9. (a) STM image (size: $50 \times 50\,\text{nm}^2$, bias: -1V) of an as-grown $MoSe_2$ on HOPG. (b) A close-up STM image (size: $13 \times 13\,\text{nm}^2$, bias: -1.46V) revealing the twin lines associated with each defect and the intensity undulations along the lines. Reproduced from H.J. Liu *et al.* (2014) with permission.

Figure 10(a) shows the atomically resolved ADF-STEM imaging of monolayer $MoSe_2$ grown by MBE, where Se2 columns are brighter than the Mo lattices [42]. Its fast Fourier transform (FFT, Fig. 10(a) inset) presents abnormally sharp lines connecting the first-order diffraction spot, indicating the existence of some ultranarrow and long line defects emerging in the monolayer $MoSe_2$. And also, these line defects have a specific directional distribution rather than in random directions. These line defects are actually IDBs, highlighted in blue (Fig. 10(b)) among those golden-colored triangular domains. Nanostripe-like IDBs are connecting with each other in a network form. Figure 10(c) provides a closer look at the atomic resolution ADF-STEM image of an IDB defect, where obvious four-membered rings are arranged in the zigzag direction and both domains are in a mirror symmetry in structure, as shown in Fig. 10(d). This DFT-relaxed IDB structure shows significant reorganization of Mo–Se bond lengths, which enlarges the horizontal Se–Se distance from 5.67 Å (d_0) to 6.16 Å (d_1). Then each IDB induces an uncommon lateral shift of 0.49 Å to the original Se–Se period of 5.67 Å, a fractional lattice translation, giving rise to diverse stacking orders if two such IDB-nested monolayers stack onto each other.

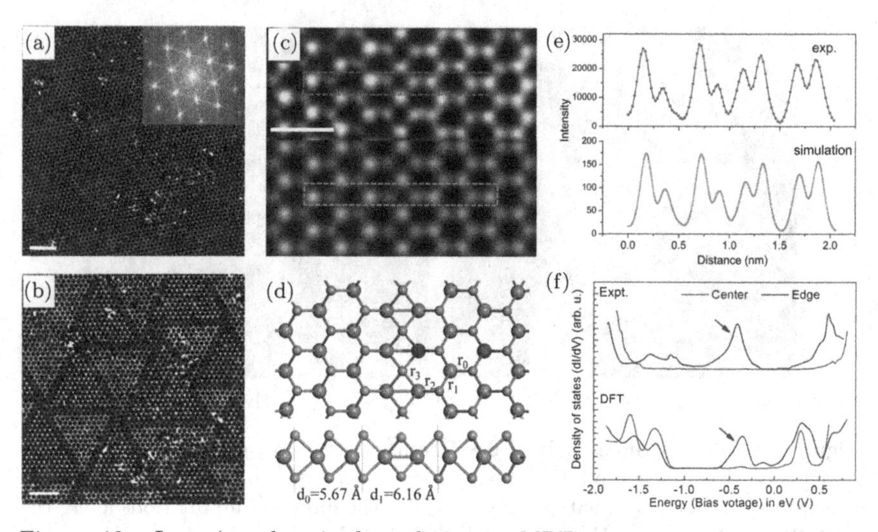

Figure 10. Inversion domain boundaries in MBE-grown monolayer $MoSe_2$. (a) Atomically resolved ADF–STEM image of monolayer $MoSe_2$. The inset FFT shows the quasi-periodicity of the ultra-narrow and long nanostructures. Scale bar: 2 nm. (b) False colored domains and boundaries. These dense inversion domain boundaries connect with each other like a wagon wheel. Scale bar: 2 nm. (c) Experimental and simulated ADF-STEM images of the boundary. Scale bar: 0.5 nm. (d) DFT relaxed atomic model of the boundary where orange balls represent Se atoms and cyan ones represent Mo atoms. (e) ADF-STEM intensity profiles along the long sides of the rectangular stripes marked in (c). (f) DFT-calculated DOS and experimental STS spectra from the domain center and the boundary. The calculated DOS were from boundary or domain Mo atoms highlighted in blue and red in (d), respectively, since Se atoms make negligible contribution to the DOS around the pristine bandgap. Reproduced from Jin *et al.* (2017) with permission.

The slight deviation of the symmetry of the ADF-STEM intensity in Fig. 10(c) is due to the presence of unintentional residual aberration such as three fold astigmatism A_2 in the focused electron probe. Considering this residual aberration in electron optics, quantitative ADF-STEM image was simulated in the lower panel of Fig. 10(c) and compared with the experimental image in the upper panel. Figure 10(e) shows the intensity line profiles extracted along the long sides of the stripes in the experimental and simulated images, both in a high consistency, confirming an Se_2-core boundary structure. In others words, neighboring triangular domains share the same line of Se_2 columns to form the Se_2-core IDBs everywhere [42].

Scanning tunneling spectra from the IDBs and domain center in Fig. 10(f) show distinctive characteristics especially around Fermi energy. The STS from the domain center is almost similar to the density of states (DOS) of a normal semiconductor with a bandgap of $\sim$2.0 eV, while that of IDBs has a remarkable midgap state at -0.41 eV and another two peaks at -1.8 eV and 0.6 eV. This metallic midgap state around Fermi level is characteristic of the IDB defects. The DFT-calculated DOS of IDB and the domain center both agree well with the experimental STS in the midgap states and bandgap characteristics, except the slight undrestimation of the bandgap.

3.4. *Stacking-band structure diversity in bilayer $MoSe_2$*

Network-like IDBs will induce fractional lattice translation to the adjacent domains. If two layers with IDBs stack together, then diverse stacking orders will inevitably appear, which occurs exactly in the MBE-grown bilayers [42]. Figure 11(a) shows the atomically resolved ADF-STEM image of a typical $MoSe_2$ bilayer without interlayer rotation. Random size of the non-periodic domains in a relatively large area rules out the possibility of Moiré patterns but confirms that they are stacking-dependent domains. This is quite different from the lattice-mismatch-induced Moiré stacking orders in hetero-bilayers [43].

After careful checking of the triangular bilayer domains, the difference in ADF-STEM imaging of different domains indicates distinctive stacking orders in the bilayer [42]. To specify the detailed stacking structure in each domain, construction of the stacking model and image simulation will be necessary. Starting from the initial high-symmetry AB–0 and AA–0 (Fig. 11(b)) configurations, the upper layer is shifted horizontally (H1, H2) or vertically (V1–V7), both parallel to the fixed bottom layer to yield various types of stacking orders, as shown in Figs. 11(b). The corresponding

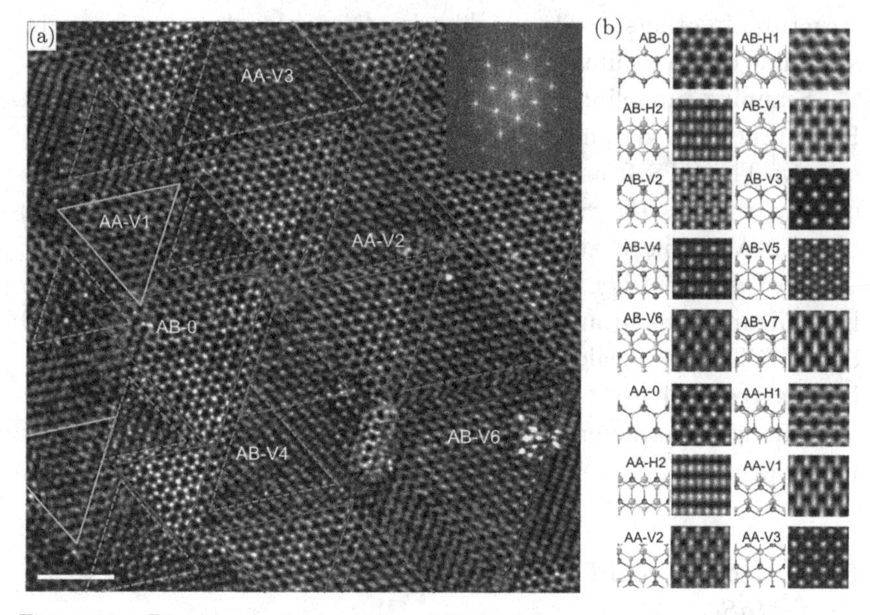

Figure 11. Diverse atomic structure of $MoSe_2$ bilayer domains. (a) Experimental HAADF image of a typically continuous and uniform bilayer $MoSe_2$. Those domains marked by triangles in the same color indicate the same stacking order. Scale bar: 2 nm. (b) Atomic model and the simulated ADF-STEM images of diverse bilayer stacking orders. The atomic structures of each domain in (a) are assigned by the comparison of experimental and simulated ADF images. Reproduced from Jin *et al.* (2017) with permission.

simulated ADF-STEM images demonstrate clearly the distinctiveness and diversity of these stacking sequences. Note that AB–0 and AA–V3 (Fig. 11(b)) stackings are actually bilayer structures in the well-known 2H and 3R phases, respectively. These simulated ADF-STEM images of each stacking structure act as fingerprints of each stacking sequence, and hence, can be directly compared with the experimental image in Fig. 11(a). Each domain can be assigned with a stacking order when the experimental images match the simulated images. As shown in Fig. 11(a), each domain is marked by triangles in different colors. Each color indicates one type of stacking order with its stacking name marked, as AB–V4, AB–V6, etc. Besides the high-symmetry configuration AB–0 and AA–V3, all the other experimentally observed stackings in low symmetry survive and get stabilized due to the confinement of the IDBs. Through this combination and comparison of the experiment/simulation results, each domain of the large-area bilayer can be unambiguously identified with one of the diverse stacking orders.

It's expected that diverse stacking bilayer structures would have different electronic structures, and distinctive electronic density of states near the fermi level. To probe the dependence of electronic structure on the stacking order, scanning tunneling spectroscopy was thus utilized to measure the electronic states from the domain centers. As shown in Fig. 12(a), experimental STS spectra collected from different domains present three types of features: olive spectra with

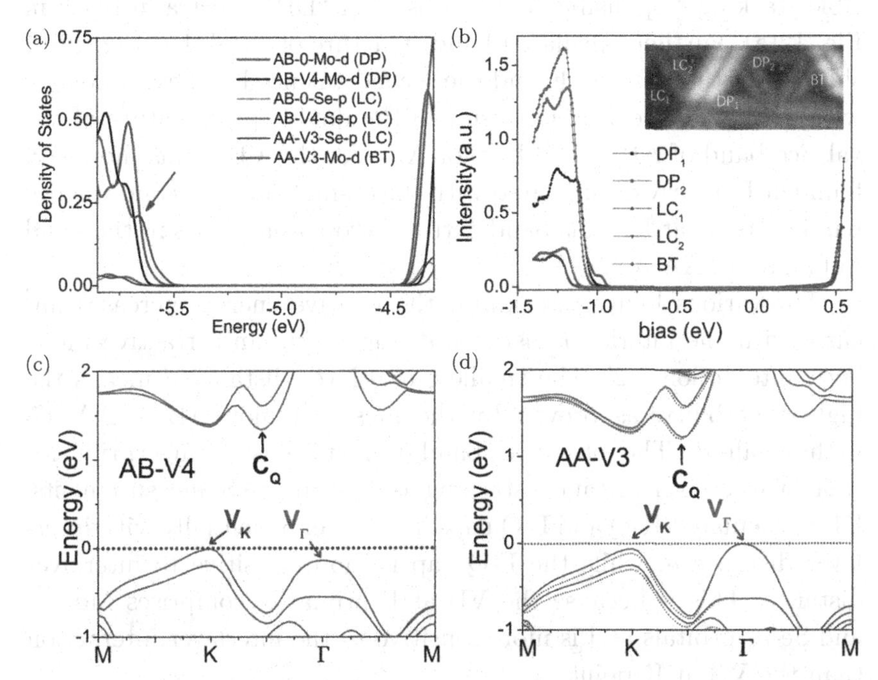

Figure 12. Distinctive electronic structures of the diverse bilayer domains. (a) DFT calculated LDOS of several typical bilayer stacking structures. (b) Experimental STS spectra measured at different domains. The inset is a STM image of the corresponding domains. The valence band edge is dependent on the stacking order. The different valence band DOS should arise from the diverse stacking orders of bilayer domains. (c)–(d) Band structures of the frequently observed stacking orders AB–V4 and AA–V3. Reproduced from Jin *et al.* (2017) with permission.

band tail (BT) states; the pronounced peak of valence band splitting into double peaks (DP) with an obvious separation; low-conductance (LC) spectra.

Three most common stackings were found in experiments: AB–0, AB–V4, AA–V3, whose electronic structure was calculated by DFT in Fig. 12. The calculated valence band DOS of AB-V4 (black curve) has an obvious double peak feature, and that of AA–V3 shows a band-tail structure and its bandgap get reduced, compared to the normal DOS of AB–0. The experimental STS in olive with band-tail feature can be assigned to the BT category, where the frequently observed AA–V3 is a typical stacking. And AB–V4 could be one possible stacking responsible for the observed "DP" spectra in black in Fig. 12(a). Further calculated band structure of AB–V4 in Fig. 12(c) shows that the VBM at K and the second valence band extreme at Γ with a 130-meV separation result in the double-peak feature of the valence band of AB–V4. While for AA–V3, the VBM was, however, found at Γ, nearly degenerated with the 59-meV-lower second valence band extreme at K. This band extreme crossover results in the band tail state, in AA–V3.

For various bilayer stackings, the relative energy increases linearly with the interlayer distance d, suggesting an attractive interlayer interaction [42]. The smallest interlayer distance d means the highest stability, as proved by the most common AB–0, AA–V3 with small d. The observed band tail state is a fingerprint for smaller-interlayer-distance stacking orders, in STS measurements. DFT-calculated K-Q and Γ-Q gaps increase exponentially with interlayer distance d, while the Γ-Q gap is more sensitive to interlayer distance. This is because the VB at Γ primarily comprises Mo d_{z2} and Se-p_z orbitals and is more sensitive to the interlayer interaction than the VB at K point.

4. Summary

In this chapter, both ADF-STEM and STM/STS demonstrate powerful atomic resolution imaging capability in the direct probing of atomic defects in 2D transition metal dichalcogendies. Point defects such as vacancy and antisite, grain/domain boundaries have been characterized by atomically resolved ADF-STEM or STM imaging, together with spectroscopy to reveal the electronic states induced by defects and low-symmetry lattice-translational stackings. Time sequential STEM to track the atomic flow also elucidate the different states involved in defects' evolution to deduce the primary kinetic pathways in the atomic migration.

In the 2D materials research, STEM/STM show their versatility in revealing the nanophysics of defects: both atomic characterization of the structures of defects and translational stackings and spectroscopic measurement of the electronic states induced.

Acknowledgments

JH and CJ acknowledge the financial support provided by the National Science Foundation of China under grant nos. 51772265, 51761165024 and 61721005, the Zhejiang Provincial Natural Science Foundation under Grant No. D19E020002, and the 111 project under no. B16042. MX acknowledges the support provided by a Collaborative Research Fund (C7036-17W) and a General Research Fund (No. 17327316) from the Research Grant Council, Hong Kong Special Administrative Region. CJ and MX also acknowledge the financial support provided by the NSFC/RGC joint research scheme (Nos. 51761165024 and N_HKU732/17). The authors acknowledge Dr. Wei Huang, Feng Jiang, and Dr. Yipu Xia for their kind assistance in preparing this chapter.

References

[1] W. Zhou, *et al.*, Intrinsic structural defects in monolayer molybdenum disulfide, *Nano. Lett.* **13**, 2615–2622 (2013).
[2] Y. Huang, *et al.*, Bandgap tunability at single-layer molybdenum disulphide grain boundaries, *Nat. Commun.* **6**, 6298 (2015).
[3] T. H. Ly, *et al.*, Misorientation-angle-dependent electrical transport across molybdenum disulfide grain boundaries, *Nat. Commun.* **7**, 10426 (2016).
[4] S. Manzeli, *et al.*, 2D transition metal dichalcogenides, *Nat. Rev. Mater.* **2**, 17033 (2017).

[5] Q. H. Wang, K. Kalantar-Zadeh, A. Kis, J. N. Coleman, M. S. Strano, Electronics and optoelectronics of two-dimensional transition metal dichalcogenides, *Nat. Nanotechnol.* **7**, 699–712 (2012).

[6] C. Ataca, H. Sahin, S. Ciraci, Stable, Single-layer MX_2 transition-metal oxides and dichalcogenides in a honeycomb-like structure. *J. Phys. Chem. C* **116**, 8983–8999 (2012).

[7] Lf. Mattheis, Band structures of transition-metal-dichalcogenide layer compounds, *Phys. Rev. B* **8**, 3719–3740 (1973).

[8] K. F. Mak, K. L. He, J. Shan, T. F. Heinz, Control of valley polarization in monolayer MoS_2 by optical helicity, *Nat. Nanotechnol.* **7**, 494–498 (2012).

[9] H. L. Zeng, J. F. Dai, W. Yao, D. Xiao, X. D. Cui, Valley polarization in MoS_2 monolayers by optical pumping, *Nat. Nanotechnol.* **7**, 490–493 (2012).

[10] T. Cao, *et al.*, Valley-selective circular dichroism of monolayer molybdenum disulphide, *Nat. Commun.* **3**, 887 (2012).

[11] S. F. Wu, *et al.*, Vapor-solid growth of high optical quality MoS_2 monolayers with near-unity valley polarization, *ACS Nano.* **7**, 2768–2772 (2013).

[12] B. Radisavljevic, A. Kis, Mobility engineering and a metal-insulator transition in monolayer MoS_2, *Nat. Mater.* **12**, 815–820 (2013).

[13] B. Radisavljevic, A. Radenovic, J. Brivio, V. Giacometti, A. Kis, Single-layer MoS_2 transistors, *Nat. Nanotechnol.* **6**, 147–150 (2011).

[14] S. B. Desai, *et al.*, MoS_2 transistors with 1-nanometer gate lengths, *Science* **354**, 99–102 (2016).

[15] S. Helveg, *et al.*, Atomic-scale structure of single-layer MoS_2 nanoclusters, *Phys. Rev. Lett.* **84**, 951–954 (2000).

[16] X. S. Wang, H. B. Feng, Y. M. Wu, L. Y. Jiao, Controlled synthesis of highly crystalline MoS_2 flakes by chemical vapor deposition, *J. Am. Chem. Soc.* **135**, 5304–5307 (2013).

[17] Y. H. Lee, *et al.*, Synthesis of large-area MoS_2 atomic layers with chemical vapor deposition, *Adv. Mater.* **24**, 2320–2325 (2012).

[18] H. P. Komsa, *et al.*, Two-dimensional transition metal dichalcogenides under electron irradiation: Defect production and doping, *Phys. Rev. Lett.* **109**, 035503 (2012).

[19] J. H. Hong, *et al.*, Exploring atomic defects in molybdenum disulphide monolayers, *Nat. Commun.* **6**, 6293 (2015).

[20] Q. Feng, *et al.*, Growth of large-area 2D $MoS_{2(1-x)}Se_{2x}$ semiconductor alloys, *Adv. Mater.* **26**, 2648–2653 (2014).

[21] Q. L. Feng, *et al.*, Growth of $MoS_{2(1-x)}Se_{2x}$ (x = 0.41–1.00) monolayer alloys with controlled morphology by physical vapor deposition, *ACS Nano.* **9**, 7450–7455 (2015).

[22] C. Gong, *et al.*, Metal contacts on physical vapor deposited monolayer MoS_2, *ACS. Nano.* **7**, 11350–11357 (2013).

[23] K. F. Mak, C. Lee, J. Hone, J. Shan, T. F. Heinz, Atomically thin MoS_2: A new direct-gap semiconductor, *Phys. Rev. Lett.* **105**, 136805 (2010).

[24] K. K. Liu, *et al.*, Growth of large-area and highly crystalline MoS_2 thin layers on insulating substrates, *Nano. Lett.* **12**, 1538–1544 (2012).
[25] Y. M. Shi, *et al.*, van der Waals epitaxy of MoS_2 layers using graphene as growth templates, *Nano. Lett.* **12**, 2784–2791 (2012).
[26] Y. C. Lin, D. O. Dumcencon, Y. S. Huang, K. Suenaga, Atomic mechanism of the semiconducting-to-metallic phase transition in single-layered MoS_2, *Nat. Nanotechnol.* **9**, 391–396 (2014).
[27] J. B. Hannon, S. Kodambaka, F. M. Ross, R. M. Tromp, The influence of the surface migration of gold on the growth of silicon nanowires, *Nature*, **440**, 69–71 (2006).
[28] S. Hofmann, G. Csanyi, A. C. Ferrari, M. C. Payne, J. Robertson, Surface diffusion: The low activation energy path for nanotube growth, *Phys. Rev. Lett.* **95**, 036101 (2005).
[29] L. E. Jensen, *et al.*, Role of surface diffusion in chemical beam epitaxy of InAs nanowires, *Nano. Lett.* **4**, 1961–1964 (2004).
[30] C. N. Satterfield, Mass Transfer in Heterogeneous Catalysis. (The MIT Press, 1970).
[31] J. Hong, Y. Pan, Z. Hu, D. Lv, C. Jin, W. Ji, J. Yuan, Z. Zhang, Direct imaging of kinetic pathways of atomic diffusion in monolayer molybdenum disulfide, *Nano Lett.* **17**, 3383–3390 (2017).
[32] M. Chhowalla, *et al.*, The chemistry of two-dimensional layered transition metal dichalcogenide nanosheets, *Nat. Chem.* **5**, 263–275 (2013).
[33] H. Liu, *et al.*, Dense network of one-dimensional midgap metallic modes in monolayer $MoSe_2$ and their spatial undulations, *Phy. Rev. Lett.* **113**, 066105 (2014).
[34] W. Jolie, *et al.*, Tomonaga-Luttinger liquid in a box: Electrons confined within MoS_2 mirror-twin boundaries, *Phy. Rev. X* **9**, 011055 (2019).
[35] Y. Xia, *et al.* Quantum confined Tomonaga-Luttinger liquid in $MoSe_2$ twin domain boundaries, arXiv: 1908.09259 (2019).
[36] O. Lehtinen, *et al.*, Atomic scale microstructure and properties of Se-deficient two-dimensional $MoSe_2$, *ACS Nano.* **9**, 3274–3283 (2015).
[37] J. V. Lauritsen, *et al.*, Size-dependent structure of MoS_2 nanocrystals, *Nat. Nanotechnol.* **2**, 53–58 (2007).
[38] Y. Wang, *et al.*, Monolayer $PtSe_2$, a new semiconducting transition-metal-dichalcogenide, epitaxially grown by direct selenization of Pt. *Nano. Lett.* **15**, 4013–4018 (2015).
[39] J. Lin, S. T. Pantelides, W. Zhou, Vacancy-induced formation and growth of inversion domains in transition-metal dichalcogenide monolayer, *ACS Nano.* **9**, 5189–5197 (2015).
[40] B. Feng, *et al.*, Experimental realization of two-dimensional boron sheets, *Nat. Chem.* **8**, 563–568 (2016).
[41] S. Barja, *et al.*, Charge density wave order in 1D mirror twin boundaries of single-layer $MoSe_2$, *Nat. Phys.* **12**, 751–756 (2016).

[42] J. Hong, C. Wang, H. Liu, X. Ren, J. Chen, G. Wang, J. Jia, M. Xie, C. Jin, W. Ji, J. Yuan, Z. Zhang, Inversion domain boundary induced stacking and bandstructure diversity in bilayer $MoSe_2$, *Nano. Lett.* **17**, 6653–6660 (2017).

[43] C. Zhang, C. P. Chuu, X. Ren, M. Li, L. J. Li, C. Jin, M. Y. Chou, C. K. Shih, Interlayer couplings, Moire patterns, and 2D electronic superlattices in MoS_2/WSe_2 hetero-bilayers, *Sci. Adv.* **3**, e1601459 (2017).

CHAPTER 3

Defects in Perovskites for Solar Cells and LEDs

F. BICCARI[*,†,¶], N. FALSINI[*,†], M. BRUZZI[*,‡], F. GABELLONI[*,†], N. CALISI[§] and A. VINATTIERI[*,†,‡]

[*]Department of Physics and Astronomy, University of Florence, Sesto Fiorentino (FI), Italy
[†]European Laboratory for Non-linear Spectroscopy (LENS), University of Florence, Sesto Fiorentino (FI), Italy
[‡]INFN Sezione di Firenze, Sesto Fiorentino (FI), Italy
[§]Department of Industrial Engineering, University of Florence, Firenze (FI), Italy
[¶]francesco.biccari@unifi.it

The properties of defects, including point defects, grain boundaries, and surfaces, play a major role in the carrier recombination dynamics in semiconductors and therefore determine the performances of semiconductor-based devices. Currently two classes of devices are of extreme interest: photovoltaic cells and light emitting diodes (LEDs). Such relevance comes from the demand of environmentally sustainable and commercially viable sources of energy, on the one hand, and low-cost, environmentally friendly and high-efficiency devices for light generation, on the other hand. The research for new materials and new heterojunctions for the implementation of innovative optoelectronic devices requires a deep knowledge of the defects, their effects on transport and optical properties, in particular radiative and non-radiative recombinations, and their relationship with the material crystallinity and stoichiometry. Here, we review the theoretical and experimental results about the nature of the defects and properties of two innovative materials for photovoltaics and LEDs: methylammonium lead iodide ($CH_3NH_3PbI_3$), a hybrid organic–inorganic perovskite, and cesium lead bromide ($CsPbBr_3$), a fully inorganic perovskite. The choice of these two compounds is dictated by their fundamental role in the development of perovskite-based solar cells ($CH_3NH_3PbI_3$) and perovskite-based LEDs ($CsPbBr_3$). Even though, especially in the case of solar cells, new perovskite compounds are being investigated, $CH_3NH_3PbI_3$ and $CsPbBr_3$ remain the starting point to understand the physics of the defects of this class of materials. Given the extremely large literature, this review is not expected to be exhaustive, covering all the aspects of defects in perovskites: the authors just aim to provide an overview of the main topics, inspiring the reader to undertake further studies.

1. Introduction to Perovskites

The global final energy consumption in 2016 was about 111 PWh while the global electrical energy consumption was about 23 PWh, and these numbers are expected to further increase in the future [1]. Fossil fuels represent the largest part of the energy production, but they are finite energy resources and their use causes environmental pollution and political instabilities. For this reason, currently, one of the most important research field considers the development of environmentally sustainable and commercially viable ways to use renewable energy sources and high-efficiency solution to reduce the power consumption of devices.

The sun is a virtually inexhaustible renewable energy source, enough to provide a quantity of energy per year much larger than the total energy consumption in the world. However, photovoltaic energy production was only 0.33 PWh in 2016, which is about 0.3% of the global energy consumption and about 1.4% of the global electrical energy consumption [1]. Moreover, this result was obtained by national financial incentives, since the initial cost of a photovoltaic plant is still high and the cost per generated unit energy is not yet competitive in many parts of the world with respect to the traditional energy sources (grid parity not reached) [2, 3]. Part of this cost is due to the high cost of production of the active material of the cell, i.e. silicon [2, 3]. Therefore, in order to solve this problem, a huge effort has been directed by research and technology experts to fabricate cells with low-cost raw materials, maintaining or increasing the conversion efficiency.

As regards the effort to increase the device efficiency, in particular electrical and electronic devices, a lot of research is devoted to lighting devices: lighting electricity consumption in 2016 was about 3 PWh, about 14% of the global electrical consumption [4–6]. Therefore, the demand for low-cost, environmentally friendly and high-efficiency devices for light generation, such as LEDs, is increasing worldwide. LED lamps (white and/or monochromatic) started to dominate the market for civil and domestic use because of the reduced power consumption compared not only to standard incandescent bulbs but also to compact fluorescent and energy-saving halogen lamps [6]. Moreover, lifetime of the extended LEDs, up to 50000 hours [6], makes such devices invaluable for applications such as traffic lights and street lightening, where the major cost comes from the maintenance related to the life cycle. In order to reach the widespread adoption of LED technology, it is necessary to reduce, in particular, the initial cost of the device and, subsequently, the cost per unit of generated light. Part of the former is due to the high production cost of the active material of the LED, e.g. III–V semiconductors [6], and therefore, researchers worldwide are devoting their efforts to find new low-cost, environmentally friendly and high-efficient materials.

In this context, a new class of low-cost semiconductor materials, called "perovskites" [7], have attracted a lot of attention from the scientific community in the last decade given their very good electrical and optical properties [8–10]. The general chemical formula of perovskites is ABX_3, where A and B are two cations of very different sizes, and X is an anion. Currently, the most investigated perovskites are hybrid organic–inorganic materials, like methylammonium lead iodide ($CH_3NH_3PbI_3$), or fully inorganic materials, like cesium lead bromide ($CsPbBr_3$). Many other perovskite materials are investigated, where the anion (I) and the inorganic (Pb or Cs) or organic ($CH_3NH_3^+$) cations are substituted with alternative elements/molecules.

These materials have a high absorption coefficient (10^4–10^5 cm^{-1} [11–13]) due to their direct band gap, long carrier diffusion lengths [14–17], low effective masses of electrons and holes [18, 19], high carrier mobility ($\geq$100 cm^2 V^{-1} s^{-1} [17, 20]), and significant tolerance toward defects [21, 22]; in addition, their emission is easily tunable from near-infrared to near-ultraviolet by varying the composition [23].

Perovskites take their name from the mineral *perovskite*, since they all share the same crystal structure. The perovskite mineral is made of calcium titanium oxide ($CaTiO_3$), and it was discovered by the mineralogist Gustav Rose in 1839 in the Ural Mountains. He named this mineral *perovskite* in honor of the Russian mineralogist Lev Alekseyevich von Perovski [7]. The first lead halide perovskites (like $CsPbBr_3$) were synthesized in 1892 by H. L. Wells, but their crystal structure was associated with the perovskite only in 1959 by C. K. Moller. The first measurement of the perovskite structure (by X-ray diffraction) was performed in 1945 by H. Megaw on barium titanate. Hybrid organic–inorganic lead halide perovskites (like $CH_3NH_3PbI_3$) were synthesized in 1978. Today, we know that perovskites show an ideal perfect cubic crystalline structure ($Pm\bar{3}m$ (221)) only at a "high temperature". On lowering the temperatures, these materials undergo to a first phase transition to a tetragonal structure (I4/mcm (140)), and then undergo a second phase transition to an orthorhombic structure ($Pna2_1$ (33)). Tetragonal and orthorhombic are just two slightly distorted versions of the cubic structure, with the orthorhombic structure being less symmetric than the tetragonal structure. For example, $CH_3NH_3PbI_3$ shows a cubic structure above 327 K, tetragonal structure between 162 K and 327 K, and orthorhombic structure below 162 K [7]. In contrast, $CsPbBr_3$ is stable in the orthorhombic phase up to 361 K [22].

Few research papers were devoted to perovskites until the first reports were published on solar cells based on $CH_3NH_3PbBr_3$ and $CH_3NH_3PbI_3$ by Kojima *et al.* in the period 2006–2009 [24–26], where the photovoltaic conversion efficiency was 3.8%. After that, the publications on perovskites increased exponentially making the study of halide perovskites the most actively researched topic on

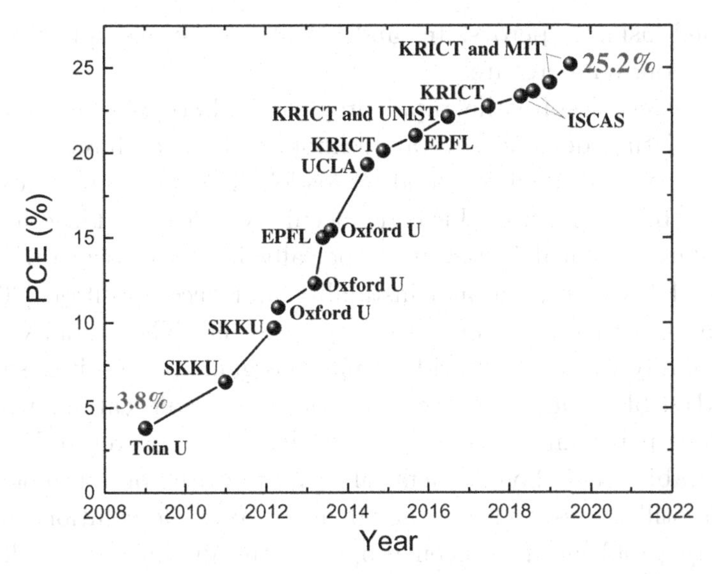

Figure 1. Best research cell efficiencies of perovskite solar cells. Re-elaborated from Ref. [28].

photovoltaics [27] and a very hot topic in many other fields. The photovoltaic conversion efficiency reached the very high value of 25.2% in 2020 [28] (see Fig. 1), making perovskite solar cells competitive to silicon-based cells, with the major advantage of low-cost fabrication techniques.

The research on perovskite-based electroluminescent LEDs is more recent [29–31]. The first report on an LED working at room temperature was published in 2014 by Tan *et al.* [32]. The highest external quantum efficiency was 0.4%. To date, the maximum external quantum efficiency that has been achieved is larger than 20% [33, 34]. Perovskites also show narrow linewidths (about 20 nm at room temperature), size-sensitive emission spectra in nanocrystals, and high photoluminescence quantum efficiency even without any surface passivation [29–31]. It is worth noting that the narrow emission and the continuous spectral tunability make perovskite nanocrystals promising candidates as light converters in backlight display applications with an ultrawide gamut [29–31]. Perovskites are interesting materials for lasing applications as well: Li *et al.* [35]

obtained lasing in perovskites under continuous-wave optical pumping at room temperature.

However, despite the rapid progress in both photovoltaic and light emitting devices, several technological issues limit the commercialization of devices based on lead halide perovskites: toxicity, bioavailability, and probable carcinogenicity of lead and lead halides; perovskite chemical instability, especially in the presence of moisture and UV light; thermal instability and decomposition [36–38]; and large area growth of perovskites [38, 39]. The stability problems mainly affect the hybrid organic–inorganic perovskites, such as $CH_3NH_3PbI_3$, due to the presence of the organic cation, whereas inorganic perovskite, such as $CsPbBr_3$ have been proven to be much more stable, even though chemical stability in the presence of moisture is still an issue. Even though many possible solutions to the stability problems have been proposed, the lifetime of the devices has not yet reached the market standards.

Recently, in order to overcome the presence of lead, investigations were carried out using Sn or Ge as a substitute for Pb, both in hybrid organic–inorganic and fully inorganic perovskites [40–42]. However, to date, the performances of these lead-free alternatives are not yet comparable with those of the lead-based perovskites.

Despite the impressive technological advancement and the increase in the number of published papers, the nature and the physical and chemical properties of the intrinsic defects in this class of materials are in large part unknown, even though point defects, grain boundaries, and surfaces play a major role in the performances of semiconductor-based devices, in particular solar cells and LEDs. The research for new materials and new heterojunctions for the implementation of innovative optoelectronic devices requires a vast knowledge of the defects, their effects on transport and optical properties, in particular radiative and non-radiative recombinations, and their relationship with the material crystallinity and stoichiometry.

In this chapter, we review the theoretical and experimental results to gain an understanding regarding the nature of the defects and properties of the hybrid organic–inorganic perovskite methylammonium lead iodide ($CH_3NH_3PbI_3$) and the fully inorganic perovskite cesium lead bromide ($CsPbBr_3$). The choice of these two compounds comes from their fundamental role in the development of perovskite-based solar cells ($CH_3NH_3PbI_3$) and perovskite-based LEDs ($CsPbBr_3$). Even though, especially in the case of solar cells, it is now known that better performances are obtained with more complex compounds, where methylammonium is mixed or substituted by formamidinium and bromine is partly introduced in place of the iodine, $CH_3NH_3PbI_3$ and $CsPbBr_3$ remain the starting point to understand the physics of the defects of this class of materials.

The first section will be devoted to $CH_3NH_3PbI_3$ (also called $MAPbI_3$ or MAPI), while the second one will be devoted to $CsPbBr_3$. Both sections are divided into two parts: one dedicated to theoretical *ab initio* calculations and the other dedicated to experimental results pertaining to the identification, properties, and effects of the defects. Indeed, to date there is no experimental technique able to accurately identify the nature and the properties of the defects of a semiconductor, and therefore a strong support of *ab initio* calculations is necessary in any serious attempt to understand the physics of the defects of a material.

2. Defects in $MAPbI_3$

2.1. *Theoretical calculations*

The identification of the nature of point defects in a semiconductor and the measurement of their physical properties cannot be done without a proper support for theoretical calculations.

The most important property obtained by *ab initio* calculations is the defect formation energy (DFE) of each defect. The DFE of a native defect X^q is defined as the energy difference between the crystal with the defect and the energy of the components in their reference state [43]:

$$\mathrm{DFE}[X^q] = E[\mathrm{bulk} + X^q] - E[\mathrm{bulk}] - \sum_i n_i \mu_i + q(E_{\mathrm{VBM}} + \mu_{\mathrm{e}}),$$

where q is the charge state of the defect (expressed in unit of the module of the electron charge), n_i represents the number of atoms of species i added to the bulk to form the defect, taken from a reservoir described by a chemical potential μ_i, and μ_e is the electron chemical potential (Fermi level) defined with respect to the maximum of the valence band E_{VBM}. Since the simulated crystals are usually composed of a very few unitary cells in order to limit the computation time and since periodic boundary conditions are employed, the interaction between the replica of a charged defect must be corrected [43].

It is intuitively clear that the DFE of a defect depends on the environmental condition in which the perovskite is immersed, that is the chemical potential of its constituting species. The thermodynamic stability region of $MAPbI_3$ is determined by the following constraints [44]:

$$\mu(\mathrm{MA}) + \mu(\mathrm{Pb}) + 3\mu(\mathrm{I}) = \mu(\mathrm{MAPbI_3}),$$
$$\mu(\mathrm{MA}) + \mu(\mathrm{I}) < \mu(\mathrm{MAI}),$$
$$\mu(\mathrm{Pb}) + 2\mu(\mathrm{I}) < \mu(\mathrm{PbI_2}).$$

The first equation expresses the equilibrium between $MAPbI_3$ and its constituting species. The two inequalities indictate that the formation of MAI and PbI_2 is not favored. The result is the red region shown in Fig. 2 [44]. According to this diagram, three points in Fig. 2 are typically chosen for the calculation of the DFEs: point A (metal

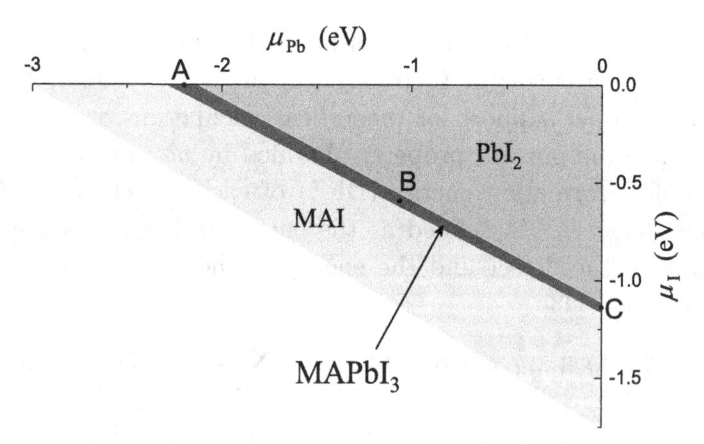

Figure 2. Thermodynamic stable range for equilibrium growth of $CH_3NH_3PbI_3$ is the narrow and long red region.

poor, halide rich), point B (intermediate conditions), and point C (metal rich, halide poor).

By fixing the chemical potentials, DFEs show a linear dependence on the Fermi level in the system, which is allowed to vary within the material band gap, whose slope is determined by the charge of the defect. The thermodynamic ionization level between two different charge states of a defect is defined as the Fermi energy (E_F) where the DFEs of the two different charge states, q and q', are equal, i.e. $\mathrm{DFE}(q', E_F) = \mathrm{DFE}(q, E_F)$ with the constraint that the Fermi energy is restricted within the band gap of the crystal. The level is usually indicated as $\varepsilon(q/q')$. The concentration of a particular defect at the thermodynamic equilibrium can be easily estimated by applying the Boltzmann law where the activation energy is the DFE and where the prefactor of the exponential is given by the concentration of possible defect sites in the lattice. Finally, from the calculated defect densities, it is possible to predict the position of the Fermi level in the system and, thus, whether the material will be n- or p-doped by solving the associated electroneutrality equations.

The first set of papers that were published on $MAPbI_3$ defect calculation were by Yin *et al.* [44–46]. The point defects considered include three vacancies (V_{MA}, V_{Pb}, V_I), three interstitial substitutions (MA_i, Pb_i, I_i), two cation substitutions (MA_{Pb}, Pb_{MA}), and four antisite substitutions ($MA_I, Pb_I, I_{MA}, I_{Pb}$). The second published paper was by Kim *et al.* [47]. They calculated formation energies of Schottky defects, such as PbI_2 and CH_3NH_3I vacancy, and Frenkel defects, such as V_{Pb}–Pb_i, V_I–I_i, and V_{MA}–MA_i. Another paper was published by Agiorgousis *et al.* [48]. The inference from all these papers is that only shallow levels are present in the $MAPbI_3$ bandgap and, moreover, that it is also possible that they aggregate or self-compensate by forming Schottky defects [49], reducing their activity even further. However, two papers were published by Buin *et al.* [50, 51], stating that the actual deep defect in MAPI is Pb_I. These contradictory results were explained by the fact they all used a GGA approach without considering the spin–orbit coupling (SOC). This approach, even though it fortuitously gives an almost correct value of the band gap, must be avoided since the SOC effects in $MAPbI_3$ are very strong. This is well explained by Du in Ref. [52].

Calculations taking into account hybrid functionals including SOC can indeed predict the presence of deep-trap states, which are not predicted by the other methods. Du [53, 54] showed that halogen vacancy is a shallow donor and, for the first time, that I_i is a deep recombination center. In [52], Du also shows the energy levels of several $MAPbI_3$ deep defects. In Fig. 3, we show the energy levels of the defects calculated by Du [52] and the microscopic configuration of the charge states of the iodine interstitial point defect ([52], refinement over [53]). The HSE–SOC calculations predict that, among

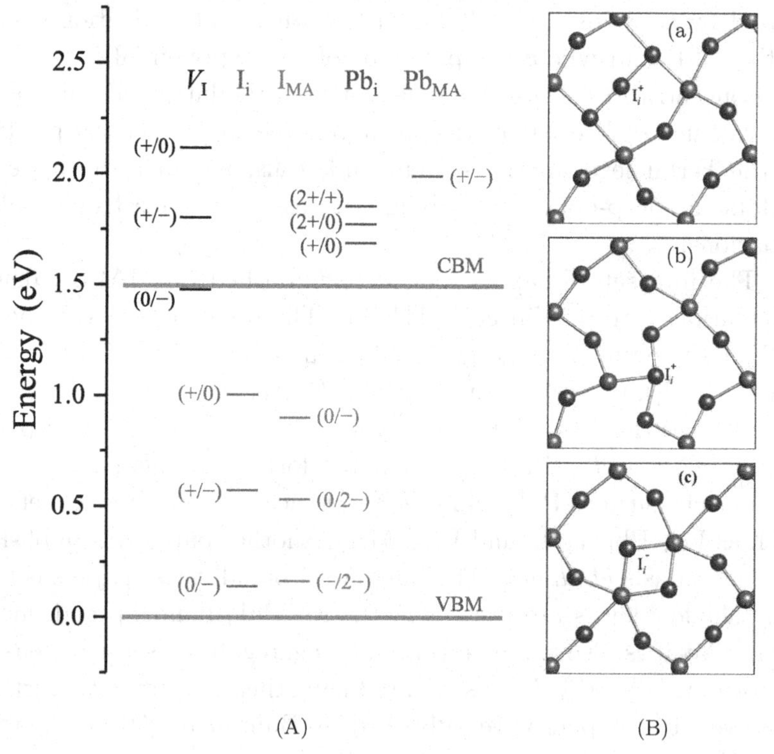

Figure 3. (A) Charge transition energy levels of several point defects in $MAPbI_3$ (orthorhombic phase). (B) Microscopic configuration of two I_i^+ structures (a,b) and the structure of I_i^- (c). Only one Pb–I layer is shown for clarity. Red and blue balls represent Pb and I atoms, respectively. Reprinted with permission from Ref. [52]. Copyright (2015) American Chemical Society.

native point defects (vacancies, interstitials, antisites, and their complexes), only I_i and its complexes (such as I_{MA}) introduce deep levels inside of the band gap of $MAPbI_3$. Other native point defects are shallow.

Let us analyze the energy levels in more detail. The strong structural relaxation upon charge trapping leads to the negative correlation energy behavior of I_i and I_{MA}. Let us consider I_i as an example. Its (+/0) level is higher than its (0/−) level, indicating strong structural relaxation upon electron trapping at the defect. The neutral charge state of I_i is metastable. The (+/−) level determines the charge transition between the two stable charge states, that is, +1 and −1. When the Fermi level is higher than (+/−), I_i^+ captures an electron, and soon after it relaxes, it captures another electron. The (+/0) level is the electron-trapping level for I_i^+, while the (0/−) level is the hole-trapping level for I_i^-. The HSE–SOC calculations show that the (+/−) and the (0/2−) transition levels for I_i and I_{MA} are 0.57 eV and 0.54 eV above the VBM, respectively, as shown in Fig. 3. The Fermi level is usually near midgap in $MAPbI_3$ planar thin film device architectures: under these conditions, I_i^- and I_{MA}^{2-} are stable, and their acceptor levels are deep, that is, the (0/−) level for I_i and the (−/2−) level for I_{MA} are 0.15 eV and 0.17 eV above the VBM, respectively.

In Fig. 3, we have also shown the microscopic configuration of I_i as calculated by Du [52]. I_i^- is a split interstitial, where two iodine ions occupy the same iodine site. The structure of I_i^- has two I^- ions lying on the ab plane, each of which binds two Pb ions (Fig. 3(c)). I_i^+ prefers to be coordinated with two other iodine ions on the ab plane in triplet structures shown in Figs. 3(a) and 3(b). The energies of these two structures differ by less than 0.01 eV.

It is interesting to note that Whalley *et al.* [55], starting from the work of Du, have suggested using spin-sensitive techniques to identify the iodine-related complexes in $MAPbI_3$.

Calculations similar to those performed by Du were performed by Meggiolaro *et al.* [56, 57]. Reference [57] is the most recent and accurate paper that discusses defect energy levels and DFEs calculations. In Fig. 4, we report their results [57].

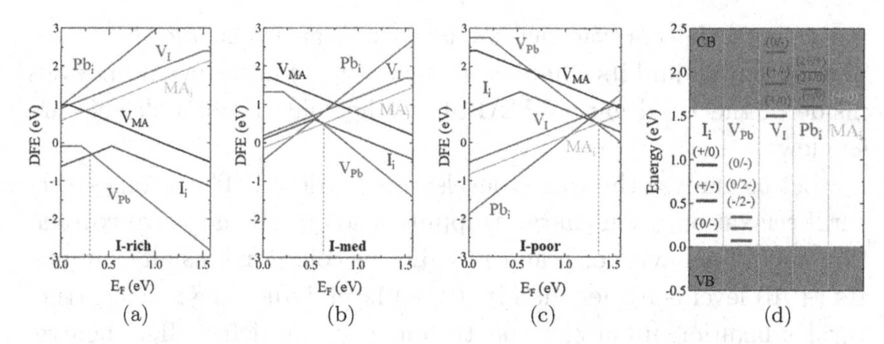

Figure 4. (a, b, c) Defect formation energies for several $MAPbI_3$ defects in three different points (A, B, C) of the stability diagram reported in Fig. 2. (d) Charge transition energy levels. Reprinted with permission from Ref. [57]. Copyright (2018) American Chemical Society.

In iodine-medium conditions, the most stable defects, i.e. those with the lowest DFEs, are MA interstitials (MA_i^+), interstitial iodine (I_i), and lead vacancies (V_{Pb}) which pin the Fermi level roughly at mid-gap, slightly in the p-type region.

Other theoretical predictions focus their attention on specific defects. For example, Vogel *et al.* [84] calculated the properties of Pb site vacancy and its impact on the charge-carrier dynamics following photoexcitation. Analogously, Li *et al.* [85].

All the papers cited thusfar discuss bulk defects ($MAPbI_3$ defect is discussed in Ref. [58, 59]). However, surface and grain boundary defects might be of huge importance for the functionality of the perovskites in optoelectronic devices. Several papers indicate that surface and grain boundary defects play a fundamental role in $MAPbI_3$ solar cells. This was proven by the effect of several surface post-growth treatments on photovoltaic conversion efficiencies. However, to date, the types of surface defects that act as carrier traps have still not been identified. As regards the defect present on the surface, the most recent and comprehensive theoretical paper is by Uratani and Yamashita [60] (some results are also reported in Ref. [56]). The authors focused on the (001) surface, which is considered as one of the major types of $MAPbI_3$ facet by first-principles calculations and X-ray powder diffraction measurements. Three types of surface termination were considered: MAI, flat, and vacant termination. The defect formation energies for these three types of terminated surfaces are calculated for three different external conditions, I-rich, moderate, and Pb-rich. In Table 1, the formation of DFEs is reported, while in Fig. 5, the defect energy levels introduced within the bandgap are shown.

The high values of the band gap and the variation of the band in the presence of defects are worth noting. Despite these inconsistencies, the results are clear: the best surface is the one with MAI termination because in that case only shallow levels have low formation energies, regardless of the growth conditions. However, it was experimentally proved [60] that, since the high volatility of MAI, the surface is not typically MAI terminated. According to their calculations, Uratani and Yamashita [60] conclude that (i) Under the I-rich condition, excessive I atoms on flat and vacant surfaces are responsible for the carrier trapping. On the contrary, under the Pb-rich condition, I atom vacancies on vacant surfaces and excessive Pb atoms on both flat and vacant surfaces act as carrier traps. (ii) The formation of

Table 1. DFEs (in eV) of several surface defects in $MAPbI_3$. Reprinted from Ref. [60].

Termination	Defect	I-rich	Moderate	Pb-rich
MAI	V_I	2.38	1.77	1.15
MAI	V_{MA}	0.32	0.91	1.54
MAI	Pb_i	4.69	3.45	2.24
MAI	Pb_{MA}	2.35	1.70	1.13
MAI	Pb_I	5.97	4.14	2.31
Flat	I_i	−0.03	0.57	1.19
Flat	V_I	4.55	3.94	3.32
Flat	Pb_i	2.94	1.71	0.50
Flat	V_{Pb}	−0.13	1.10	2.31
Vacant	I_i	−0.12	0.48	1.10
Vacant	V_I	1.76	1.15	0.54
Vacant	Pb_i	2.71	1.48	0.26
Vacant	V_{Pb}	−1.40	−0.17	1.04

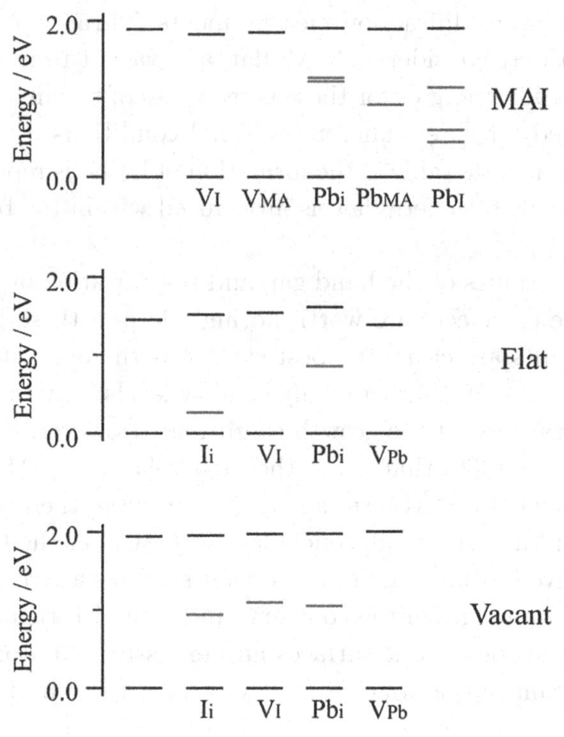

Figure 5. Charge transition energy levels of several surface defects in $MAPbI_3$. Reprinted from Ref. [60].

carrier-trapping surface defects under the Pb-rich condition is thermodynamically more unfavorable than that under the I-rich condition. (iii) Under the moderate condition, any surface defects that act as carrier traps have a high formation energy, that is, cannot easily form. From the above discussion, to reduce carrier trapping on surfaces or grain boundaries so as to improve the carrier lifetime and avoid hysteresis, we conclude that the Pb-rich condition is better than the I-rich condition, consistent with the superiority of the two-step method over the one-step method in terms of photovoltaic performance.

Besides the paper by Uratani and Yamashita, another paper specifically dedicated to the theoretical prediction of grain boundary defects in $MAPbI_3$ is that by Yin *et al.* [86]. Even though they used the same approach as that used in their previous paper on bulk defects, i.e. without spin-orbit coupling, in this case, they tested the results obtained with other simulations with spin–orbit coupling. They simulated two typical grain boundaries, experimentally observed in inorganic perovskites, $\Sigma3(111)$ GB and $\Sigma5(310)$. They found that structural disorders from both thermal fluctuation and grain boundaries introduce no deep defect states within the bandgap.

2.2. *Effects and properties of $MAPbI_3$ defects*

In this section, we will review the experimental facts related to the $MAPbI_3$ defects. In the first part, we will review all the direct measurements of the defect properties, obtained by the typical techniques used for deep defect characterization: DLTS, TSC, TAS, and so on. In the second part, we will try to briefly describe all the most important macroscopic effects related to the $MAPbI_3$ defects (defect tolerance in $MAPbI_3$ devices, hysteresis of current–voltage solar cell characteristics, and so on).

A first insight into the effect of defects in $MAPbI_3$ and a check of the theoretical calculations is given by the p-type nature or n-type nature of the $MAPbI_3$ samples. Considering the predictions shown above, the elongated shape of the red region in Fig. 2 indicates that it is possible to tune $MAPbI_3$ from n-type to p-type. For example, when the chemical potential of Pb is higher than that of I (point A in the red region of Fig. 2), the defects with the lowest formation energies in the structure, such as V_{Pb} and I_i, are predominant, and since they are acceptors, the hole carrier density is increased, thus leading to the formation of p-type $MAPbI_3$. Vice versa, when the chemical potential of Pb is lower than that of I (point C in the red region of Fig. 2), the predominant defects are the donors such as V_I, Pb_i, and MA_i, and the $MAPbI_3$ is n-type. This was proven experimentally in Ref. [61] where the authors changed the ratio of methylammonium halide (MAI) and lead iodine (PbI_2), finding that MAI-rich and PbI_2-rich perovskite films are p- and n-type, respectively. They also found that thermal annealing can convert the p-type perovskite to n-type perovskite by removing MAI. The carrier concentration varied by as much as six orders of magnitude.

The experimental study of point defects is usually related to the availability of single crystals, since the quality of the material is higher and controlled with respect to nano- or micro-crystalline thin films and spurious effects like grain boundaries are not present. Fabrication of large $MAPbI_3$ single crystals has been possible since 2015 thanks to the work of Shi *et al.* [62]. Several measurements were carried out in order to measure the properties of $MAPbI_3$ intrinsic defects, both on single crystals and thin films. We will discuss them in the order of technique.

Adinolfi *et al.* [63] measured a $MAPbI_3$ single crystal using Space Charge Limited Current (SCLC) as a function of temperature. The crystal is nearly intrinsic but n-type in the bulk with a concentration of carriers of about $4 \times 10^9\,\mathrm{cm}^{-3}$. This value is similar to other works on $MAPbI_3$ single crystals. It is interesting to note, as we will see in the surface defects section, that the surface has a strong n-type behavior, due to the pinning of the Fermi level (0.12 eV from the conduction band) caused by a high concentration of defects, which are also responsible for the fast PL decay component typically observed. The mobility is similar for electrons and holes and is about $50\,\mathrm{cm}^2\,\mathrm{V}^{-1}\,\mathrm{s}^{-1}$ for electrons (Hall effect) and about $60\,\mathrm{cm}^2\,\mathrm{V}^{-1}\,\mathrm{s}^{-1}$ for holes (SCLC). The conductivity is about $3 \times 10^{-8}\,\Omega^{-1}\,\mathrm{cm}^{-1}$. The diffusion length is similar for electrons and holes, about $12\,\mu\mathrm{m}$, as obtained using the Einstein relation for the charge diffusion coefficient and the lifetime found by the long component of the PL-decay measurements.

Using SCLC measurements as a function of temperature on two dedicated $MAPbI_3$-based devices, one only hole-injecting and the other one only electron-injecting, Adinolfi *et al.* [63] could measure the density of states of the traps, as reported in Fig. 6. The donor concentration is about one order of magnitude larger than the acceptor concentration ($3 \times 10^{11}\,\mathrm{cm}^{-3}$ versus $3 \times 10^{10}\,\mathrm{cm}^{-3}$). The donor DOS is located about 0.2 eV from the conduction band while the acceptor DOS is located around 0.1 eV above the valence band. The

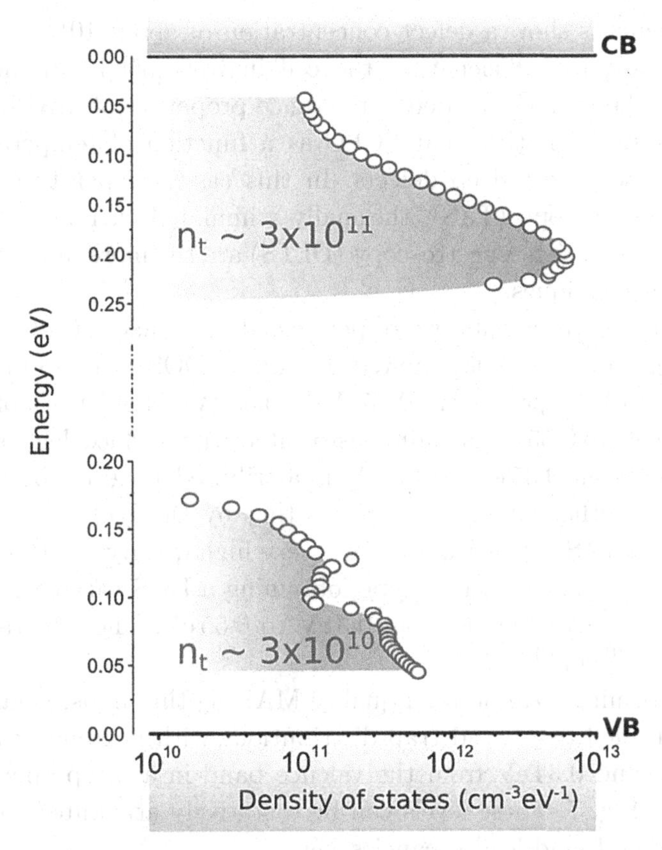

Figure 6. Density of the trap states within the bandgap of a $MAPbI_3$ single crystal. Reprinted with permission from Ref. [63]. Copyright (2016) John Wiley and Sons.

values are consistent with the theoretical calculations, with the possible identification of the acceptor states with lead vacancies V_{Pb} and the donor states with iodine vacancies V_I.

The concentration of traps found by Adinolfi *et al.* [63] is consistent with other works [62, 64], where the SCLC is used only at room temperature to estimate the total defect concentration and not their density of states as a function of the energy. Instead, as expected, $MAPbI_3$ thin films show larger defect concentration, due to the presence of grain boundaries [65]. However, even though SCLC measurements show a defect concentration of about $10^{16}\,\mathrm{cm}^{-3}$ [65], surprisingly, the efficiency of the cell in Ref. [65] is around 20%, showing the remarkable defect tolerance properties of $\mathrm{MAPbI_3}$.

It is worth noting that SCLC as a function of temperature is not able to detect deep defects. In this case, temperature admittance spectroscopy (TAS), thermally stimulated current (TSC), or deep level transient spectroscopy (DLTS) are the more indicated and common techniques.

TAS measurements were performed by Shao *et al.* [66] on $\mathrm{MAPbI_3}$ thin films. They showed that defect DOS varies from 10^{11} to $10^{13}\,\mathrm{cm}^{-3}\,\mathrm{eV}^{-1}$ going from 0.35 eV to 0.55 eV. The density of states from 0.4 eV to 0.55 eV is mainly associated with surface defects, while the DOS from 0.35 eV to 0.4 eV is attributed to grain boundaries defects. Another paper on TAS was that by Dong *et al.* [64]. They performed TAS measurements on a very high quality $\mathrm{MAPbI_3}$ single crystal, that was slightly p-type, obtaining a DOS varying from 10^9 to $10^{10}\,\mathrm{cm}^{-3}\,\mathrm{eV}^{-1}$ going from 0.3 eV to 0.55 eV. The integral value is $3.6 \times 10^{10}\,\mathrm{cm}^{-3}$.

Performing TAS on high-quality $\mathrm{MAPbI_3}$ thin films, Duan *et al.* [67] identified two broad trap distributions with maxima at around 0.165 eV and 0.34 eV from the valence band in a p-type perovskite film. See Fig. 7. These levels can be tentatively attributed to iodine interstitials $\mathrm{I_i}$ and lead vacancies V_{Pb}.

Heo *et al.* [68], using TAS on $\mathrm{MAPbI_3}$ solar cells (thin films), found a trap level at 0.27 eV with a concentration that can be reduced by one order of magnitude after iodic acid treatment. After the treatment, the trap density of states shift at 0.28 eV indicating, maybe, the presence of two close but different trap levels.

Another technique able to measure defect density and concentration is the Thermally Stimulated Current (TSC). Baumann *et al.* [69] performed TSC on a $\mathrm{MAPbI_3}$ solar cell (thin film) with an efficiency of 10%. They found a lower limit for the defect density of about $10^{15}\,\mathrm{cm}^{-3}$ and an energy level of this trap located at about 500 meV from VB or CB. We can tentatively assign it to iodine interstitials $\mathrm{I_i}$ or lead vacancies V_{Pb}. The only other study to date using TSC is that by Gordillo *et al.* [70] on $\mathrm{MAPbI_3}$ thin films. They found

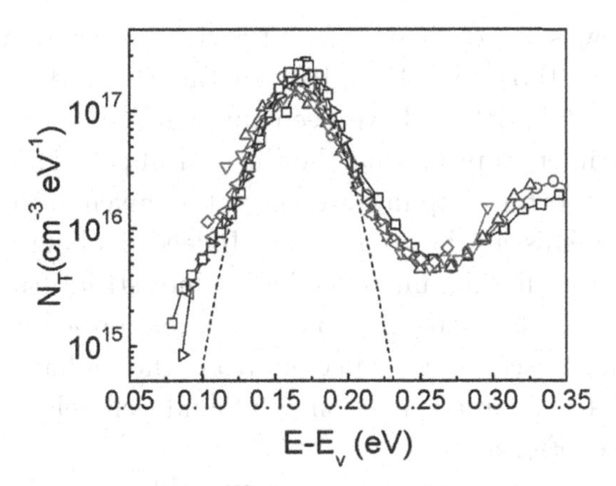

Figure 7. Density of trap states measured on a high-quality $MAPbI_3$ thin film by TAS. Reprinted with permission from Ref. [67]. Copyright (2015) Royal Society of Chemistry.

that oxygen plays a fundamental role in the defect formation on the $MAPbI_3$ samples. At low oxygen pressure, the only trap levels found are at 0.18 eV and 0.49 eV (from VB or CB cannot be determined by TSC). The 0.49-eV level is the same as that found by Baumann *et al.*, while the other is similar to that found by Duan *et al.* [67] by TAS.

DLTS measurements on pure $MAPbI_3$ were performed only by Heo *et al.* [71] on p-type thin films, finding two deep levels at 0.62 eV and 0.75 eV from the conduction band, using solar cells with an efficiency around 17%. The authors attribute these levels to I_{Pb} and I_{MA}, respectively, according to the *ab initio* calculations of Yin *et al.* [46], which, however, do not take into account, as described above, the spin–orbit coupling. It is important to note that, as pointed out by Futscher *et al.* [72], DLTS measurements can be misinterpreted if the ion migration is very fast. Indeed, as we will see below, fast ion migration is a fundamental property of $MAPbI_3$.

An alternative approach by Landi *et al.* [73], based on the noise spectroscopy of the random current fluctuations measured at different temperatures and for different illumination levels in $MAPbI_3$ solar cells, found a level of about 0.8 eV below the conduction band.

Up to now, some of the important review papers on the $MAPbI_3$ trap defects are those by [74, 76, 101]. In Table 2, all the experimental works on $MAPbI_3$ defects have been summarized.

An experimental proof that the theoretical calculations based on HSE and spin-orbit coupling are correct is given by the expected luminescence emission below the gap. Indeed, looking at the calculated configuration diagram of the iodine interstitial (see Fig. 8(A)), as explained in Ref. [56], for low excitation power a broad non-negligible luminescence is expected from the radiative transition $\mathrm{I}_\mathrm{i}^0 + e^- \rightarrow \mathrm{I}_\mathrm{i}^- + \gamma$. A broad emission was found well below the bandgap transition (see Fig. 8(B)).

Other less direct experiments are available, giving some insights about the nature of the $MAPbI_3$ defects [75, 76]. In particular, the exposure of $MAPbI_3$ to I_2 gas induces the following [75]: (i) irreversible p-type doping of the perovskite with an increased hole conductivity; (ii) formation of metallic Pb on the surface, as revealed by XPS experiments; and (iii) PL quenching. On the basis of calculations reported above for indium-rich conditions, it is clear that the exposure to I_2 gas generates a larger concentration of iodine interstitials and lead vacancies. The increase of iodine interstitials explains the PL quenching while the increase of lead vacancies generates a larger p-type doping and a segregation of Pb on the surface.

It is clear from this brief review that the results about direct measurements of $MAPbI_3$ defects are few and quite scattered. Therefore, more accurate analyses are necessary to have a better view of the defects physics of $MAPbI_3$ and perovskites in general. This effort is made more difficult by the low stability of $MAPbI_3$, its phase transitions, the several forms in which it is used (single crystal, nanocrystals, and polycrystalline thin films), and also by the many possible defects. However, this challenge must be dealt with if it is believed that a deeper understanding of the physics and chemistry of $MAPbI_3$ defects can bring about an improvement in the efficiency and stability of perovskite solar cells and devices in general. A thorough study on defect properties is necessary, clearly comparing results obtained by several experimental techniques (DLTS, TSC, TAS, etc.) performed on the same set of samples, including single crystals and high-quality

Table 2. Trap density and trap energy levels for $MAPbI_3$ defects reported in the literature.

Experimental method	Reference	Type of sample	Defect energy level (eV)	Trap density (cm^{-3})	Notes
SCLC vs T	Adinolfi *et al.* [63]	SC — n-type	0.2 (CB)	3×10^{11}	
			0.1 (VB)	3×10^{10}	
TAS	Shao *et al.* [66]	TF — p-type	0.4–0.55 (VB)	10^{10}–10^{12}	Surface traps
			0.35–0.4 (VB)	10^{9}–10^{10}	Grain bound
	Dong *et al.* [64]	SC — p-type	0.3–0.55 (VB)	3.6×10^{10}	
	Duan *et al.* [67]	TF — p-type	0.165 (VB)	10^{16}	
			0.34 (VB)	3×10^{15}	
	Heo *et al.* [68]	TF	0.275 (VB)	10^{16}–10^{17}	
TSC	Baumann *et al.* [69]	TF	0.5	$>10^{15}$	
	Gordillo *et al.* [70]	TF	0.18	9.1×10^{16}	
			0.49	4.8×10^{16}	
DLTS	Heo *et al.* [71]	TF — p-type	0.62 (CB)	1.3×10^{15}	$\sigma = 2.5 \times 10^{-15}$ cm^2
			0.75 (CB)	3.9×10^{14}	$\sigma = 2.7 \times 10^{-15}$ cm^2
Noise Spectr.	Landi *et al.* [73]	TF	0.8 (CB)	2×10^{15}	

Notes: SC: single crystal, TF: thin film. VB/CB: value referenced to the valence band maximum/conduction band minimum (TSC cannot determine if the activation energy is relative to the VB or CB). σ is the carrier capture cross-section.

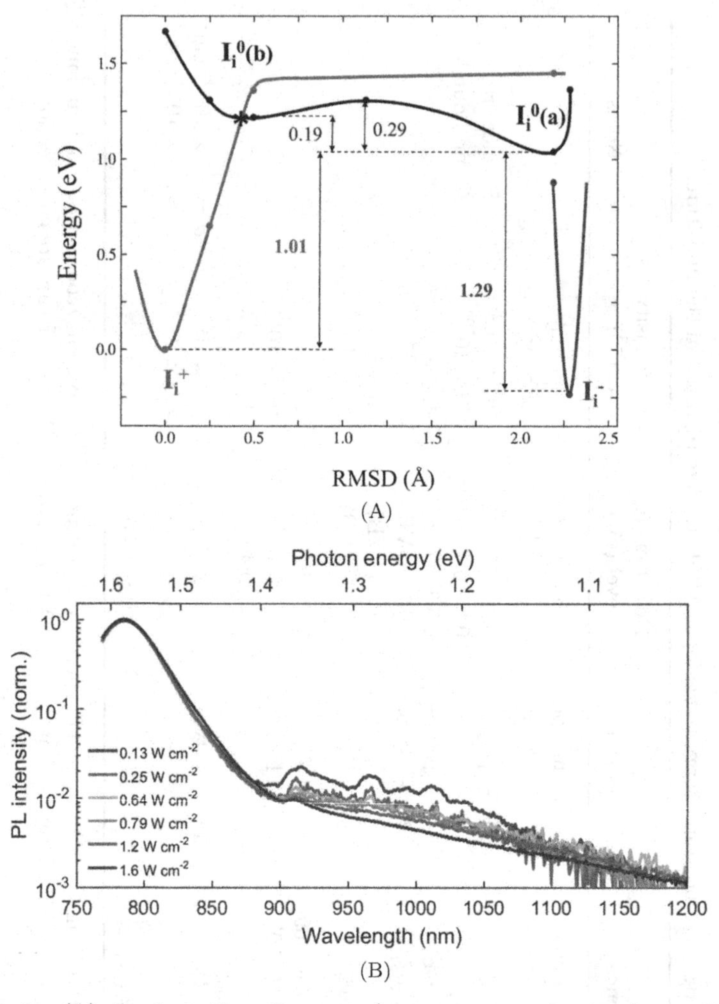

Figure 8. (A) Configuration diagrams (energy versus structural coordinates) of interstitial iodine in its positive (red), neutral (black), and negative (blue) charge states versus the root mean square displacement (RMSD) of the defect iodine atoms with respect to their position in I_i^+. The energies for the positive species are referenced to the VB maximum, while those of the negative species are referred to the CB minimum, simulating the energetics after photoexcitation (before recombination). Two possible radiative transitions between I_i^0 and I_i^- are shown by yellow arrows. (B) PL emission where the emission below the band gap energy is clearly visible. Reprinted, after some modifications, with permission from Ref. [56]. Copyright (2018) Royal Society of Chemistry.

thin films, as a function of growth conditions (stoichiometry, p- or n-type) and post-growth treatments.

After this brief review of the direct identification of trap defects in $MAPbI_3$, we want to discuss how the most commonly observed phenomena related to $MAPbI_3$ defects are explained in the literature.

The first one, in order of importance, is the defect tolerance property shown by $MAPbI_3$ (high V_{OC} and high PL efficiency despite the defect concentration). Shallow defects may be abundant, in particular due to off-stoichiometric conditions during film formation [61] or due to strain [77]; however, it is known that shallow defects may trap electrons or holes but they are not efficient at trapping both, as explained by the Shockley–Read–Hall theory. On the contrary, deep defects can be very detrimental but we have shown above that very few of them have low formation energies (mainly only the iodine interstitials). However, the trap density in thin films is usually similar to that of other semiconductors for solar cells, without a significant impact on the photovoltaic conversion efficiencies. A first factor, limiting the efficacy of deep defects, is the high dielectric constant of $MAPbI_3$, about 24, which leads to a polaron screening of the charged defects and therefore to a reduction of the non-radiative recombination [78]. As explained above, another factor limiting the detrimental effect of iodine interstitials (the only deep defects present), in particular the I_i^+, is their natural behavior to become inactive after electron capture [79]. In addition, Meggiolaro *et al.* [75, 79] suggested that moderate quantities of dry oxygen can effectively passivate the I_i^- iodine interstitials (whereas too much oxygen is detrimental [75]). The possible defect complexes that can be formed between iodine and oxygen lower the hole-trap level of I_i^- closer to the valence band (up to 0.25 eV), reducing its non-radiative efficiency. While all these effects limit the detrimental effect of the only deep defect in $MAPbI_3$, we have to point out that $MAPbI_3$ is also tolerant against other defects. For example, as predicted by Walsh *et al.* [49] and experimentally demonstrated by Steirer *et al.* [80], $MAPbI_3$ turns out to be very tolerant to MA and I loss by the self-compensating mechanism of V_{MA} and V_I with the concurrent formation of paired complexes, Schottky defects, with no effect on the structure and electronic properties of $MAPbI_3$ up to a very large stoichiometry deviation.

Another very important aspect of the $MAPbI_3$ defects is their very high diffusion coefficient and therefore their high mobility. This has several detrimental effects for proper working of the devices, such as hysteretic *I–V* characteristics. Several explanations were proposed for the hysteresis observed in *I–V* characteristics of $MAPbI_3$ solar cells: ferroelectricity, trapping–detrapping from bulk and surface defects, ion migration [81]. Now it is accepted in the literature that only ion migration is considered the cause of the hysteresis. Ion migration is also considered as a main factor in many other interesting phenomena [76] observed in $MAPbI_3$: giant switchable photovoltaic effect, giant dielectric constant, diminished transistor behavior at room temperature, photo-induced phase separation and the self-poling effect, and electrical field-driven reversible conversion between PbI_2 and $MAPbI_3$ phases. Updated reviews on ion migration can be found in Ref. [72, 76, 81]. In Fig. 9, a typical *I–V* showing the hysteresis is presented [82]. The hysteresis is due to

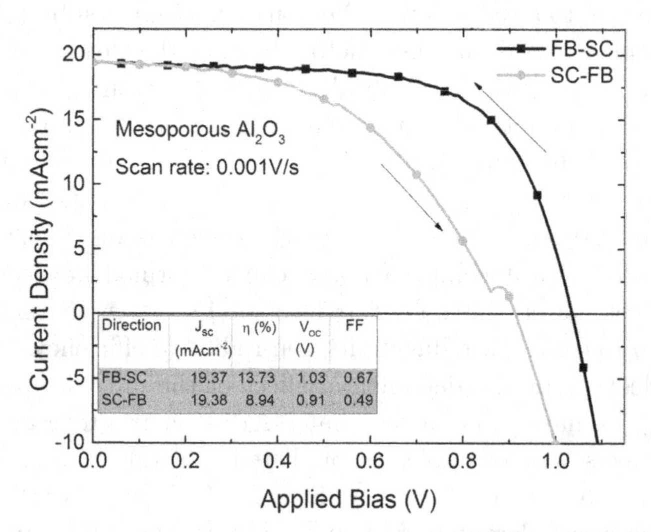

Direction	J_{sc} (mAcm^{-2})	η (%)	V_{oc} (V)	FF
FB-SC	19.37	13.73	1.03	0.67
SC-FB	19.38	8.94	0.91	0.49

Figure 9. Current–voltage, *I–V*, characteristics showing the typical hysteresis observed in perovskite based solar cells. Reprinted with permission from Ref. [82]. Copyright (2014) American Chemical Society.

the migration of ions under an external electrical field toward the opposite interfaces (perovskite/ETL and perovskite/HTL, respectively). These accumulated ions result in both a change of the internal field and a modulation of the interfacial barriers, thus giving rise to a hysteretic behavior [81]. Since the ion migration proceeds through defects, a material without defects does not show any hysteresis. This can be obtained with several surface treatments or specific growth approach and it also brings out the highest photovoltaic efficiencies [72, 81]. Moreover, once the defects accumulate at the transport layer interfaces, irreversible chemical processes can take place, with detrimental effects. This situation can be avoided with specific blocking layers between the transport layers and the perovskite [72, 81].

Among the various possible ions migrating throughout the perovskite layer (I^-, Pb^{2+}, MA^+), it is now generally accepted, both by theoretical predictions and experimental direct evidences, that I^- ions are the most mobile and dominate ion migration [57, 72, 81]. The movement of ions proceeds by a hopping mechanism between the atomic lattices via defect states (i.e. vacancies, interstitials, etc.) [81] and this hopping process to the neighboring sites requires energy to overcome the barriers (activated process). There is a large scattering of both the data and the predictions about the migration activation energies of these defects. The reader is referred to the most recent review articles on this topic [72, 81] for a comparison of the values found in the literature. However, the general consensus is that the fastest ion species is that of iodine, followed by the MA and finally by Pb.

As a final remark, we want to point out that, as reviewed by Ran *et al.* [76], the presence of point defects influences the $MAPbI_3$ chemical stability as well, for example, the presence of some particular defects can increase the instability in air, in particular with moisture, under light irradiation, and under thermal treatments.

In the last part of this section, we will briefly discuss the current experimental knowledge about grain boundaries (GBs) and surfaces in $MAPbI_3$. A general review paper on the effect of grain boundaries in perovskite solar cells is that by Lee *et al.* [83], while a general review paper on $MAPbI_3$ surfaces and interfaces is that by Schulz *et al.* [102].

In solar cells, LEDs, and other devices, $MAPbI_3$ is typically used in the form of thin films, and therefore, GBs are obviously present. Even though the grain size affects the performances of solar cells, this detrimental effect is less pronounced in $MAPbI_3$ with respect to the other materials (CdTe, Si, etc.). This can be explained by the work of Yin *et al.* [86] described above, where no deep defects are introduced by grain boundaries but only shallow levels that can increase the effective mass of the holes, thus reducing the diffusion length of these carriers. However, several experimental studies were performed: Kelvin probe force microscopy and AFM/conductive-AFM proved that in $MAPbI_3$ the grain boundaries, thanks to the charged state of their defects, are characterized by a band bending with a built-in potential that helps the carrier collection and reduces the non-radiative recombination [87, 88], similar to what happens in $CuIn_xGa_{1-x}Se_2$ (CIGS) and other chalcogenide compounds. Similar analyses also pointed out that at the nanoscale, the photovoltaic parameters also exhibit an intra-grain change, not only between grains but also on the grain boundaries, ascribing this difference to the different defect densities in each different facet, thus suggesting that the control of facet orientation may allow to optimize polycrystalline films for photovoltaic and lighting applications [89]. Time-resolved micro-PL investigations have shown instead a faster recombination at the grain boundaries with respect to the internal part of the grains [90, 91], even though these results are strongly affected by the stoichiometric deviation of the $MAPbI_3$ [92]. Finally, it was experimentally shown that, as expected, ion migration is faster along grain boundaries and their presence enhances all the effects related to ion migration [83].

In $MAPbI_3$, numerous studies have addressed how surface passivation can modify the carrier diffusion length [59, 83, 93] and therefore the solar cell. Many studies by XPS were conducted, trying to understand the chemistry of the exposed surfaces of $MAPbI_3$ [94], their degradation [95], and therefore their relation with the performance of the solar cells.

3. Defects in $CsPbBr_3$

3.1. *Theoretical calculations*

Kang and Wang [96] calculated the defect formation energies (DFE) and thermodynamic transition energy levels using an HSE–SOC approach.

In Fig. 10(A), we report the stability diagram of $CsPbBr_3$ taken from Ref. [96]. It is calculated based on the following conditions:

$$\mu(\mathrm{Cs}) + \mu(\mathrm{Pb}) + 3\mu(\mathrm{Br}) = \mu(\mathrm{CsPbBr_3}),$$
$$\mu(\mathrm{Cs}) + \mu(\mathrm{Br}) < \mu(\mathrm{CsBr}),$$
$$\mu(\mathrm{Pb}) + 2\mu(\mathrm{Br}) < \mu(\mathrm{PbBr_2}).$$

As usual, three points are typically selected for DFE calculations: A, B, and C. A corresponds to Br-rich Pb-poor, C corresponds to Br-poor Pb-rich, and B corresponds to intermediate conditions between A and C.

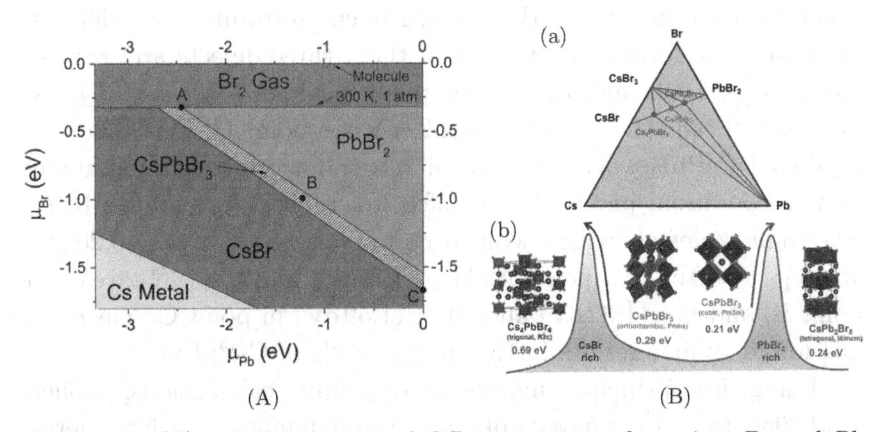

Figure 10. (A) stability regions of different compounds against Br and Pb chemical potentials. Reprinted with permission from Ref. [96]. Copyright (2017) American Chemical Society. (B) (a) Schematic representation of the ternary phase together with paths between precursors (CsBr and $PbBr_2$) and highlighted compounds ($CsPbBr_3$, $CsPb_2Br_5$, and Cs_4PbBr_6). (b) Crystal structures and decomposition energies of orthorhombic- and cubic-phase $CsPbBr_3$, tetragonal-phase $CsPb_2Br_5$, and trigonal-phase Cs_4PbBr_6. Reprinted with permission from Ref. [97]. Copyright (2018) American Chemical Society.

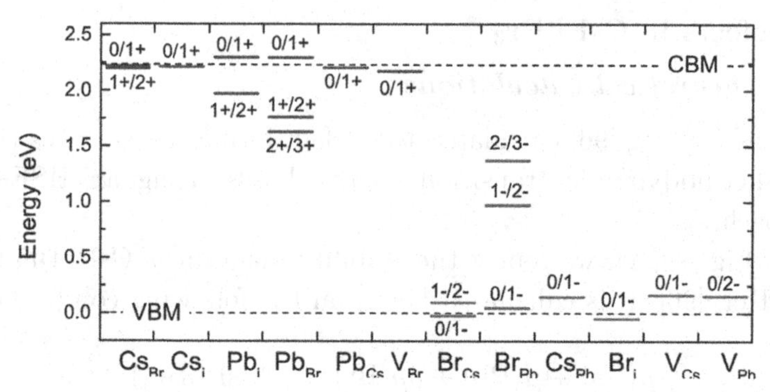

Figure 11. Defect charge-transition levels calculated by HSE–SOC. Reprinted with permission from Ref. [96]. Copyright (2017) American Chemical Society.

Actually, as summarized by Yin *et al.* [97], other phases can be formed. This is clear from the ternary diagram reported in Fig. 10(B). We will not consider these spurious phases here, but we will limit ourselves to $CsPbBr_3$ and a small deviation from its stoichiometry.

The results of Kang and Wang [96] are reported in Fig. 11, where it is clear that only Pb_{Br}, Br_{Pb}, and Pb_i can introduce deep defects. Moreover, the formation energies of these three defects are, respectively, 6.28, 1.40, and 4.66 eV for point A, 4.38, 3.30, and 3.44 eV for point B, and 2.48, 5.20, and 2.22 eV for point C. It is therefore clear that $CsPbBr_3$ deep defects will have a low concentration in any growth condition, justifying the defect tolerance nature of $CsPbBr_3$. The shallow defects with lowest formation energies are V_{Cs} (0.20 eV) and V_{Pb} (0.49 eV) in point A, V_{Cs} (0.98 eV) and Cs_{Pb} (1.26 eV) in point B, and V_{Br} (1.32 eV) and Pb_{Cs} (1.59 eV) in point C. Yin *et al.* [97] found similar results in the energy levels of $CsPbBr_3$.

Using first-principle molecular dynamics calculations, Cohen *et al.* [98] studied the impact of structural dynamics on defect energy levels. Usually, these energy levels have always been considered static. The authors focus on V_{Br} energy level and they found that this level is not fixed in time as considered usually, but it oscillates by as much as 1 eV on the picosecond time scale. It is likely that this can be extended to many other defect levels and different perovskites, and it could have an important impact on the effects of the defects in these materials.

Ab initio calculations on $CsPbBr_3$ grain boundaries were performed by Guo *et al.* [99]. The authors simulated the $\Sigma3(111)$, $\Sigma3(112)$, $\Sigma5(210)$, and $\Sigma5(310)$ grain boundaries. The calculated GB energy shows that the most stable configuration is $\Sigma5(210)$ (0.4,0), where (0.4,0) indicates the shifting vector in unit of lattice constant of one of the two grains with respect to the other. The electronic properties confirm that all the GBs in halide perovskites do not create midgap states in fundamental bandgap, because of the large distance between unsaturated atoms and the atomic relaxation in forming the stable GB configurations.

Thind *et al.* [100] found, by STEM experiments and DFT calculations, that two types of prevalent planar defects from atomic resolution imaging are observed: Br-rich [001](210) $\Sigma5$ GBs and Ruddlesden–Popper (RP) planar faults. The *ab initio* calculations reveal that neither of these planar faults induce deep defect levels, but their Br-deficient counterparts do. It is found that the $\Sigma5$ GB repels electrons and attracts holes, similar to an n–p–n junction, and the RP planar defects repel both electrons and holes, similar to a semiconductor–insulator–semiconductor junction.

We are not aware of any study relating to *ab initio* calculations of $CsPbBr_3$ surfaces and defect surfaces.

3.2. *Effects and properties of $CsPbBr_3$ defects*

$CsPbX_3$ (X = I, Br, Cl) perovskites are relevant not only for photovoltaics but also for LED and lasers: therefore, the theoretical and experimental investigations mostly address the quenching of radiative recombination related to non-radiative defects. In particular, in nanocrystals, these studies focus on the surface since it plays a fundamental role. Single crystals are, instead, fundamental to investigate bulk defects. Single crystals of $CsPbBr_3$, being completely inorganic, are easier to grow than single crystals of $MAPbI_3$. Single crystals of $CsPbBr_3$ can be prepared with different methods [103, 104]: Bridgman, low-temperature solution (inverse-temperature crystallization, antisolvent vapor-assisted crystallization), and electronic dynamic gradient.

Very few measurements to identify the defect density and defect energy levels have been performed to date. Vitale *et al.*

[105] employed thermal admittance spectroscopy on a nanocrystalline $CsPbBr_3$ thin-film obtained by co-evaporation. However, the authors focused on only transport mechanism and dielectric relaxation processes.

The next known report is a conference proceeding of Chung *et al.* [106]. They performed DLTS, thermally stimulated luminescence (TSL), and photoluminescence (PL) on $CsPbBr_3$ single crystals. The data have been published as an online video of the conference and are not accessible.

He *et al.* [107] used TSC on a very high quality $CsPbBr_3$ p-type single crystal. The authors found the following trap energy levels and concentrations: $1.25 \times 10^{15}\,cm^{-3}$ at 0.18 eV, $2.10 \times 10^{15}\,cm^{-3}$ at 0.21 eV, $5.29 \times 10^{15}\,cm^{-3}$ at 0.23 eV, and $0.41 \times 10^{15}\,cm^{-3}$ at 0.28 eV. We want to recall the fact that TSC cannot determine whether the measured energy levels are with respect to the valence or conduction band. The authors do not try to attribute these levels to any known calculated level or to any impurity found in the sample (they also performed a very detailed and sensitive glow discharge mass spectrometry), but they simply conclude that no deep levels are found in their sample.

Zhang *et al.* [108] used TSC on $CsPbBr_3$ single crystals, grown in three different conditions: 1% Pb-poor, stoichiometric, and 1% Pb-rich. They investigated the temperature range 20–300 K. Five possible intrinsic point defects, namely, V_{Cs} and V_{Br} vacancies, Cs_i and Pb_i interstitials, and Pb_{Br} antisites, were found in the $CsPbBr_3$ crystal. A broad emission, in the range 20–160 K, was attributed to the presence of several energy levels in the range from 0.02 eV to 0.31 eV, characterized by cross sections in the range 10^{-17}–$10^{-14}\,cm^2$ and concentrations of the order of 10^{12}–$10^{14}\,cm^{-3}$. Another set of peaks, in the range 200–300 K, were attributed to energy levels at 0.48 eV–0.53 eV that were probably lead related.

The most recent study on microcrystalline $CsPbBr_3$ thin films by Bruzzi *et al.* [109] analyzed the TSC emissions with beta-variation and delayed heating methods, which were able to evidence that such broad emissions are a signature of the very low capture cross sections of the involved defects, estimated down to $10^{-22}\,cm^2$. Even though the trap concentrations are high, up to $10^{16}\,cm^{-3}$, these small capture cross sections significantly reduce the impact of defects on the optoelectronic performance of $CsPbBr_3$, justifying its high defect tolerance. Below room temperature, TSC emissions are dominated by a quasi-continuum distribution of energy levels in the range, 0.11–0.27 eV, with almost a constant density of states. In the range 300–400 K, of interest for solar cells and LEDs applications, TSC is mainly due to energy levels at 0.40–0.45 eV (they too are characterized by very low capture cross sections, as mentioned above). Such defects appear to dominate room temperature photoconductivity and the background current of $CsPbBr_3$ samples. In Table 3, all the experimental works on $CsPbBr_3$ defects have been summarized.

Bruzzi *et al.* [109] provided for this chapter some unpublished figures related to their paper about TSC measurements on $CsPbBr_3$ microcrystalline thin film to highlight how these kind of measurements are carried out (see Fig. 12). In Fig. 12(a) the temperature profile is reported. The figure shows three regions: filling, heating and cooling. First, the current (see Fig. 12(b)) is measured during the filling process, obtained by shining on the sample, biased with a constant voltage of 5 V, a 0.8 mW, 400 nm LED, for a filling time of 640 s at a fixed temperature of about 300 K. Then, the sample is placed in dark and the temperature is increased with a fixed constant rate, about 0.08 K/s, up to 390 K (heating region). During heating the current first increases (see inset of Fig. 12(b)), then decreases, showing a peak of emission at about 360 K. The decrease of the current is due to the exhaustion of traps. Peak parameters as maximum temperature, temperature at the maximum, peak intensity, intensity at the maximum, and peak FWHM, can be used to estimate the energy level position in the forbidden gap, the trap capture cross section, and the concentration of the trap. In this case it has been shown [109] that a single trap, with energy level about 0.45 eV, and a very low cross section, about $10^{-22}\,cm^2$, is dominating the TSC emission. During cooling, the current decrease, almost exponential with temperature, is thermally activated by an energy around 0.4 eV. Nonetheless, due to the complexity often encountered in TSC spectra, various methods of analysis frequently need to be applied. Some

Table 3. Trap density and trap energy levels for $CsPbBr_3$ defects reported in the literature.

Experimental method	Reference	Type of sample	Defect energy level (eV)	Trap density (cm^{-3})	Notes
DLTS	Chung *et al.* [106]	SC			Not accessible
TSC	He *et al.* [107]	SC — p-type	0.18	1.25×10^{15}	
			0.21	2.10×10^{15}	
			0.23	5.29×10^{15}	
			0.28	0.41×10^{15}	
	Zhang *et al.* [108]	SC	0.028	0.63×10^{13}	Cs_i
			0.055	1.53×10^{13}	Pb_{Br}
			0.070	1.48×10^{13}	V_{Br}
			0.138	9.45×10^{13}	Cs_{Pb}
			0.160	10.7×10^{13}	V_{Cs}
			0.217	3.92×10^{13}	Pb_i
	Bruzzi *et al.* [109]	TF	0.4		$\sigma \approx 10^{-22}$ cm^2

Notes: SC: single crystal, TF: thin film. (TSC cannot determine if the activation energy is relative to the VB or CB). σ is the carrier capture cross section. The concentrations related to the paper of Zhang *et al.* [108] refers to the stoichiometric condition except for the Cs_{Pb} defect.

of them are purely based on summing the emission of a set of energy levels independent of each other in a SIMultaneous-Peak Analysis (SIMPA). More accurately, a set of experiments is needed (as, e.g. in beta-variation and delayed-heating methods [109]), where selected parameters such as heating rate, filling time, and filling temperature are opportunely changed, in view to cross-correlating the resulting TSC spectra in an attempt to separate as much as possible emissions coming from overlapping peaks.

Grain boundaries in $CsPbBr_3$ were studied only in few papers [100, 110, 111]. As already explained in the previous section, Thind *et al.* [100] found, by STEM experiments and DFT calculations, that two types of prevalent planar defects from atomic resolution imaging

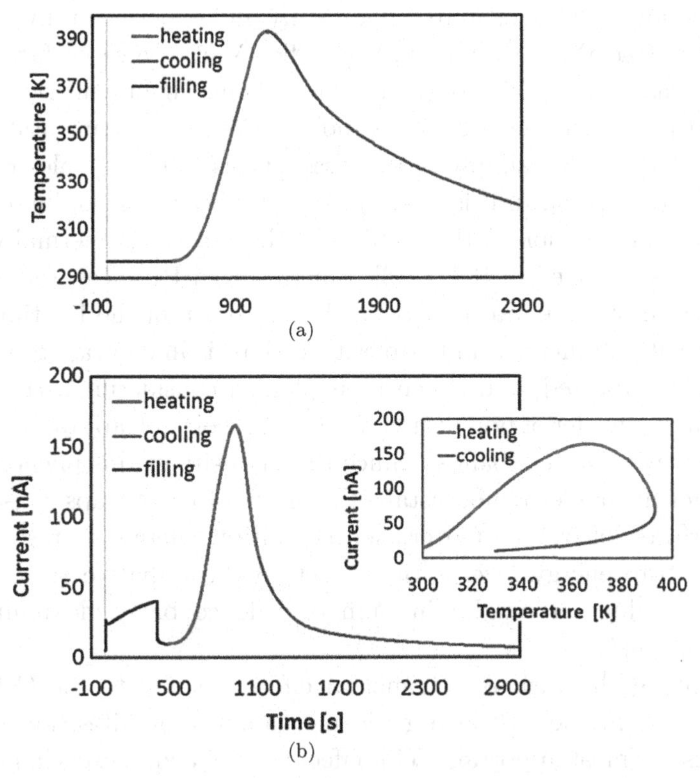

Figure 12. TSC measurement on a $CsPbBr_3$ polycrystalline thin film. Kindly provided by Bruzzi *et al.* (unpublished graphs whose results have been published in [109]). See text for explanations.

are observed: Br-rich [001](210) $\Sigma5$ GBs and Ruddlesden–Popper (RP) planar faults. The *ab initio* calculations reveal that neither of these planar faults induce deep defect levels, but their Br-deficient counterparts do. It is found that the $\Sigma5$ GB repels electrons and attracts holes, similar to an n–p–n junction, and the RP planar defects repel both electrons and holes, similar to a semiconductor–insulator–semiconductor junction.

Surface defects and their effects have been studied by several authors. This topic is particularly important since $CsPbBr_3$ nanoparticles or quantum dots are often used directly into the devices or as starting point for thin film formation [112, 122]. Many surface passivation techniques were theoretically and experimentally investigated in $CsPbX_3$ [113–116] and the efficacy of oxygen surface treatment was recently demonstrated to enhance light emission [117]. Nevertheless, surface states are not necessarily detrimental, since some of them can simply act as trap states only for holes or only for electrons [118]. In this case, the captured carriers can be released by thermal emission in the bands and the process is thermally activated. As reported in Ref. [118], time-resolved PL measurements as a function of temperature show a slowing down of the PL time evolution with an increase in temperature. This is in contrast to what is typically expected, and it can be explained considering a thermally activated transfer of population from a reservoir of trap states toward the radiative states (bands). Since the intensity of this phenomenon is larger in smaller nanocrystals, the origin of these traps is ascribed to surface defects. Moreover, the activation energy, corresponding to the trap energy level with respect to the radiative state (most likely conduction band minimum or valence band maximum), is about 40 meV.

Among the various detrimental effects related to the $CsPbBr_3$ defects, as in the hybrid organic–inorganic perovskites, ion migrations is the most important. This effect is detrimental especially when mixed inorganic perovskites are used, for example, in order to tune the emission wavelength: the field-driven anion exchange can separate the two phases and therefore completely change the emission color [119–121].

Acknowledgments

This work was partially supported by Fondazione Ente Cassa di Risparmio di Firenze within the project PERBACCO ("Stabilized perovskites for high-efficiency eco-friendly light emitters and solar cells"; no. 2016.1084), within the project EPICO ("Eco-Perovskiti Inorganiche: Crescita e proprietà Optoelettroniche"; no. 2018.0950), and within the project no. 2017.0756 "Advanced optical spectroscopy for interface control in perovskite-based solar cells". The authors acknowledge Prof. Bogani and Dr. Caporali for critical reading of the manuscript and suggestions.

References

[1] International Energy Agency (IEA), Key World Energy Statistics (2018).
[2] International Energy Agency (IEA), Photovoltaic Power Systems Programme. Annual Report 2017 (2017).
[3] A. Luque, S. Hegedus (eds.), *Handbook of Photovoltaic Science and Engineering* (2011).
[4] International Energy Agency (IEA), Light's Labour's Lost (2006).
[5] International Energy Agency (IEA), World Energy Outlook 2018 (2018).
[6] T. Q. Khanh, P. Bodrogi, Q. T. Vinh, H. Winkler (eds.), LED Lighting: Technology and Perception (2015).
[7] N.-G. Park, M. Grätzel, T. Miyasaka (eds.), Organic–Inorganic Halide Perovskite Photovoltaics (2016).
[8] J. S. Manser, J. A. Christians, P. V. Kamat, Intriguing optoelectronic properties of metal halide perovskites, *Chem. Rev.* **116**, 12956 (2016).
[9] Q. Chen, N. D. Marco, Y. M. Yang, T.-B. Song, C.-C. Chen, H. Zhao, Z. Hong, H. Zhou, Y. Yang, Under the spotlight: The organic–inorganic hybrid halide perovskite for optoelectronic applications, *Nano Today* **10**, 355 (2015).
[10] Z. Xiao, Y. Yuan, Q. Wang, Y. Shao, Y. Bai, Y. Deng, Q. Dong, M. Hu, C. Bi, J. Huang, Thin-film semiconductor perspective of organometal trihalide perovskite materials for high-efficiency solar cells, *Mater. Sci. Eng. R* **101**, 1 (2016).
[11] S. D. Wolf, J. Holovsky, S.-J. Moon, P. Löper, B. Niesen, M. Ledinsky, F.-J. Haug, J.-H. Yum, C. Ballif, Organometallic halide perovskites: Sharp optical absorption edge and its relation to photovoltaic performance, *J. Phys. Chem. Lett.* **5**, 1035 (2014).
[12] P. Löper, M. Stuckelberger, B. Niesen, J. Werner, M. Filipič, S.-J. Moon, J.-H. Yum, M. Topič, S. D. Wolf, C. Ballif, Complex refractive index spectra of $CH_3NH_3PbI_3$ perovskite thin films determined by spectroscopic ellipsometry and spectrophotometry, *J. Phys. Chem. Lett.* **6**, 66 (2015).
[13] W.-J. Yin, T. Shi, Y. Yan, Unique properties of halide perovskites as possible origins of the superior solar cell performance, *Adv. Mater.* **26**, 4653 (2014).
[14] Q. Dong, Y. Fang, Y. Shao, P. Mulligan, J. Qiu, L. Cao, J. Huang, Electron-hole diffusion lengths >175 μm in solution-grown $CH_3NH_3PbI_3$ single crystals, *Science* **347**, 967 (2015).
[15] J. Song, Q. Cui, J. Li, J. Xu, Y. Wang, L. Xu, J. Xue, Y. Dong, T. Tian, H. Sun, H. Zeng, Ultralarge all-inorganic perovskite bulk single crystal for high-performance visible-infrared dual-modal photodetectors, *Adv. Opt. Mater.* **5**, 1700157 (2017).
[16] S. D. Stranks, G. E. Eperon, G. Grancini, C. Menelaou, M. J. P. Alcocer, T. Leijtens, L. M. Herz, A. Petrozza, H. J. Snaith, Electron-hole diffusion lengths exceeding 1 micrometer in an organometal trihalide perovskite absorber, *Science* **342**, 341 (2013).

[17] G. R. Yettapu, D. Talukdar, S. Sarkar, A. Swarnkar, A. Nag, P. Ghosh, P. Mandal, Terahertz conductivity within colloidal $CsPbBr_3$ perovskite nanocrystals: Remarkably high carrier mobilities and large diffusion lengths, *Nano Lett.* **16**, 4838 (2016).

[18] G. Giorgi, J.-I. Fujisawa, H. Segawa, K. Yamashita, Small photocarrier effective masses featuring ambipolar transport in methylammonium lead iodide perovskite: A density functional analysis, *J. Phys. Chem. Lett.* **4**, 4213 (2013).

[19] A. Miyata, A. Mitioglu, P. Plochocka, O. Portugall, J. T.-W. Wang, S. D. Stranks, H. J. Snaith, R. J. Nicholas, Direct measurement of the exciton binding energy and effective masses for charge carriers in organic–inorganic tri-halide perovskites, *Nat. Phys.* **11**, 582 (2015).

[20] D. Shi, V. Adinolfi, R. Comin, M. Yuan, E. Alarousu, A. Buin, Y. Chen, S. Hoogland, A. Rothenberger, K. Katsiev, Y. Losovyj, X. Zhang, P. A. Dowben, O. F. Mohammed, E. H. Sargent, O. M. Bakr, Low trap-state density and long carrier diffusion in organolead trihalide perovskite single crystals, *Science* **347**, 519 (2015).

[21] W.-J. Yin, T. Shi, Y. Yan, Superior photovoltaic properties of lead halide perovskites: Insights from first-principles theory, *J. Phys. Chem. C* **119**, 5253 (2015).

[22] J. Kang, L.-W. Wang, High defect tolerance in lead halide perovskite $CsPbBr_3$, *J. Phys. Chem. Lett.* **8**, 489 (2017).

[23] L. Protesescu, S. Yakunin, M. I. Bodnarchuk, F. Krieg, R. Caputo, C. H. Hendon, R. X. Yang, A. Walsh, M. V. Kovalenko, Nanocrystals of cesium lead halide perovskites ($CsPbX_3$, X = Cl, Br, and I): Novel optoelectronic materials showing bright emission with wide color gamut, *Nano Lett.* **15**, 3692 (2015).

[24] A. Kojima, K. Teshima, T. Miyasaka, Y. Shirai, Novel photoelectrochemical cell with mesoscopic electrodes sensitized by lead-halide compounds (2), in *210th ECS Meeting*, Abstract #397 (2006).

[25] A. Kojima, K. Teshima, Y. Shirai, T. Miyasaka, Novel photoelectrochemical cell with mesoscopic electrodes sensitized by lead-halide compounds (5), in *212th ECS Meeting*, Abstract #352 (2007).

[26] A. Kojima, K. Teshima, Y. Shirai, T. Miyasaka, Organometal halide perovskites as visible-light sensitizers for photovoltaic cells, *J. Am. Chem. Soc.* **131**, 6050 (2009).

[27] J.-P. Correa-Baena, M. Saliba, T. Buonassisi, M. Grätzel, A. Abate, W. Tress, A. Hagfeldt, Promises and challenges of perovskite solar cells, *Science* **358**, 739 (2017).

[28] National Renewable Energy Laboratory (NREL), Best research-cell efficiencies, www.nrel.gov/pv/.

[29] Y.-H. Kim *et al.*, Metal halide perovskite light emitters, *PNAS* **113**, 11694 (2016).

[30] S. A. Veldhuis *et al.*, Perovskite materials for light-emitting diodes and lasers, *Adv. Mater.* **28**, 6804 (2016).

[31] Q. V. Le, H. W. Jang, S. Y. Kim, Recent advances toward high-efficiency halide perovskite light-emitting diodes: Review and perspective, *Small Meth.* **2**, 1700419 (2018).
[32] Z.-K. Tan, R. S. Moghaddam, M. L. Lai, P. Docampo, R. Higler, F. Deschler, M. Price, A. Sadhanala, L. M. Pazos, D. Credgington, F. Hanusch, T. Bein, H. J. Snaith, R. H. Friend, Bright light-emitting diodes based on organometal halide perovskite, *Nat. Nanotechnol.* **9**, 687 (2014).
[33] Y. Cao, N. Wang, H. Tian, J. Guo, Y. Wei, H. Chen, Y. Miao, W. Zou, K. Pan, Y. He, H. Cao, Y. Ke, M. Xu, Y. Wang, M. Yang, K. Du, Z. Fu, D. Kong, D. Dai, Y. Jin, G. Li, H. Li, Q. Peng, J. Wang, W. Huang, Perovskite light-emitting diodes based on spontaneously formed submicrometre-scale structures, *Nature* **562**, 249 (2018).
[34] K. Lin, J. Xing, L. N. Quan, F. P. G. de Arquer, X. Gong, J. Lu, L. Xie, W. Zhao, D. Zhang, C. Yan, W. Li, X. Liu, Y. Lu, J. Kirman, E. H. Sargent, Q. Xiong, Z. Wei, Perovskite light-emitting diodes with external quantum efficiency exceeding 20 per cent, *Nature* **562**, 245 (2018).
[35] Z. Li, J. Moon, A. Gharajeh, R. Haroldson, R. Hawkins, W. Hu, A. Zakhidov, Q. Gu, Room-temperature continuous-wave operation of organometal halide perovskite lasers, *ACS Nano* **12**, 10968 (2018).
[36] M. I. Asghar *et al.*, Device stability of perovskite solar cells — A review, *Renew. Sustain. Energy Rev.* **77**, 131 (2017).
[37] B. Salhi, Y. S. Wudil, M. K. Hossain, A. Al-Ahmed, F. A. Al-Sulaiman, Review of recent developments and persistent challenges in stability of perovskite solar cells, *Renew. Sustain. Energy Rev.* **90**, 210 (2018).
[38] H. J. Snaith, P. Hacke, Enabling reliability assessments of pre-commercial perovskite photovoltaics with lessons learned from industrial standards, *Nat. Energy* **3**, 459 (2018).
[39] Y. Chen, L. Zhang, Y. Zhang, H. Gao, H. Yan, Large-area perovskite solar cells — A review of recent progress and issues, *RSC Adv.* **8**, 10489 (2018).
[40] S. F. Hoefler, G. Trimmel, T. Rath, Progress on lead-free metal halide perovskites for photovoltaic applications: A review, *Monatshefte für Chemie* **148**, 795 (2017).
[41] M. Konstantakou, T. Stergiopoulos, A critical review on tin halide perovskite solar cells, *J. Mater. Chem. A* **5**, 11518 (2017).
[42] Z. Li, W. Yin, Recent progress in Pb-free stable inorganic double halide perovskites, *J. Semiconduc.* **39**, 071003 (2018).
[43] C. Freysoldt, B. Grabowski, T. Hickel, J. Neugebauer, G. Kresse, A. Janotti, C. G. Van de Walle, First-principles calculations for point defects in solids, *Rev. Mod. Phys.* **86**, 253 (2014).
[44] W.-J. Yin, T. Shi, Y. Yan, Unusual defect physics in $CH_3NH_3PbI_3$ perovskite solar cell absorber, *Appl. Phys. Lett.* **104**, 063903 (2014).
[45] W.-J. Yin, T. Shi, Y. Yan, Unique properties of halide perovskites as possible origins of the superior solar cell performance, *Adv. Mater.* **26**, 4653 (2014).

[46] W.-J. Yin, T. Shi, Y. Yan, Superior photovoltaic properties of lead halide perovskites: Insights from first-principles theory, *J. Phys. Chem. C* **119**, 5253 (2015).

[47] J. Kim, S.-H. Lee, J. H. Lee, K.-H. Hong, The role of intrinsic defects in methylammonium lead iodide perovskite, *J. Phys. Chem. Lett.* **5**, 1312 (2014).

[48] M. L. Agiorgousis, Y. Y. Sun, H. Zeng, S. B. Zhang, Strong covalency-induced recombination centers in perovskite solar cell material $CH_3NH_3PbI_3$, *J. Am. Chem. Soc.* **136**, 14570 (2014).

[49] A. Walsh, D. O. Scanlon, S. Chen, X. G. Gong, S.-H. Wei, Self-regulation mechanism for charged point defects in hybrid halide perovskites, *Angew. Chem. Int. Ed.* **54**, 1791 (2015).

[50] A. Buin, P. Pietsch, J. Xu, O. Voznyy, A. H. Ip, R. Comin, E. H. Sargent, Materials processing routes to trap-free halide perovskites, *Nano Lett.* **14**, 6281 (2014).

[51] A. Buin, R. Comin, J. Xu, A. H. Ip, E. H. Sargent, Halide-dependent electronic structure of organolead perovskite materials, *Chem. Mater.* **27**, 4405 (2015).

[52] M.-H. Du, Density functional calculations of native defects in $CH_3NH_3PbI_3$: Effects of spin–orbit coupling and self-interaction error, *J. Phys. Chem. Lett.* **6**, 1461 (2015).

[53] M. H. Du, Efficient carrier transport in halide perovskites: Theoretical perspectives, *J. Mater. Chem. A* **2**, 9091 (2014).

[54] H. Shi, M.-H. Du, Shallow halogen vacancies in halide optoelectronic materials, *Phys. Rev. B* **90**, 174103 (2014).

[55] L. D. Whalley, R. Crespo-Otero, A. Walsh, H-Center and V-Center defects in hybrid halide perovskites, *ACS Energy Lett.* **2**, 2713 (2017).

[56] D. Meggiolaro, S. G. Motti, E. Mosconi, A. J. Barker, J. Ball, C. A. R. Perini, F. Deschler, A. Petrozza, F. De Angelis, Iodine chemistry determines the defect tolerance of lead-halide perovskites, *Energy Environ. Sci.* **11**, 702 (2018).

[57] D. Meggiolaro, F. De Angelis, First-principles modeling of defects in lead halide perovskites: Best practices and open issues, *ACS Energy Lett.* **3**, 2206 (2018).

[58] D. Han, C. Dai, S. Chen, Calculation studies on point defects in perovskite solar cells, *J. Semicond.* **38**, 011006 (2017).

[59] C. Ran, J. Xu, W. Gao, C. Huang, S. Dou, Defects in metal triiodide perovskite materials towards high-performance solar cells: Origin, impact, characterization, and engineering, *Chem. Soc. Rev.* **47**, 4581 (2018).

[60] H. Uratani, K. Yamashita, Charge carrier trapping at surface defects of perovskite solar cell absorbers: A first-principles study, *J. Phys. Chem. Lett.* **8**, 742 (2017).

[61] Q. Wang, Y. Shao, H. Xie, L. Lyu, X. Liu, Y. Gao, J. Huang, Qualifying composition dependent p and n self-doping in $CH_3NH_3PbI_3$, *Appl. Phys. Lett.* **105**, 163508 (2014).

[62] D. Shi, V. Adinolfi, R. Comin, M. Yuan, E. Alarousu, A. Buin, Y. Chen, S. Hoogland, A. Rothenberger, K. Katsiev, Y. Losovyj, X. Zhang, P. A. Dowben, O. F. Mohammed, E. H. Sargent, O. M. Bakr, Low trap-state density and long carrier diffusion in organolead trihalide perovskite single crystals, *Science* **347**, 519 (2015).
[63] V. Adinolfi, M. Yuan, R. Comin, E. S. Thibau, D. Shi, M. I. Saidaminov, P. Kanjanaboos, D. Kopilovic, S. Hoogland, Z.-H. Lu, O. M. Bakr, E. H. Sargent, The in-gap electronic state spectrum of methylammonium lead iodide single-crystal perovskites, *Adv. Mater.* **28**, 3406 (2016).
[64] Q. Dong, Y. Fang, Y. Shao, P. Mulligan, J. Qiu, L. Cao, J. Huang, Electron-hole diffusion lengths >175 μm in solution-grown $CH_3NH_3PbI_3$ single crystals, *Science* **347**, 967 (2015).
[65] D. Yang, X. Zhou, R. Yang, Z. Yang, W. Yu, X. Wang, C. Li, S. F. Liu, R. P. H. Chang, Surface optimization to eliminate hysteresis for record efficiency planar perovskite solar cells, *Energy Environ. Sci.* **9**, 3071 (2016).
[66] Y. Shao, Z. Xiao, C. Bi, Y. Yuan, J. Huang, Origin and elimination of photocurrent hysteresis by fullerene passivation in $CH_3NH_3PbI_3$ planar heterojunction solar cells, *Nat. Comm.* **5**, 5784 (2014).
[67] H.-S. Duan, H. Zhou, Q. Chen, P. Sun, S. Luo, T.-B. Song, B. Bob, Y. Yang, The identification and characterization of defect states in hybrid organic–inorganic perovskite photovoltaics, *Phys. Chem. Chem. Phys.* **17**, 112 (2015).
[68] J. H. Heo, D. H. Song, H. J. Han, S. Y. Kim, J. H. Kim, D. Kim, H. W. Shin, T. K. Ahn, C. Wolf, T.-W. Lee, S. H. Im, Planar $CH_3NH_3PbI_3$ Perovskite solar cells with constant 17.2% average power conversion efficiency irrespective of the scan rate, *Adv. Mater.* 27, 3424 (2015).
[69] A. Baumann, S. Väth, P. Rieder, M. C. Heiber, K. Tvingstedt, V. Dyakonov, Identification of trap states in perovskite solar cells, *J. Phys. Chem. Lett.* **6**, 2350 (2015).
[70] G. Gordillo, C. A. Otálora, M. A. Reinoso. Trap center study in hybrid organic–inorganic perovskite using thermally stimulated current (TSC) analysis, *J. Appl. Phys.* **122**, 075304 (2017).
[71] S. Heo, G. Seo, Y. Lee, D. Lee, M. Seol, J. Lee, J.-B. Park, K. Kim, D.-J. Yun, Y. S. Kim, J. K. Shin, T. K. Ahn, M. K. Nazeeruddin, Deep level trapped defect analysis in $CH_3NH_3PbI_3$ perovskite solar cells by deep level transient spectroscopy, *Energy Environ. Sci.* **10**, 1128 (2017).
[72] M. H. Futscher, J. M. Lee, T. Wang, A. Fakharuddin, L. Schmidt-Mende, B. Ehrler, Quantification of ion migration in $CH_3NH_3PbI_3$ perovskite solar cells by transient capacitance measurements, (2018), arXiv:1801.08519.
[73] G. Landi, H. C. Neitzert, C. Barone, C. Mauro, F. Lang, S. Albrecht, B. Rech, S. Pagano, Correlation between electronic defect states distribution and device performance of perovskite solar cells, *Adv. Sci.* **4**, 1700183 (2017).

[74] T. Kirchartz, L. Krückemeier, E. L. Unger, Research update: Recombination and open-circuit voltage in lead-halide perovskites, *APL Materials* **6**, 100702 (2018).
[75] D. Meggiolaro, F. De Angelis, First-principles modeling of defects in lead halide perovskites: Best practices and open issues, *ACS Energy Lett.* **3**, 2206 (2018).
[76] C. Ran, J. Xu, W. Gao, C. Huang, S. Dou, Defects in metal triiodide perovskite materials towards high-performance solar cells: Origin, impact, characterization, and engineering, *Chem. Soc. Rev.* **47**, 4581 (2018).
[77] M. I. Saidaminov, J. Kim, A. Jain, R. Quintero-Bermudez, H. Tan, G. Long, F. Tan, A. Johnston, Y. Zhao, O. Voznyy, E. H. Sargent, Suppression of atomic vacancies via incorporation of isovalent small ions to increase the stability of halide perovskite solar cells in ambient air, *Nature Energy* **3**, 648 (2018).
[78] F. Ambrosio, J. Wiktor, F. De Angelis, A. Pasquarello, Origin of low electron–hole recombination rate in metal halide perovskites, *Energy Environ. Sci.* **11**, 101 (2018).
[79] D. Meggiolaro, S. G. Motti, E. Mosconi, A. J. Barker, J. Ball, C. A. R. Perini, F. Deschler, A. Petrozza, F. De Angelis, Iodine chemistry determines the defect tolerance of lead-halide perovskites, *Energy Environ. Sci.* **11**, 702 (2018).
[80] K. X. Steirer, P. Schulz, G. Teeter, V. Stevanovic, M. Yang, K. Zhu, J. J. Berry, Defect tolerance in methylammonium lead triiodide perovskite, *ACS Energy Lett.* **1**, 360 (2016).
[81] C. Li, A. Guerrero, Y. Zhong, S. Huettner, Origins and mechanisms of hysteresis in organometal halide perovskites, *J. Phys. Condens. Matter* **29**, 193001 (2017).
[82] H. J. Snaith, A. Abate, J. M. Ball, G. E. Eperon, T. Leijtens, N. K. Noel, S. D. Stranks, J. T.-W. Wang, K. Wojciechowski, W. Zhang, Anomalous hysteresis in perovskite solar cells, *J. Phys. Chem. Lett.* **5**, 1511 (2014).
[83] J.-W. Lee, S.-H. Bae, N. De Marco, Y.-T. Hsieh, Z. Dai, Y. Yang, The role of grain boundaries in perovskite solar cells, *Mater. Today Energy* **7**, 149 (2018).
[84] D. J. Vogel, T. M. Inerbaev, D. S. Kilin, Role of lead vacancies for optoelectronic properties of lead-halide perovskites, *J. Phys. Chem. C* **122**, 5216 (2018).
[85] W. Li, Y.-Y. Sun, L. Li, Z. Zhou, J. Tang, O. V. Prezhdo, Control of charge recombination in perovskites by oxidation state of halide vacancy, *J. Am. Chem. Soc.* **140**, 15753 (2018).
[86] W.-J. Yin, H. Chen, T. Shi, S.-H. Wei, Y. Yan, Origin of high electronic quality in structurally disordered $CH_3NH_3PbI_3$ and the passivation effect of Cl and O at grain boundaries, *Adv. Electron. Mater.* **1**, 1500044 (2015).
[87] J. S. Yun, A. Ho-Baillie, S. Huang, S. H. Woo, Y. Heo, J. Seidel, F. Huang, Y.-B. Cheng, M. A. Green, Benefit of grain boundaries in organic–inorganic halide planar perovskite solar cells, *J. Phys. Chem. Lett.* **6**, 875 (2015).

[88] Q. Chen, H. Zhou, T.-B. Song, S. Luo, Z. Hong, H.-S. Duan, L. Dou, Y. Liu, Y. Yang, Controllable self-induced passivation of hybrid lead iodide perovskites toward high performance solar cells, *Nano Lett.* **14**, 4158 (2014).
[89] S. Y. Leblebici, L. Leppert, Y. Li, S. E. Reyes-Lillo, S. Wickenburg, E. Wong, J. Lee, M. Melli, D. Ziegler, D. K. Angell, D. F. Ogletree, P. D. Ashby, F. M. Toma, J. B. Neaton, I. D. Sharp, A. Weber-Bargioni, Facet-dependent photovoltaic efficiency variations in single grains of hybrid halide perovskite, *Nat. Energy* **1**, 16093 (2016).
[90] D. W. de Quilettes, S. M. Vorpahl, S. D. Stranks, H. Nagaoka, G. E. Eperon, M. E. Ziffer, H. J. Snaith, D. S. Ginger, Impact of microstructure on local carrier lifetime in perovskite solar cells, *Science* **348**, 683 (2015).
[91] S. Ham, Y. J. Choi, J.-W. Lee, N.-G. Park, D. Kim, Impact of excess CH_3NH_3I on free carrier dynamics in high-performance non-stoichiometric perovskites, *J. Phys. Chem. C* **121**, 3143 (2017).
[92] M. Yang, Y. Zeng, Z. Li, D. H. Kim, C.-S. Jiang, J. van de Lagemaat, K. Zhu, Do grain boundaries dominate non-radiative recombination in $CH_3NH_3PbI_3$ perovskite thin films? *Phys. Chem. Chem. Phys.* **19**, 5043 (2017).
[93] W. Konga, T. Dinga, G. Bib, H. Wu, Optical characterizations of surface states in hybrid lead halide perovskites, *Phys. Chem. Chem. Phys.* **18**, 12626 (2016).
[94] C. Rocks, V. Svrcek, P. Maguirea, D. Mariotti, Understanding surface chemistry during $MAPbI_3$ spray deposition and its effect on photovoltaic performance, *J. Mater. Chem. C* **5**, 902 (2017).
[95] N.-K. Kim, Y. H. Min, S. Noh, E. Cho, G. Jeong, M. Joo, S.-W. Ahn, J. S. Lee, S. Kim, K. Ihm, H. Ahn, Y. Kang, H.-S. Lee, D. Kim, Investigation of thermally induced degradation in $CH_3NH_3PbI_3$ perovskite solar cells using *in situ* synchrotron radiation analysis, *Sci. Reports* **7**, 4645 (2017).
[96] J. Kang, L.-W. Wang, High defect tolerance in lead halide perovskite $CsPbBr_3$, *J. Phys. Chem. Lett.* **8**, 489 (2017).
[97] J. Yin, H. Yang, K. Song, A. M. El-Zohry, Y. Han, O. M. Bakr, J.-L. Brédas, O. F. Mohammed, Point defects and green emission in zero-dimensional perovskites, *J. Phys. Chem. Lett.* **9**, 5490 (2018).
[98] A. V. Cohen, D. A. Egger, A. M. Rappe, L. Kronik, Breakdown of the static picture of defect energetics in halide perovskites: The case of the Br vacancy in $CsPbBr_3$, (2018), arXiv:1810.04462.
[99] Y. Guo, Q. Wang, W. A. Saidi, Structural stabilities and electronic properties of high-angle grain boundaries in perovskite cesium lead halides, *J. Phys. Chem. C* **121**, 1715 (2017).
[100] A. S. Thind, G. Luo, J. A. Hachtel, M. V. Morrell, S. B. Cho, A. Y. Borisevich, J.-C. Idrobo, Y. Xing, R. Mishra, Atomic structure and electrical activity of grain boundaries and Ruddlesden–Popper faults in cesium lead bromide perovskite, *Adv. Mater.* 1805047 (2018).
[101] J. Chen, N.-G. Park, Causes and solutions of recombination in perovskite solar cells, *Adv. Mater.* 1803019 (2018).

[102] P. Schulz, D. Cahen, A. Kahn, Halide perovskites: Is it all about the interfaces? (2018), arXiv:1812.04908.
[103] H. Zhang, X. Liu, J. Dong, H. Yu, C. Zhou, B. Zhang, Y. Xu, W. Jie, Centimeter-sized inorganic lead halide perovskite $CsPbBr_3$ crystals grown by an improved solution method, *Cryst. Growth Des.* **17**, 6426 (2017).
[104] M. Zhang, Z. Zheng, Q. Fu, Z. Chen, J. He, S. Zhang, C. Chen, W. Luo, Synthesis and single crystal growth of perovskite semiconductor $CsPbBr_3$, *J. Crystal Growth* **484**, 37 (2018).
[105] G. Vitale, G. Conte, P. Aloe, F. Somma, An impedance spectroscopy investigation of nanocrystalline $CsPbBr_3$ films, *Mater. Sci. Eng. C* **25**, 766 (2005).
[106] D. Y. Chung, M. G. Kanatzidis, F. Meng, C. D. Malliakas, Synthesis, purification, and characterization of perovskite semiconductor $CsPbBr_3$ as a new candidate for y-ray detector (Conference Presentation), in *Proc. SPIE 9968, Hard X-Ray, Gamma-Ray, and Neutron Detector Physics XVIII*, 996819, 2 November 2016, doi: 10.1117/12.2238221.
[107] Y. He, L. Matei, H. J. Jung, K. M. McCall, M. Chen, C. C. Stoumpos, Z. Liu, J. A. Peters, D. Y. Chung, B. W. Wessels, M. R. Wasielewski, V. P. Dravid, A. Burger, M. G. Kanatzidis, High spectral resolution of gamma-rays at room temperature by perovskite $CsPbBr_3$ single crystals, *Nat. Comm.* **9**, 1609 (2018).
[108] M. Zhang, Z. Zheng, Q. Fu, P. Guo, S. Zhang, C. Chen, H. Chen, M. Wang, W. Luo, Y. Tian, Determination of defect levels in melt-grown all-inorganic perovskite $CsPbBr_3$ crystals by thermally stimulated current spectra, *J. Phys. Chem. C* **122**, 10309 (2018).
[109] M. Bruzzi, F. Gabelloni, N. Calisi, S. Caporali, A. Vinattieri, Defective states in micro-crystalline $CsPbBr_3$ and their role on photoconductivity, *Nanomaterials* **9**, 177 (2019).
[110] Y. Yu, D. Zhang, P. Yang, Ruddlesden–Popper phase in two-dimensional inorganic halide perovskites: A plausible model and the supporting observations, *Nano Lett.* **17**, 5489 (2017).
[111] L. Gomez, J. Lin, C. de Weerd, L. Poirier, S. C. Boehme, E. von Hauff, Y. Fujiwara, K. Suenaga, T. Gregorkiewicz, Extraordinary interfacial stitching between single all-inorganic perovskite nanocrystals, *ACS Appl. Mater. Interfa.* **10**, 5984 (2018).
[112] S. Yuan, Z.-K. Wang, M.-P. Zhuo, Q.-S. Tian, Y. Jin, L.-S. Liao, Self-assembled high quality $CsPbBr_3$ quantum dot films toward highly efficient light-emitting diodes, *ACS Nano* **12**, 9541 (2018).
[113] D. Yang, X. Li, H. Zeng, Surface chemistry of all inorganic halide perovskite nanocrystals: Passivation mechanism and stability, *Adv. Mater. Interfa.* **5**, 1701662 (2018).
[114] D. P. Nenon, K. Pressler, J. Kang, B. A. Koscher, J. H. Olshansky, W. T. Osowiecki, M. A. Koc, L.-W. Wang, A. P. Alivisatos, Design principles for trap-free $CsPbX_3$ nanocrystals: Enumerating and eliminating surface halide vacancies with softer Lewis bases, *J. Am. Chem. Soc.* **140**, 17760 (2018).

[115] L. Song, X. Guo, Y. Hu, Y. Lv, J. Lin, Y. Fan, N. Zhang, X. Liu, Improved performance of $CsPbBr_3$ perovskite light-emitting devices by both boundary and interface defects passivation, *Nanoscale* **10**, 18315 (2018).

[116] X. Liu, X. Guo, Y. Lv, Y. Hu, Y. Fan, J. Lin, X. Liu, X. Liu, High brightness and enhanced stability of $CsPbBr_3$-based perovskite light-emitting diodes by morphology and interface engineering, *Adv. Optical Mater.* **6**, 1801245 (2018).

[117] D. Lu, Y. Zhang, M. Lai, A. Lee, C. Xie, J. Lin, T. Lei, Z. Lin, C. S. Kley, J. Huang, E. Rabani, P. Yang, Giant light-emission enhancement in lead halide perovskites by surface oxygen passivation, *Nano Lett.* **18**, 6967 (2018).

[118] F. Gabelloni, F. Biccari, G. Andreotti, D. Balestri, S. Checcucci, A. Milanesi, N. Calisi, S. Caporali, A. Vinattieri, Recombination dynamics in $CsPbBr_3$ nanocrystals: Role of surface states, *Opt. Mater. Expr.* **7**, 4367 (2017).

[119] P. Vashishtha, J. E. Halpert, Field-driven ion migration and color instability in red-emitting mixed halide perovskite nanocrystal light-emitting diodes, *Chem. Mater.* **29**, 5965 (2017).

[120] D. Pan, Y. Fu, J. Chen, K. J. Czech, J. C. Wright, S. Jin, Visualization and studies of ion-diffusion kinetics in cesium lead bromide perovskite nanowires, *Nano Lett.* **18**, 1807 (2018).

[121] D. R. Ceratti, Y. Rakita, L. Cremonesi, R. Tenne, V. Kalchenko, M. Elbaum, D. Oron, M. A. C. Potenza, G. Hodes, D. Cahen, Self-healing inside $APbBr_3$ halide perovskite crystals, *Adv. Mater.* **30**, 1706273 (2018).

[122] H.-C. Wang, Z. Bao, H.-Y. Tsai, A.-C. Tang, R.-S. Liu, Perovskite quantum dots and their application in light-emitting diodes, *Small* **14**, 1702433 (2018).

CHAPTER 4

Color Centers in Wide-Gap Semiconductors for Quantum Technology

Y. YAMAZAKI, S. ONODA and T. OHSHIMA

National Institutes for Quantum and Radiological Science and Technology, 1233 Watanuki, Takasaki, Gunma 370-1292, Japan

In this chapter, color centers that act as qubits and quantum sensors in diamond and silicon carbide will be introduced. Then, their creation methodologies using particle irradiation such as protons, heavy ions and electrons will be discussed.

1. Introduction

Wide-gap semiconductors such as silicon carbide (SiC), gallium nitride (GaN) and diamond are well known as promising materials for next-generation power electronics due to various superior physical properties such as high electrical breakdown strength and thermal conductivity [1–8]. Recently, these materials have received wide attention for their power electronics properties because specific color centers in these materials show the unique properties, e.g. long spin coherence time and optical addressability of electron spin. These properties are the most essential for quantum applications such as quantum computing, quantum communication, and quantum sensing. Until now, some candidates, such as superconducting circuit, laser-trapped ions, rare-earth ions embedded in a crystal, quantum dots, and color centers in semiconductor materials, have been demonstrated as a quantum bit (qubit) in proof-of-concept experiments. From the viewpoint of industrialization, practical ways for multi-qubit states, quantum manipulation (initialization, preservation, operation, and readout), and entanglement (led by precise position control of qubits) should be established. Considering the compatibility of the semiconductor device fabrication processes, which is the at present most sophisticated technology, a solid-state quantum bit is promising. In this sense, silicon (Si) and gallium arsenide (GaAs) quantum dot is a prime candidate because it is possible to control each dot position precisely, which leads to entanglement [9–11]. However, these quantum dots can be operated only at very low temperature (~cryogenic temperature) due to their bandgap (1.11 eV for Si and 1.43 eV for GaAs). In contrast, the energy levels of color centers in wide-gap semiconductors (2.23–3.26 eV for 4H–, 6H–, and 3C–SiC, 3.39 eV for GaN and 5.47 eV for diamond) are usually adequately isolated from both valence and conduction bands. Therefore, the color centers act well as single-photon sources (SPSs) even at room temperature. This is of considerable advantage for realizing quantum devices.

To observe color centers, a laser scanning confocal fluorescent microscopy (CFM) is widely used. Figure 1 shows a schematic image of a CFM system. A wavelength of a continuous-wave (CW) laser should be appropriately chosen for highly efficient excitation, for example 532 nm for nitrogen-vacancy composite defect in diamond (known as an NV center) and 730 nm for silicon vacancy in SiC. For efficient photon collection from a color center, an objective lens with a large numerical aperture is commonly used. Excitation light is rejected by a dichroic mirror (DM) and a long-pass filter (LPF). Fluorescent light from the color center passes through DM and LPF. A pair of convex lenses are used for passing the

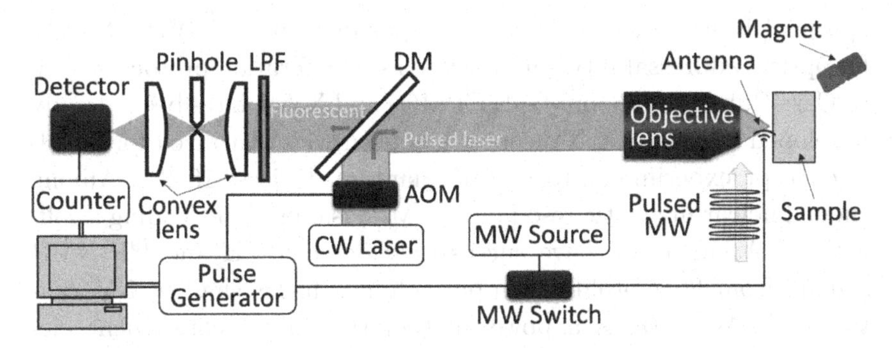

Figure 1. Schematic image of a CFM system.

light through the pinhole. Fluorescent light from the color center is detected by a highly sensitive photodetector such as an avalanche photo diode and a fast counter. Pulse-mode measurement is also frequently employed for improving not only signal-to-noise ratio but also quantum manipulation. Acousto-optic modulator (AOM), microwave (MW) switch, pulse generator, antenna and magnet are used for quantum manipulation.

In this chapter, we focus on color centers in diamond and SiC. The two crystals have advantage as a host material for color centers as a qubit owing to stable isotopes of group IV element, such as ^{12}C, ^{28}Si, and ^{74}Ge, having no nuclear spin compared with III–V compounds because nuclear spin interference against electron spins of solid-state qubits results in degrading quantum properties. In addition, it enables us to utilize the isotopic enrichment technique to prolong spin coherence time, which has been already demonstrated in the case of color center of diamond [12].

1.1. *Diamond*

Nitrogen-vacancy composite defect (NV center) is one of the most famous color centers in diamond [13, 14], as shown in Fig. 2. Its spin and optical properties have been intensively investigated since the fluorescence from individual NV center can be observed with CFM at room temperature [13]. In addition, NV center shows a very long spin coherent time even at room temperature. Especially, spin coherent time reaches up to 1.8 ms for the ^{12}C-enriched diamond epitaxial layer grown by chemical vapor deposition (CVD) [12]. As for optical addressability, sophisticated spin operation protocols, such as Carr-Purcell-Meiboom-Gill (CPMG) and XY-8, have been already developed [15]. Hence, NV center has been widely utilized for proof-of-concept experiments to realize quantum devices [16–21]. Among them, nuclear magnetic resonance (NMR) is one of the leading applications. Aslam *et al.* have successfully detected ^{1}H and ^{19}F NMR signals from 20-zeptoliter sample volumes using shallow NV centers [21]. As NMR is a powerful tool to analyze constituent elements in a matter, it must be widely utilized not only in various disciplines but also in various industries, for example, biology and medicine/healthcare, in future. Figure 2 also shows the new color centers in diamond found recently. SiV [22, 23], GeV [24] and SnV [25] consist of an interstitial foreign atom located between two neighboring vacancies. An NE8 has a more complex structure, with one nickel atom surrounded by four nitrogen atoms [26].

Table 1 shows the optical properties of color centers in diamond shown in Fig. 2. The vacancy-related color centers have a similar wavelength in the red light region in terms of zero-phonon line (ZPL), optical transition energy of color center, even though constituent element is widely varies from second to fifth period elements. Figure 3 summarizes photoluminescence (PL) spectra of these vacancy-related color centers observed at room temperature. A prominent broad emission at higher wavelength of ZPL for the NV center is attributed to the phonon sideband, whereas for other vacancy-related color centers, lower phonon sideband emission was observed. These suggest

Table 1. The properties of color centers in diamond [13, 23, 25, 26].

	NV	SiV	GeV	SnV	NE8
λ_{ZPL} (mm)	637	738	602	~619	802
Spin value	1				
ZFS (GHz)	2.870	48	~170	~850	
ODMR	○ (RT)	×	×	×	×

Note: All ZFSs are the value for the ground state.

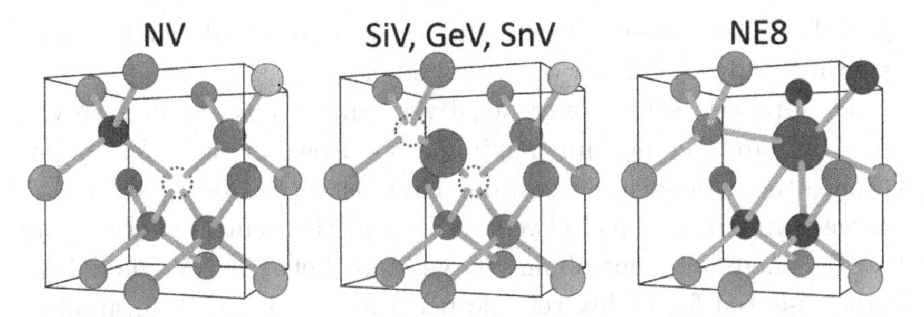

Figure 2. Schematic configurations of color centers in diamond [14, 23–26]. The gray, blue and brown balls denote carbon, nitrogen, and nickel atoms, respectively. The dot-line balls are a vacancy. For SiV, GeV and SnV, the red ball represents a corresponding foreign atom, which is located in an interstitial site.

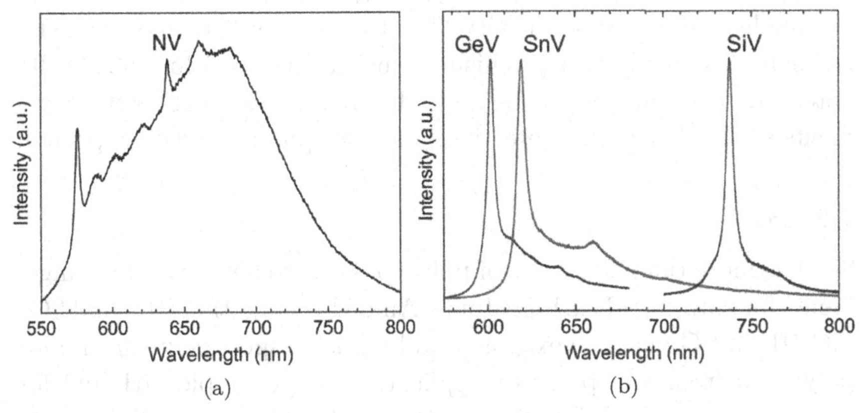

Figure 3. PL spectra of the vacancy-related color centers in diamond [25] (a) NV and (b) SiV, GeV, and SnV.

that phonon coupling with SiV, GeV, and SnV centers is weaker than that with NV center. This feature is one of advantages for a stable single-photon source. The NE8 shows near-infrared ZPL. Electron paramagnetic resonance (EPR) has revealed zero-field splitting (ZFS) of each color centers, except for NE8. In contrast to ZPL, ZFS increases with atomic number for the vacancy-related color centers. Optical addressability of electron spin, which is a key function for quantum applications, can be checked by optically detected magnetic resonance (ODMR), a standard technique to manipulate electron spins. For color centers in diamond, optical addressability has been experimentally proven for only NV center.

In a precise sense, only negatively charged NV center, NV^-, has a feature of quantum manipulation. However, an NV^- population is not stable under various circumstances due to stochastic charge-state transitions between NV^- and the neutral charge state (NV^0). Therefore, controlling charge state between NV^- and NV^0 is the essential for highly reliable performance of quantum applications. Some approaches using optical and electrical methods have been demonstrated [28–31]. Especially, electrical methods play an important role in realizing integrated quantum devices for various applications. On this point, electrical detection of spin state is gathering much attention for not only integrated quantum devices but also higher detection sensitivity [32]. Electrically detected magnetic resonance (EDMR) is a promising fundamental tool for electrically detecting the spin state instead of ODMR [32, 33]. Further improvements such as electrical detection of single spin are widely expected.

1.2. *SiC*

SiC has more than 200 types of polytypes due to a variety of arrangements for silicon and carbon atoms. Among them, 4H–, 6H–, and 3C–SiC (H and C denote hexagonal and cubic, respectively) are major polytypes from viewpoint of application, for example, 4H–SiC for power device and 6H-SiC for a substrate for GaN-based light emitting diode and 3C-SiC for heteroepitaxy of Si and SiC aiming for a large-size epitaxial layer. Unlike diamond, color centers of SiC have various configurations because of two possible lattice sites (Si and C) for a vacancy and a substitutional atom, polytypes (4H, 6H, 3C, and others), and their combination. Recently, it has been reported that specific defects of SiC act as a SPS [34–43]. Although the dependence of polytype has been investigated, we mainly focus on color centers in 4H-SiC in this chapter. Details of 3C– and 6H–SiC color centers can be found in various studies.

Figure 4 shows the schematic configuration of color centers in 4H–SiC. The configurations have been identified by EPR and PL

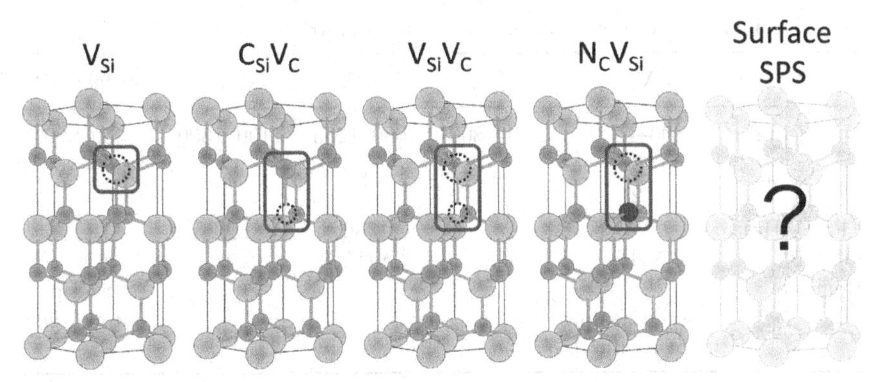

Figure 4. Schematic configuration of SPSs in 4H–SiC. One representative configuration of each color center is shown here. Gray and green balls denote carbon and silicon atoms, respectively. Dot-line balls represent vacancies. Nitrogen atom is displayed as a blue ball for $N_C V_{Si}$. A structure of surface SPS has not been clarified yet.

for silicon-vacancy (V_{Si}), carbon–antisite and carbon–vacancy pair ($C_{Si}V_C$), divacancy ($V_{Si}V_C$), and silicon-vacancy and nitrogen pair (N_CV_{Si}). There are two nonequivalent atomic sites of Si and C in 4H–SiC, a quasi-hexagonal *h* site and a quasi-cubic *k* site. Hence, for example, V_{Si} is of two different types: V1 (*h* site) and V2 (*k* site), and there are four inequivalent forms of divacancy ($V_{Si}V_C$, neighboring Si and C vacancies pair): *hh*, *kk*, *hk*, and *kh* sites. Unlike diamond (diamond does not have oxide), bright color centers can be found at an interface between SiO_2 and SiC, which is called "surface SPS". Creation and annihilation of the surface SPS can be reproducibly repeated by oxidation and removal of the oxide layer. Although the experimental results imply that the surface SPS is an oxygen-related defect, the configuration has not yet been clarified.

The optical properties of 4H-SiC color centers are summarized in Table 2. The unique point of these centers is a wide distribution of ZPL ranging from the visible to the near-infrared region. This can allow us to select an appropriate color center to meet the demand of individual application. For example, longer wavelength is suitable for biological application due to lower absorption by body tissue, meaning that one can directly detect information in a deeper region of the

Table 2. The properties of color centers in 4H–SiC [34, 36, 41–45].

	V_{Si}	$C_{Si}V_C$	$V_{Si}V_C$	N_CV_{Si}	Surface SPS
λ_{ZPL} (mm)	800–1000	600–800	1000–1200	1100–1300	550–800
Spin value	$\frac{3}{2}$		1	1	
ZFS (MHz)	70		1350	1270	
ODMR	○ (RT)	×	○ (LT)	×	×
Remarks	Post-annealing not required				Structure not clarified

Note: All ZFSs are the value for the ground state.

human body from outside. Optical communication requires narrow bandwidth at near-infrared light region because transmission loss of optical fiber represents the minimum value in the wavelength range. V_{Si} has half-integer spin $S = \frac{3}{2}$. $V_{Si}V_C$, and N_CV_{Si} are spin $S = 1$ centers. ZFS for V_{Si} has been determined to be 70 MHz. This value is considerably smaller than that for $V_{Si}V_C$, N_CV_{Si} in SiC and NV center in diamond, which have almost the same ZFS lying in the GHz range. ODMR spectra have been confirmed for V_{Si} and $V_{Si}V_C$ so far, and have been especially detectable for V_{Si} even at room temperature. Concerning creation process of color centers, high-temperature post-annealing is usually needed in order to form a complex defect. For V_{Si}, only ion irradiation process is required because of a single vacancy. This is one of the advantages for combining V_{Si} and SiC devices, which usually has a limited thermal budget. It should be noted that V_{Si} is abruptly annealed out above 600°C [34].

As for applications using SiC color centers, magnetometry [46–48] and thermometry [49] have been demonstrated using V_{Si}. Details of the two applications are summarized in the review paper [50]. For SiC, electrical control and detection of the spin state are a more promising compared with diamond color centers because various practical devices and sophisticated fabrication processes such as growth and doping are ready to utilize for examining capability of electrical functions. There are some reports regarding electrical controlling of V_{Si} [51], $V_{Si}V_C$ [52] and surface SPS [53, 54]. As SiC devices have been already put to practical use, there are plenty of room for early realization of quantum applications. Therefore, both fundamental and application researches are strongly required.

2. Creation of Color Centers in Wide-Gap Semiconductors

There are two types of techniques widely in use for creating color centers in wide-gap semiconductors, one is crystal growth, and the other is particle irradiation such as electron and ion beam. The introduction of some impurities is possible during the crystal growth; however, there is still room to improve the growth techniques for introducing color centers. On the contrary, the ion implantation easily introduces various ions at aiming position and concentration.

In this chapter, the creation of NV centers in diamond and V_{Si} in SiC using particle irradiation is reviewed.

2.1. *Overview of creation method*

Figure 5 shows the schematic diagram for the creation of NV center. Type Ib and type IIa diamonds are commonly used for the creation of color centers. As shown in Fig. 5(a), type IIa diamonds do not

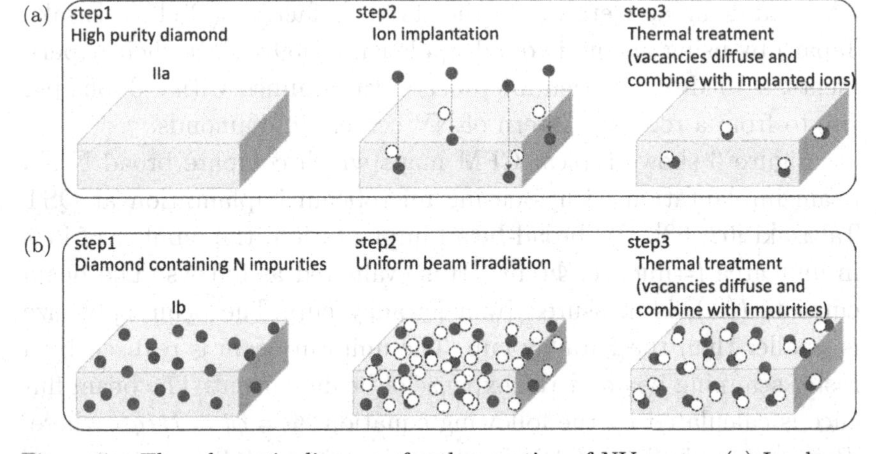

Figure 5. The schematic diagrams for the creation of NV center. (a) Ion beam is implanted into type IIa diamond. (b) Ion or electron beams are irradiated into type Ib diamond.

contain nitrogen atoms or other impurity atoms. When nitrogen ions are implanted, the vacancies are created along its track at the same time. After annealing at high temperature (typically 800–1000°C), where vacancies easily diffuse, vacancies combine with the implanted nitrogen atoms, and as a result, NV centers are formed. As shown in Fig. 5(b), type Ib diamonds contain nitrogen atoms as their main impurity. Ion or electron beams are irradiated into Ib diamonds to introduce vacancies by the knock-on process of host carbon atoms. Then, the vacancies are combined with substitutional nitrogen atoms by subsequent annealing at high temperature.

2.2. *MeV Focusing microbeam implantation*

The heavy ion microbeam systems were utilized for the creation of NV center in 2005 [55–57]. Meijer *et al.* demonstrated the generation of single-NV centers by 2 MeV nitrogen implantation into type IIa diamonds. They focused nitrogen beams provided from the dynamitron tandem accelerator in DTL (Ruhr University Bochum, Germany) by using a 15 T superconducting solenoid lens [58]. The regular pattern of implantation sites was generated by stepwise raster scanning beams over the sample. In much the same manner, Yamamoto *et al.* performed N ion microbeam implantation provided from the 3 MV tandem accelerator in the TIARA facility (QST Takasaki, Japan) by using magnetic quadrupole lenses [59–61]. In their experiments, a 10 MeV nitrogen microbeam was scanned with ~3 ions per site to from a regular pattern of NV centers in diamonds.

Figure 6 shows typical CFM maps which compare broad N ion beam implantation with focusing microbeam implantation at QST Takasaki [62, 63]. For broad-beam implantation, the number of ions in unit area (=fluence, $\Phi(\mathrm{cm}^{-2})$) is evaluated as follows: The beam current, $I(\mathrm{A})$, is measured by a Faraday cup. The beam spot size is smaller than the Faraday cup. The uniform beam is realized by a raster scanning beam at the expense of beam current. The beam fluence is calculated by the following equation, $\Phi = \phi t = It/qS$ where, $\phi(\mathrm{cm}^{-2}/\mathrm{s}^{-1})$ is flux, $t(\mathrm{s})$ is implantation time, $q(\mathrm{C})$ is unit charge, and $S(\mathrm{cm}^2)$ is the area of the Faraday cup, respectively. Figure 6(a)

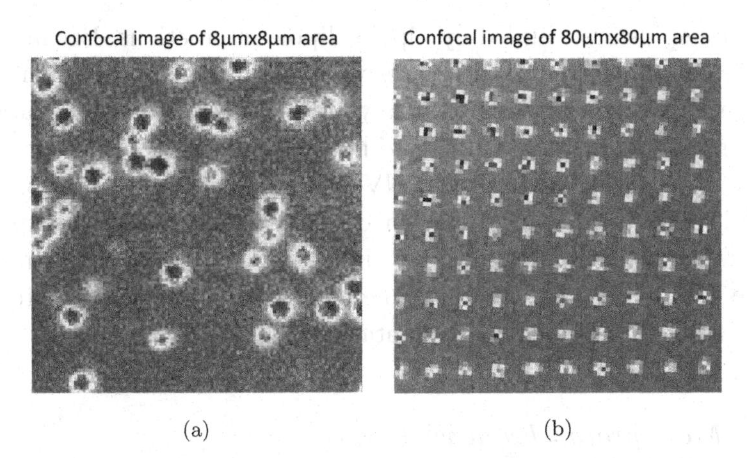

(a) (b)

Figure 6. The typical CFM images created by nitrogen ion implantations. (a) Broad 10 MeV nitrogen beam implantation with the fluence of $10^8\,\mathrm{cm}^{-2}$. (b) Focusing 10 MeV nitrogen microbeam implantation.

shows the CFM map when a 10 MeV nitrogen beam with the fluence of $10^8\,\mathrm{cm}^{-2}$ is uniformly implanted. As seen in Fig. 6(a), 33 NV centers are found in $64\,\mu\mathrm{m}^2$, which correspond to $\sim 5 \times 10^7\,\mathrm{NV/cm}^2$. This result means that ~50% of the implanted nitrogen atoms is converted to NV centers. The conversion ratio is called creation yield. It is well known that the creation yield increases with an increase in the energy [56, 57, 64]. In contrast to broad-beam implantation, the regular pattern of NV centers can be created by microbeam, as shown in Fig. 6(b). Before implantation, the beam flux is reduced to tens of ions per second. The combination of an aperture a fast beam chopping system enable the single-ion implantation. Then the number of ions per spot is varied from one to several thousands. As stated above, this method is applicable for any ion species. The microbeam of various ion species (C, N, O, F, Si, P, Fe, Ni, Cu, I, and Au) is available at QST Takasaki and is used for exploring new color centers.

One of the applications of NV centers in diamond is solid-state spin quantum bit at room temperature. One possible architecture of quantum register, which is a key technology for quantum information processing (QIP), is an array of NV centers in which the neighboring electron spins are coupled by dipole–dipole interactions. Because the dipole coupling strength decreases with an increase in the separation distance, NV centers need to satisfy both shorter separation distance (lesser than a few tens of nm) and better spin properties at the same time. To realize closely placed NV–NV pairs, ion implantation is a useful technique; however, the inaccuracy originates from the varying results in the extremely low probability of NV–NV creation. In fact, there is only one case in which the coupled NV–NV was created by MeV focusing microbeam implantation [65].

2.3. *MeV proton beam writing*

There are a large number of MeV-range proton microbeam lines available in the world. Most of these are optimized for particle-induced X-ray emission (PIXE) application, which is useful to analyze wide varieties of matters from inorganic to organic (including biological sample) materials [66]. Using the etching feature of the proton beam and a precise position-controlling system, the designed structures can be fabricated on soft matter such as organic materials [67]. This technique is the so-called proton beam writing (PBW).

PBW has been applied for the creation of V_{Si} of SiC [68–70]. Figure 7(a) shows a CFM image of V_{Si} array created into a bulk substrate. Proton fluence was changed line by line (higher fluence on the left side in the figure), and emission intensity changed in the same manner as that for fluence. This result clearly indicates that V_{Si} can be created by controlling the position and density (number of V_{Si} per spot) simultaneously by PBW. From the result, V_{Si} creation yield can be calculated. As reported by Kraus *et al.* [68] the yield was determined to be 0.1 V_{Si}/proton for 1–3 MeV proton with a fluence range of 10^{10}–10^{12} cm^{-2}. Utilizing precise position controllability of PBW, V_{Si} can be introduced into a particular area of an SiC device as shown in Fig. 7(b). Electroluminescence (EL) of V_{Si} was successfully observed by applying forward bias to the pn junction in addition to PL, which suggests electrical control of the color center.

The feature of V_{Si} creation in the desired area is preferable for the introduction of V_{Si} into an SiC device because the device

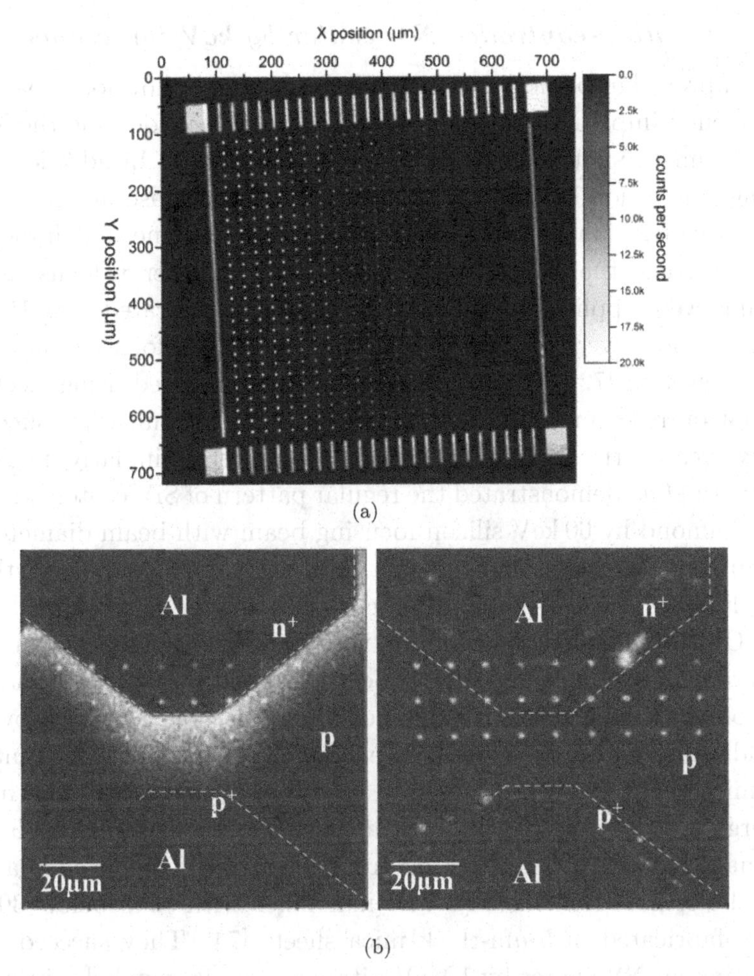

Figure 7. The CFM image of V_{Si} array created by PBW. (a) PL mapping pattern with different fluence on a bulk substrate [69]. (b) EL (left) and PL (right) mapping patterns on in-plane *pn* junction diode [70]. A pronounced luminescence observed along the p–n^+ junction for EL was emitted not from V_{Si} but from D1 defects [71].

usually has a three-dimensionally complicated structure. The depth of V_{Si} is readily controlled by changing the proton energy. The fact that no post-annealing is needed after PBW is also advantageous for transforming V_{Si} into an SiC device having limited thermal budget.

2.4. *Position-controlled NV center by keV ion beams*

An improved in-plane positioning accuracy might allow for a spatially controlled implantation of color centers. However, focusing the MeV ion beam to spot sizes below 1 μm is challenging. In addition, the straggling of ion in matter increases with the increase in ion energy, which causes inaccurate targeting for both in-plane and in-depth. To overcome the poor in-plane resolution, low-energy focusing ion beams were applied. In 2013, Lesik *et al.* demonstrated that 15 keV nitrogen ions were focused to ~100 nm by using a focused ion beam (FIB) system [72]. The nitrogen beam was produced using electron cyclotron resonance (ECR) plasma source. As a result, they successfully created the regular pattern of NV centers. Similarly, in 2014, Tamura *et al.* demonstrated the regular pattern of SiV centers in type IIa diamond by 60 keV silicon focusing beam with beam diameter of 60 nm [73]. Although a significant improvement was reported, there is a limit to accurate targeting focusing beam.

On the contrary, the combination of low-energy broad-beam and a variety of nanohole masks have been proposed [64, 74–82]. First, the nanohole in mica sheets is explained. The nanohole is created by the irradiation of heavy ions with an energy of ~GeV. The disadvantage of mica method is the randomly distributed nanoholes. The high-energy heavy ion passes through the mica sheets, resulting in a latent damaged channel along the ion track. The latent damaged channel can be removed chemically. The nanoholes with sizes below 30 nm were fabricated in 5-μm-thick mica sheets [74]. They succeeded in creating an NV center by 1 MeV nitrogen ion via nanoholes in mica. In addition, Dale *et al.* demonstrated that the coupled NV–NV was created by 1 MeV nitrogen ion implantation via nanoholes in mica sheet [83]. They found NV–NV pair with the distance of 25 nm in an implantation site. Second, a system combining a low-energy nitrogen beam and a pierced atomic force microscopy (AFM) tip is explained [75, 76]. The nitrogen beam is irradiated on the backside of the AFM tip that has a nanohole with a diameter of 30 nm, and the resolution of the set-up thus directly depends on the size of the nanohole. Moller *et al.* demonstrated that 5 keV nitrogen ion beam through the AFM tip was implanted into the photonic cavity [84]. This method is quite useful for the creation of color centers at the desired positions even if the sample has a structure. Third, the implantation via nanohole in a resist mask fabricated by electron beam lithography is explained [64, 77–82]. The original work of nanohole in resist mask was performed by Toyli *et al.* in 2010 [77]. The in-plane accuracy is determined by the nanohole diameter as well as straggling, in the same manner as that followed for mica and AFM. Jakobi *et al.* have investigated the implantation conditions to fabricate coherently coupled NV pairs at a reasonable probability [64]. The optimal energy of 30 keV was applied for nitrogen implantation through nanoholes of diameter 50 nm in PMMA resist mask. Ten coupled NV–NV pairs were found among 6,000 implantation sites. Scarabelli *et al.* succeeded in creating small nanoholes (8–20 nm) on a thin gold film under PMMA by a combination of metal etching technique and electron beam lithography [79]. Because the electron beam lithography technique is highly developed, the nanohole in resist mask is of primary importance in the engineered accurate patterning.

For creating a coupled NV–NV center, nitrogen molecular (N_2) ion implantation was proposed [85, 86]. When an N_2 ion hits the diamond, it decomposes into two individual nitrogen atoms. Because two nitrogen atoms hit at the same position, the targeting accuracy is solely determined by straggling. In short, the molecular ion implantation is regarded as an ideal point source. Gaebel *et al.* reported NV–N pair fabrication by 14 keV N_2 implantation in 2006 [85]. They reported that the estimated separation distance and dipole coupling strength between NV and N were 1.5 nm and 14 MHz, respectively. Similarly, Yamamoto *et al.* reported the fabrication of an NV–NV pair by 20 keV N_2 implantation in 2013 [86]. It is obvious that ion implantation with higher number of nitrogen atoms in single molecules such as N_3 or more could lead to further integration of NV centers. However, there has been no report of coupled NV centers produced by nitrogen cluster ions as the formation of a pure nitrogen cluster has been hardly implemented.

Here, we take up a lot of space for “in-plane” position-controlled NV center from the viewpoint of quantum information processing (QIP), as described above. However, “in-depth” accuracy is also important, especially for quantum sensing. One of the most famous applications is a nanoscaled NMR spectroscopy [87–94]. The detection of NMR signals from ^{1}H, ^{19}F, ^{29}Si, and ^{31}P in small detection-volume samples, using NV centers implanted near the surface, has been reported recently. The sensing volume depends strongly on the depth of NV center from the surface, if the target nuclear spins are located on the diamond surface. To create an NV center near the surface, lower energy is required. Most studies involved ion implantation with an acceleration energy of 2–6 keV.

2.5. *Several hundreds of MeV ion irradiation*

keV and MeV ions are mainly used to create NV centers for quantum applications, such as quantum information processing and quantum sensing, etc. The several hundreds of MeV is not adequate for such purposes. In this section, the novel application of NV center is introduced.

When the heavy ion with several hundreds of MeV hits diamond containing nitrogen impurity, vacancies along ion track are created. The vacancies are captured by nitrogen impurity after annealing, as shown in Fig. 5(b). By observing CFM, the trail-like emissions from NV centers along ion track are detected, which can be used for fluorescent nuclear track detector (FNTD) [95, 96]. Figure 8(a) shows the schematic diagram of 490 MeV osmium ion irradiation into a diamond. The 490 MeV Os ions provided by the azimuthally varying field (AVF) cyclotron accelerator at QST Takasaki, are irradiated at

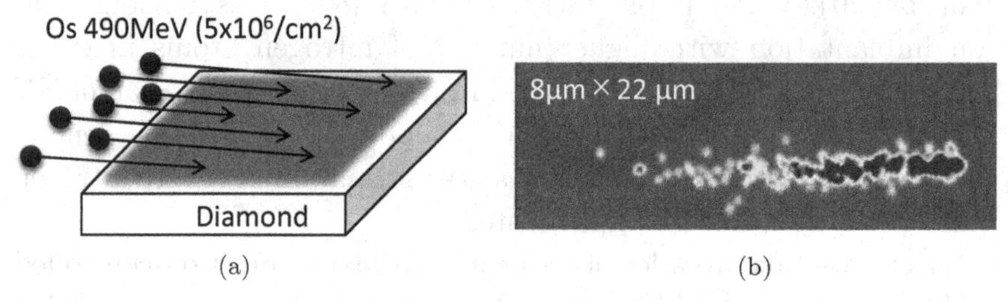

Figure 8. (a) The schematic diagram of irradiation, and (b) CFM image of ion track.

a fluence of $5 \times 10^6\,\mathrm{cm}^{-2}$. Samples are slightly tilted at an angle of $\sim 10°$. After annealing at 1000°C for 2 hours in vacuum, the trail-like emission is observed by CFM. As a result, the ion tracks are clearly seen as shown in Fig. 8(b). However, there is still room to improve the sensitivity because the measured track length is shorter than the calculated value. The most likely reason for the underestimation of track length is due to the insufficient nitrogen concentration. The advantage of diamond-FNTD is high spatial resolution of ion track. In addition, this technique has considerable potential when using a super-high-resolution imaging technique, such as stimulated emission depletion (STED) microscopy [97].

2.6. *MeV electron irradiation*

The uniform electron beam is effective for the introduction of vacancies in wide-gap semiconductors. In the case of type Ib diamond, vacancy combines with nitrogen impurity after annealing, as shown in Fig. 5(b). For SiC, the subsequent annealing is not necessary to form V_{Si}. In this section, effectiveness of MeV electron irradiation into diamond is introduced.

When electron beams are irradiated into a material, the orbital electrons and its host nuclide receive the energy by coulomb force. If the energy exceeds the threshold energy, the orbital electrons are ionized, and the nuclides are displaced. As a result, the primary-knock-on-atoms (PKAs) and vacancies are created. The maximum scattered energy of the lattice atom, $E_{p,\max}$, is described by following equation: $E_{p,\max} = 2E(E + 2mc^2)/mc^2$, where m and M are the mass of the incident electron and target material, c is velocity of photons, respectively. The threshold energy of diamond is assumed to be 37.5 eV [98], and the maximum energy is calculated to be 175 keV. Figure 9 shows the ray trace of electron beams with energies of 200 keV and 2 MeV. CASINO (monte Carlo SImulation of electroN trajectory in sOlids) code is used for the calculation [99]. The sample is diamond with the size of 5 mm $\times$ 5 mm $\times$ 0.5 mm. When the energy of incident electrons is 200 keV, the electrons stop inside the diamond and the vacancies are created near the surface.

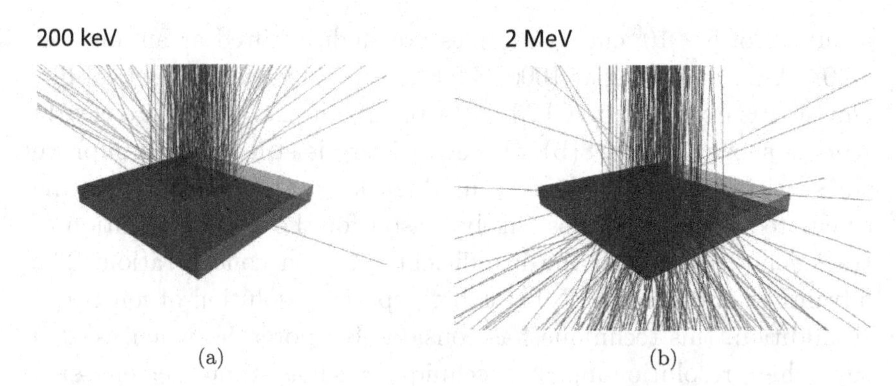

Figure 9. The ray trace of electron beam calculated by CASINO code. (a) 200-keV electrons stop inside the diamond. Solid blue and red lines represent the incident and back-scattered electrons. (b) All 2 MeV electrons pass through the diamond.

On the contrary, the electrons with energy of 2 MeV pass through the whole diamond and vacancies are created in the whole diamond. It is concluded that MeV electron beams are more suitable than ion beams to create color centers with uniform concentration in whole substrate.

For several decades, behavior of vacancies in diamonds created by electron irradiation has been systematically studied by optical absorption/luminescence spectroscopy and EPR [100–102]. The production rates of vacancies at 100 K and 350 K were reported to be 1.53 cm^{-1} for 2 MeV electron irradiation [103]. After annealing at high temperature (typically 800–1000° C) where vacancies easily diffuse, the vacancies combine with impurities such as substitutional nitrogen atoms to form NV centers in diamonds. Vacancies can also recombine with self-interstitials to form extended defects, or they can become trapped at dislocations or at the surface. Some of the extended defects by the aggregation of vacancies have electron spins, and these paramagnetic vacancy clusters disturb the spin properties of NV centers. To reduce these paramagnetic vacancy clusters, subsequent annealing at high temperature is typically applied. Here, subsequent annealing is called *ex situ* annealing. In addition, *in situ* annealing is one of the most promising technique to reduce the paramagnetic vacancy clusters [59]. In fact, 2 MeV. electron irradiation combined with *ex situ* and *in situ* annealing helped in realizing the highest concentration of NV centers, such as ~45 ppm [104].

References

[1] B. J. Baliga, *IEEE Elect. Dev. Lett.* **10**, 455 (1989).
[2] M. Bhatnagar, B. J. Baliga, *IEEE Trans. Elect. Dev.* **40**, 645 (1993).
[3] A. Itoh, H. Matsunami, *Phys. Status Solidi a* **162**, 389 (1997).
[4] T. Kimoto, *Jpn. J. Appl. Phys.* **54**, 040103 (2015).
[5] Y. Ando, Y. Okamoto, H. Miyamoto, T. Nakayama, T. Inoue, M. Kuzuhara, *IEEE Elect. Dev. Lett.* **24**, 289 (2003).
[6] M. Kodama, M. Sugimoto, E. Hayashi, N. Soejima, O. Ishiguro, M. Kanechika, K. Itoh, H. Ueda, T. Uesugi, T. Kachi, *Appl. Phys. Exp.* **1**, 021104 (2008).
[7] J. E. Butler, M. W. Geis, K. E. Krohn, J. Lawless Jr, S. Deneault, T. M. Lyszczarz, D. Flechtner, R. Wright, *Semicond. Sci. Tech.* **18**, S67 (2003).
[8] T. Iwasaki, J. Yaita, H. Kato, T. Makino, M. Ogura, D. Takeuchi, H. Okushi, S. Yamasaki, M. Hatano, *IEEE Elect. Dev. Lett.* **35**, 241 (2014).
[9] B. M. Maune, M. G. Borselli, B. Huang, T. D. Ladd, P. W. Deelman, K. S. Holabird, A. A. Kiselev, I. Alvarado-Rodriguez, R. S. Ross, A. E. Schmitz, M. Sokolich, C. A.Watson, M. F. Gyure, A. T. Hunter, *Nature* **481**, 344 (2012).
[10] M. Veldhorst, C. H. Yang, J. C. C. Hwang, W. Huang, J. P. Dehollain, J. T. Muhonen, S. Simmons, A. Laucht, F. E. Hudson, K. M. Itoh, A. Morello, A. S. Dzurak, *Nature* **526**, 410 (2015).
[11] J. R. Petta, A. C. Johnson, J. M. Taylor, E. A. Laird, A. Yacoby, M. D. Lukin, C. M. Marcus, M. P. Hanson, A. C. Gossard, *Science* **309**, 2180 (2005).
[12] G. Balasubramanian, P. Neumann, D. Twitchen, M. Markham, R. Kolesov, N. Mizuochi, J. Isoya, J. Achard, J. Beck, J. Tissler, V. Jacques, P. R. Hemmer, F. Jelezko, J. Wrachtrup, *Nat. Mater.* **8**, 383 (2009).
[13] A. Gruber, A. Dräbenstedt, C. Tietz, L. Fleury, J. Wrachtrup, C. von Borczyskowski, *Science* **276**, 2012 (1997).
[14] N. Mizuochi, T. Makino, H. Kato, D. Takeuchi, M. Ogura, H. Okushi, M. Nothaft, P. Neumann, A. Gali, F. Jelezko, J. Wrachtrup, S. Yamasaki, *Nat. Photon.* **6**, 299 (2012).
[15] G. de Lange, Z. H. Wang, D. Ristè, V. V. Dobrovitski, R. Hanson, *Science* **330**, 60 (2010).
[16] B. Hensen, H. Bernien, A. E. Dréau, A. Reiserer, N. Kalb, M. S. Blok, J. Ruitenberg, R. F. L. Vermeulen, R. N. Schouten, C. Abellán, W. Amaya, V. Pruneri, M. W. Mitchell, M. Markham, D. J. Twitchen, D. Elkouss, S. Wehner, T. H. Taminiau, R. Hanson, *Nature* **526**, 682 (2015).

[17] G. Kucsko, P. C. Maurer, N. Y. Yao, M. Kubo, H. J. Noh, P. K. Lo, H. Park, M. D. Lukin, *Nature* **500**, 54 (2013).
[18] V. M. Acosta, E. Bauch, M. P. Ledbetter, A. Waxman, L.-S. Bouchard, D. Budker, *Phys. Rev. Lett.* **104**, 070801 (2010).
[19] F. Dolde, H. Fedder, M. W. Doherty, T. Nöbauer, F. Rempp, G. Balasubramanian, T. Wolf, F. Reinhard, L. C. L. Hollenberg, F. Jelezko, J. Wrachtrup, *Nat. Phys.* **7**, 459 (2011).
[20] T. Iwasaki, W. Naruki, K. Tahara, T. Makino, H. Kato, M. Ogura, D. Takeuchi, S. Yamasaki, M. Hatano, *ACS Nano* **11**, 1238 (2017).
[21] N. Aslam, M. Pfender, P. Neumann, R. Reuter, A. Zappe, F. F. de Oliveira, A. Denisenko, H. Sumiya, S. Onoda, J. Isoya, J. Wrachtrup, *Science* **357**, 67 (2017).
[22] C. Wang, C. Kurtsiefer, H. Weinfurter, B. Burchard, *J. Phys. B: At. Mol. Opt. Phys.* **39**, 37 (2006).
[23] E. Neu, D. Steinmetz, J. Riedrich-Möller, S. Gsell, M. Fischer, M. Schreck, C. Becher, *New J. Phys.* **13**, 025012 (2011).
[24] T. Iwasaki, F. Ishibashi, Y. Miyamoto, Y. Doi, S. Kobayashi, T. Miyazaki, K. Tahara, K. D. Jahnke, L. J. Rogers, B. Naydenov, F. Jelezko, S. Yamasaki, S. Nagamachi, T. Inubushi, N. Mizuochi, M. Hatano, *Sci. Rep.* **5**, 12882 (2015).
[25] T. Iwasaki, Y. Miyamoto, T. Taniguchi, P. Siyushev, M. H. Metsch, F. Jelezko, M. Hatano, *Phys. Rev. Lett.* **119**, 253601 (2017).
[26] T. Gaebel, I. Popa, A. Gruber, M. Domhan, F. Jelezko, J. Wrachtrup, *New J. Phys.* **6**, 98 (2004).
[27] I. Aharonovich, S. Castelletto, D. A. Simpson, A. Stacey, J. McCallum, A. D. Greentree, S. Prawer, *Nano Lett.* **9**, 3191 (2009).
[28] Y. Doi, T. Fukui, H. Kato, T. Makino, S. Yamasaki, T. Tashima, H. Morishita, S. Miwa, F. Jelezko, Y. Suzuki, N. Mizuochi, *Phys. Rev. B* **93**, 081203(R) (2016).
[29] B. Grotz, M. V. Hauf, M. Dankerl, B. Naydenov, S. Pezzagna, J. Meijer, F. Jelezko, J. Wrachtrup, M. Stutzmann, F. Reinhard, J. A. Garrido, *Nat. Commun.* **3**, 729 (2012).
[30] H. Kato, M. Wolfer, C. Schreyvogel, M. Kunzer, W. Müller-Sebert, H. Obloh, S. Yamasaki, C. Nebel, *Appl. Phys. Lett.* **102**, 151101 (2013).
[31] Y. Doi, T. Makino, H. Kato, D. Takeuchi, M. Ogura, H. Okushi, H. Morishita, T. Tashima, S. Miwa, S. Yamasaki, P. Neumann, J. Wrachtrup, Y. Suzuki, N. Mizuochi, *Phys. Rev. X* **4**, 011057 (2014).
[32] F. M. Hrubesch, G. Braunbeck, M. Stutzmann, F. Reinhard, M. S. Brandt, *Phys. Rev. Lett.* **118**, 037601 (2017).
[33] M. Gulka, E. Bourgeois, J. Hruby, P. Siyushev, G. Wachter, F. Aumayr, P. R. Hemmer, A. Gali, F. Jelezko, M. Trupke, M. Nesladek, *Phys. Rev. Appl.* **7**, 044032 (2017).
[34] F. Fuchs, B. Stender, M. Trupke, D. Simin, J. Pflaum, V. Dyakonov, G. V. Astakhov, *Nat. Commun.* **6**, 7578 (2015).

[35] M. Widmann, S. Y. Lee, T. Rendler, N. T. Son, H. Fedder, S. Paik, L. P. Yang, N. Zhao, S. Yang, I. Booker, A. Denisenko, M. Jamali, S. A. Momenzadeh, I. Gerhardt, T. Ohshima, A. Gali, E. Janzén, J. Wrachtrup, *Nat. Mater.* **14**, 164 (2015).
[36] S. Castelletto, B. C. Johnson, V. Ivády, N. Stavrias, T. Umeda, A. Gali, T. Ohshima, *Nat. Mater.* **13**, 151 (2014).
[37] D. J. Christle, A. L. Falk, P. Andrich, P. V. Klimov, J. Ul Hassan, N. T. Son, E. Janzén, T. Ohshima, D. D. Awschalom, *Nat. Mater.* **14**, 160 (2015).
[38] W. F. Koehl, B. B. Buckley, F. J. Heremans, G. Calusine, D. D. Awschalom, *Nature* **479**, 84 (2011).
[39] A. Lohrmann, N. Iwamoto, Z. Bodrog, S. Castelletto, T. Ohshima, T. J. Karle, A. Gali, S. Prawer, J. C. McCallum, B. C. Johnson, *Nat. Commun.* **6**, 7783 (2015).
[40] A. Lohrmann, S. Castelletto, J. R. Klein, T. Ohshima, M. Bosi, M. Negri, D. W. M. Lau, B. C. Gibson, S. Prawer, J. C. McCallum, B. C. Johnson, *Appl. Phys. Lett.* **108**, 021107 (2016).
[41] Y. Abe, T. Umeda, M. Okamoto, R. Kosugi, S. Harada, M. Haruyama, W. Kada, O. Hanaizumi, S. Onoda, T. Ohshima, *Appl. Phys. Lett.* **112**, 031105 (2018).
[42] H. J. von Bardeleben, J. L. Cantin, E. Rauls, U. Gerstmann, *Phys. Rev. B* **92**, 064104 (2015).
[43] S. A. Zargaleh B. Eble, S. Hameau, J.-L. Cantin, L. Legrand, M. Bernard, F. Margaillan, J.-S. Lauret, J.-F. Roch, H. J. von Bardeleben, E. Rauls, U. Gerstmann, F. Treussart, *Phys. Rev. B* **94**, 060102(R) (2016).
[44] A. L. Falk, B. B. Buckley, G. Calusine, W. F. Koehl, V. V. Dobrovitski, A. Politi, C. A. Zorman, P. X.-L. Feng, D. D. Awschalom, *Nat. Commun.* **4**, 1819 (2013).
[45] H. J. von Bardeleben, J. L. Cantin, A. Csóré, A. Gali, E. Rauls, U. Gerstmann, *Phys. Rev. B* **94**, 121202(R) (2016).
[46] D. Simin, F. Fuchs, H. Kraus, A. Sperlich, P. G. Baranov, G. V. Astakhov, V. Dyakonov, *Phys. Rev. Appl.* **4**, 014009 (2015).
[47] C. J. Cochrane, J. Blacksberg, M. A. Anders, P. M. Lenahan, *Sci. Rep.* **6**, 37077 (2016).
[48] D. Simin, V. A. Soltamov, A. V. Poshakinskiy, A. N. Anisimov, R. A. Babunts, D. O. Tolmachev, E. N. Mokhov, M. Trupke, S. A. Tarasenko, A. Sperlich, P. G. Baranov, V. Dyakonov, G. V. Astakhov, *Phys. Rev. X* **6**, 031014 (2016).
[49] A. N. Anisimov, D. Simin, V. A. Soltamov, S. P. Lebedev, P. G. Baranov, G. V. Astakhov, V. Dyakonov, *Sci. Rep.* **6**, 33301 (2016).
[50] T. Ohshima, T. Satoh, H. Kraus, G. V. Astakhov, V. Dyakonov, P. G. Baranov, *J. Phys. D: Appl. Phys.* **51**, 333002 (2018).
[51] F. Fuchs, V. A. Soltamov, S. Väth, P. G. Baranov, E. N. Mokhov, G. V. Astakhov, V. Dyakonov, *Sci. Rep.* **3**, 1637 (2013).
[52] C. F. de las Casas, D. J. Christle, J. U. Hassan, T. Ohshima, N. T. Son, D. D. Awschalom, *Appl. Phys. Lett.* **111**, 262403 (2017).

[53] S.-I. Sato, T. Honda, T. Makino, Y. Hijikata, S.-Y. Lee, T. Ohshima, *ACS Photon.* **5**, 3159 (2018).

[54] M. Widmann, M. Niethammer, T. Makino, T. Rendler, S. Lasse, T. Ohshima, J. Ul Hassan, N. T. Son, S.-Y. Lee, J. Wrachtrup, *Appl. Phys. Lett.* **112**, 231103 (2018).

[55] J. Meijer, B. Burchard, M. Domhan, C. Wittmann, T. Gaebel, I. Popa, F. Jelezko, J. Wrachtrup, *Appl. Phys. Lett.* **87**, 261909 (2005).

[56] S. Pezzagna, B. Naydenov, F. Jelezko, J. Wrachtrup, J. Meijer, *New J. Phys.* **12**, 065017 (2010).

[57] S. Pezzagna, D. Rogalla, D. Wildanger, J. Meijer, A. Zaitsev, *New J. Phys.* **13**, 035024 (2011).

[58] J. Meijer, A. Stephan, *Nucl. Instrum. Metho. Phys. Res. B* **188**, 9 (2002).

[59] T. Yamamoto, T. Umeda, K. Watanabe, S. Onoda, M. L. Markham, D. J. Twitchen, B. Naydenov, L. P. McGuinness, T. Teraji, S. Koizumi, F. Dolde, H. Fedder, J. Honert, J. Wrachtrup, T. Ohshima, F. Jelezko, J. Isoya, *Phys. Rev. B* **88**, 075206 (2013).

[60] T. Yamamoto, S. Onoda, T. Ohshima, T. Teraji, K. Watanabe, S. Koizumi, T. Umeda, L. P. McGuinness, C. Muller, B. Naydenov, F. Dolde, H. Fedder, J. Honert, M. L. Markham, D. J. Twitchen, J. Wrachtrup, F. Jelezko, J. Isoya, *Phys. Rev. B* **90**, 081117(R) (2014).

[61] T. Teraji, T. Yamamoto, K. Watanabe, Y. Koide, J. Isoya, S. Onoda, T. Ohshima, L. J. Rogers, F. Jelezko, P. Neumann, J. Wrachtrup, S. Koizumi, *Phys. Status Solidi a* **212**, 2365 (2015).

[62] T. Sakai, T. Hamano, T. Suda, T. Hirao, T. Kamiya, *Nucl. Instrum. Methods Phys. Res. B* **130**, 498 (1997).

[63] T. Kamiya, T. Sakai, Y. Naitoh, T. Hamano, T. Hirao, *Nucl. Instrum. Methods Phys. Res. B* **158**, 255 (1999).

[64] I. Jakobi, S. A. Momenzadeh, F. F-Oliveira, J. Michil, F. Ziem, M. Schreck, P. Neumann, A. Denisenko, J. Wrachtrup, *J. Phys.: Conf. Series* **752**, 012001 (2016).

[65] P. Neumann, R. Kolesov, B. Naydenov, J. Beck, F. Rempp, M. Steiner, V. Jacques, G. Balasubramanian, M. L. Markham, D. J. Twitchen, S. Pezzagna, J. Meijer, J. Twamley, F. Jelezko, J. Wrachtrup, *Nat. Phys.* **6**, 249 (2010).

[66] S. A. E. Johansson, J. L. Campbell, K. G. Malmqvist, eds., *Particle-induced X-Ray Emission Spectrometry (PIXE)*, John Wiley & Sons Inc. (1995).

[67] F. Watt, M. B. H. Breese, A. A. Bettiol, J. A. van Kan, *Materialstoday* **10**, 20 (2007).

[68] H. Kraus, D. Simin, C. Kasper, Y. Suda, S. Kawabata, W. Kada, T. Honda, Y. Hijikata, T. Ohshima, V. Dyakonov, G.V. Astakhov, *Nano Lett.* **17**, 2865 (2017).

[69] T. Ohshima, T. Honda, S. Onoda, T. Makino, M. Haruyama, T. Kamiya, T. Satoh, Y. Hijikata, W. Kada, O. Hanaizumi, A. Lohrmann, J. R. Klein, B. C. Johnson, J. C. McCallum, S. Castelletto, B. C. Gibson, H. Kraus, V. Dyakonov, G. V. Astakhov, *Mater. Sci. Forum* **897**, 233 (2017).

[70] Y. Yamazaki, Y. Chiba, T. Makino, S.-I. Sato, N. Yamada, T. Satoh, Y. Hijikata, K. Kojima, S.-Y. Lee, T. Ohshima, *J. Mater. Res.* **33**, 3355 (2018).
[71] L. Patrick, W. J. Choyke, *Phys. Rev. B* **5**, 3253 (1972).
[72] M. Lesik, P. Spinicelli, S. Pezzagna, P. Happel, V. Jacques, O. Salord, B. Rasser, A. Delobbe, P. Sudraud, A. Tallaire, J. Meijer, J-F. Roch, *Phys. Status Solidi a* **210**, 2055 (2013).
[73] S. Tamura, G. Koike, A. Komatsubara, T. Teraji, S. Onoda, L. P. McGuinness, L. Rogers, B. Naydenov, E. Wu, L. Yan, F. Jelezko, T. Ohshima, J. Isoya, T. Shinada, T. Tanii, *Appl. Phys. Exp.* **7**, 115201 (2014).
[74] S. Pezzagna, D. Rogalla, H.-W. Becker, I. Jakobi, F. Dolde, B. Naydenov, J. Wrachtrup, F. Jelezko, C. Trautmann, J. Meijer, *Phys. Stat. Soli. A* **208**, 2017 (2011).
[75] J. Meijer, S. Pezzagna, T. Vogel, B. Burchard, H. H. Bukow, I. W. Rangelow, Y. Sarov, H. Wiggers, I. Plümel, F. Jelezko, J. Wrachtrup, F. S. Kaler, W. Schnitzler, K. Singer, *Appl. Phys. A* **91**, 567 (2008).
[76] S. Pezzagna, D. Wildanger, P. Mazarov, A. D. Wieck, Y. Sarov, I. Rangelow, B. Naydenov, F. Jelezko, S. W. Hell, J. Meijer, *Small* **6**, 2117 (2010).
[77] D. M. Toyli, C. D. Weis, G. D. Fuchs, T. Schenkel, D. D. Awschalom, *Nano Lett.* **10**, 3168 (2010).
[78] S. Sangtawesin, T. O. Brundage, Z. J. Atkins, J. R. Petta, *Appl. Phys. Lett.* **105**, 063107 (2014).
[79] D. Scarabelli, M. Trusheim, O. Gaathon, D. Englund, S. J. Wind, *Nano Lett.* **16**, 4982 (2016).
[80] P. Spinicelli, A. Dréau, L. Rondin, F. Silva, J. Achard, S. Xavier, S. Bansropun, T. Debuisschert, S. Pezzagna, J. Meijer, V. Jacques, J-F. Roch, *New J. Phys.* **13**, 025014 (2011).
[81] J. Wang, F. Feng, J. Zhang, J. Chen, Z. Zheng, L. Guo, W. Zhang, X. Song, G. Guo, L. Fan, C. Zou, L. Lou, W. Zhu, G. Wang, *Phys. Rev. B* **91**, 155404 (2015).
[82] R. Fukuda, P. Balasubramanian, I. Higashimata, G. Koike, T. Okada, R. Kagami, T. Teraji, S. Onoda, M. Haruyama, K. Yamada, M. Inaba, H. Yamano, F. M Stürner, S. Schmitt, L. P. McGuinness, F. Jelezko, T. Ohshima, T. Shinada, H. Kawarada, W. Kada, O. Hanaizumi, T. Tanii, J. Isoya, *New J. Phys.* **20**, 083029 (2018).
[83] F. Dolde, I. Jakobi, B. Naydenov, N. Zhao, S. Pezzagna, C. Trautmann, J. Meijer, P. Neumann, F. Jelezko, J. Wrachtrup, *Nat. Phys.* **9**, 139 (2013).
[84] J. R. Moller, S. Pezzagna, J. Meijer, C. Pauly, F. Mucklich, M. Markham, A. M. Edmonds, C. Becher, *Appl. Phys. Lett.* **106**, 221103 (2015).
[85] T. Gaebel, M. Domhan, I. Popa, C. Wittmann, P. Neumann, F. Jelezko, J. R. Rabeau, N. Stavrias, A. D. Greentree, S. Prawer, J. Meijer, J. Twamley, P. R. Hemmer, J. Wrachtrup, *Nat. Phys.* **2**, 408 (2006).
[86] T. Yamamoto, C. Muller, L. P. McGuinness, T. Teraji, B. Naydenov, S. Onoda, T. Ohshima, J. Wrachtrup, F. Jelezko, J. Isoya, *Phys. Rev. B* **88**, 201201(R) (2013).

[87] T. Staudacher, F. Shi, S. Pezzagna, J. Meijer, J. Du, C. A. Meriles, F. Reinhard, J. Wrachtrup, *Science* **339**, 561 (2013).
[88] H. J. Mamin, M. Kim, M. H. Sherwood, C. T. Rettner, K. Ohno, D. D. Awschalom, D. Rugar, *Science* **339**, 557 (2013).
[89] C. Müller, X. Kong, J.-M. Cai, K. Melentijevic, A. Stacey, M. Markham, D. Twitchen, J. Isoya, S. Pezzagna, J. Meijer, J. F. Du, M. B. Plenio, B. Naydenov, L. P. McGuinness, F. Jelezko, *Nat. Commun.* **5**, 4703 (2014).
[90] M. Loretz, S. Pezzagna, J. Meijer, C. L. Degen, *Appl. Phys. Lett.* **104**, 033102 (2014).
[91] S. J. DeVience, L. M. Pham, I. Lovchinsky, A. O. Sushkov, N. Bar-Gill, C. Belthangady, F. Casola, M. Corbett, H. Zhang, M. Lukin, H. Park, A. Yacoby, R. L. Walsworth, *Nat. Nanotechnol.* **10**, 129 (2015).
[92] D. Rugar, H. J. Mamin, M. H. Sherwood, M. Kim, C. T. Rettner, K. Ohno, D. D. Awschalom, *Nat. Nanotechnol.* **10**, 120 (2015).
[93] I. Lovchinsky, A. O. Sushkov, E. Urbach, N. P, Leon, S. Choi, K. Greve, R. Evans, R. Gertner, E. Bersin, C. Muller, L. McGuinness, F. Jelezko, R. L. Walsworth, H. Park, M. D. Lukin, *Science* **351**, 836 (2016).
[94] L. M. Pham, S. J. DeVience, F. Casola, I. Lovchinsky, A. O. Sushkov, E. Bersin, J. Lee, E. Urbach, P. Cappellaro, H. Park, A. Yacoby, M. Lukin, R. L. Walsworth, *Phys. Rev. B* **93**, 045425 (2016).
[95] S. Onoda, M. Haruyama, T. Teraji, J. Isoya, W. Kada, O. Hanaizumi, T. Ohshima, *Phys. Status Solidi a* **212**, 2641 (2015).
[96] S. Onoda, K. Tatsumi, M. Haruyama, T. Teraji. J. Isoya, W. Kada, T. Ohshima, O. Hanaizumi, *Phys. Status Solidi a* **214**, 1700160 (2017).
[97] S. W. Hell, J. Wichmann, *Opt. Lett.* **19**, 780 (1994).
[98] J. Koike, D. M. Parkin, T. E. Mitchell, *Appl. Phys. Lett.* **60**, 1450 (1992).
[99] http://www.gel.usherbrooke.ca/casino/.
[100] G. Davies, S. C. Lawson, A. T. Collins, A. Mainwood, S. J. Sharp, *Phys. Rev. B* **46**, 13157 (1992).
[101] J. N. Lomer, A. M. A. Wild, *Radi. Eff.* **17**, 37 (1973).
[102] J. Loubser, J. Wyk, *Rep. Prog. Phys.* **41**, 1201 (1978).
[103] D. C. Hunt, D. J. Twitchen, M. E. Newton, J. M. Baker, T. R. Anthony, W. F. Banholzer, S. S. Vagarali, *Phys. Rev. B* **61**, 3863 (2000).
[104] S. Choi, J. Choi, R. Landig, G. Kucsko, H. Zhou, J. Isoya, F. Jelezko, S. Onoda, H. Sumiya, V. Khemani, C. von Keyserlingk, N. Y. Yao, E. Demler, M. D. Lukin, *Nature* **543**, 221 (2017).

CHAPTER 5

Point Defects in InN

XINQIANG WANG* **and HUAPENG LIU**

State Key Laboratory of Artificial Microstructure and Mesoscopic Physics, School of Physics, Peking University, Beijing 100871, China
*wangshi@pku.edu.cn

In this chapter, we review the theory of native point defects, impurities, and complexes defects in InN, focusing on the results of first-principles calculations based on density functional theory and pseudopotentials. We discuss the structural and electronic properties of the most relevant native defects, impurities, and complexes defects and their impact on the electrical properties of InN. In particular, we discuss the possible causes of unintentional n-type conductivity in as-grown InN and the prospects for p-type doping. In Section 2, we describe native point defects in InN, and in Section 3, we discuss the impurities in InN. In Section 4, we present the results of complexes in InN. Section 5 summarizes the chapter.

1. Introduction

Indium nitride has attracted considerable attention as it possesses a series of outstanding properties [1–3], such as a small effective mass [4] leading to a large electron mobility and a high carrier saturation velocity. Therefore, it is expected to be one of the most promising materials for high-speed and high-frequency electronic devices [5].

However, it is inevitable that native defects and impurity incorporation will exist during crystal growth. This unintentional introduction of impurities and native defects can affect the physical and electronic properties of InN and thus may degrade the performance of devices. Thus, understanding the role of native defects, impurities and dopants and the interaction between them, in altering the material properties and eventually controlling the concentration, are of crucial importance for the growth and fabrication of group III nitride-based electronic and optoelectronic devices.

Although an integration of many different experimental techniques, such as electrical, optical, magnetic, and mass spectroscopy techniques, is applied [6–8], the identification and characterization of defects and impurities in semiconductors is a formidable task through experiment. Thus, computational studies have been playing an increasingly important role in the investigation of defects in semiconductors. In particular, first-principles calculations based on the density functional theory [9–11] have proved to be a powerful tool for studying the structural and electronic properties of native defects and impurities in semiconductors [12]. Stable and metastable positions of native defects and impurities in the crystal lattice and the surrounding local structural relaxations can be determined by minimizing total energies with respect to the atomic coordinates [10, 11]. A formalism has been developed that allows for calculations of formation energies as a function of Fermi-level position and chemical potentials [12, 13]. In addition, frequencies of the local vibrational modes can be calculated by diagonalizing the calculated force-constant matrices [12] and can be directly compared with infrared or Raman spectroscopy measurements.

2. Native Point Defects in InN

In compound semiconductors such as InN, there are six possible native point defects: vacancies, which consist of missing atoms in the regular sublattices of N or In (V_N and V_{In}); self-interstitials, which consist of extra nitrogen or indium atoms occupying interstitial sites in the InN lattice (N_i and In_i); and antisites, where nitrogen or indium atoms occupy sites in the "wrong" sublattice (N_{In} and In_N), that is, an extra nitrogen atom replacing an indium atom or vice versa. Here, we show several classical results about six native point defects by using different kinds of calculation methods.

2.1. *Tight-binding calculations*

Defects in InN were previously studied by Jenkins and Dow [14], who used parametrized tight-binding calculations to investigate native defects and doping. They assumed a bandgap of 1.9 eV for InN. Jenkins and Dow suggested that the nitrogen vacancy would be responsible for the observed n-type conductivity in as-grown InN and that the In antisite would explain an optical absorption peak attributed to a defect level located in the gap. But their results for nitrogen vacancies are inconsistent with results from first-principles methods, described in the next section. For instance, nitrogen vacancies are known to induce large local lattice relaxations that greatly affect the position of transition levels, as described below. This feature was not captured in the early calculations [14]. Based on the Jenkins and Dow calculations, Tansley and Egan [15] discussed their results of optical measurements on radio-frequency reactively sputtered polycrystalline films of InN. Tansley and Egan obtained a band gap of 1.94 eV and attributed the defect levels in the upper half of the band gap to nitrogen vacancies and nitrogen antisites. Since the currently accepted value of the InN fundamental band gap is 0.7 eV, we believe that the results of Ref. [15] are artifacts related to the poor crystalline quality and/or high impurity content of their samples. In an early review on nitrides, Strite and Morkoc [16] carefully pointed out that it was believed, but not proven, that the observed n-type conductivity in InN was caused by nitrogen vacancies.

2.2. *DFT-LDA calculations*

More recently, first-principles DFT-LDA calculations of native point defects were performed by Stampfl *et al.* [17] in zinc-blende InN. Although zinc-blende is not the ground-state phase of InN, the local tetrahedral structure and the electronic properties of zinc-blende InN are very similar to those of the ground-state wurtzite InN. Therefore, it is expected that the defect physics in wurtzite InN can be quantitatively described by calculations for the more symmetric zinc-blende InN. Besides, calculations have indeed confirmed that formation energies of defects and impurities in the wurtzite phase are similar to those in the zinc-blende phase, except for interstitial configurations, for which the atomic environments are quite different: the geometry of the wurtzite structure allows for octahedral and tetrahedral interstitial sites, whereas the zinc-blende structure allows for tetrahedral and hexagonal interstitial sites.

The results obtained by Stampfl *et al.* are shown in Fig. 1, where the formation energy of each point defect is plotted as a function of the Fermi level, for In-rich conditions. Figure 1 follows the standard convention in which only Fermi-level values within the bandgap are shown, but of course, the formation energies also apply to Fermi-level positions above the CBM. The results show that the defects with lowest formation energies for any Fermi-level position in the bandgap are nitrogen vacancies and indium interstitials. Although V_N and In_i are the most favorable to be formed in the In-rich limit, V_N and In_i are also the defects with the lowest formation energies under N-rich conditions due to the small formation enthalpy of InN ($-0.4\,eV$). In detail, under N-rich conditions, the formation energy of V_N and In_i increases by an amount given by the absolute value of the formation enthalpy of InN ($\Delta H_f = -0.4\,eV$); the formation energy of V_{In} and N_i drops by ΔH_f; the formation energy of InN increases by twice the absolute value of ΔH_f; and the formation energy of N_{In} drops by twice ΔH_f.

Regardless of the conditions (In-rich or N-rich), nitrogen interstitials (N_i), indium vacancies (V_{In}), and both nitrogen and indium antisites (N_{In} and In_N) were found to have high formation energies. Therefore, these defects will be present with insignificant concentrations in as-grown InN. Due to the exceedingly high formation energies of N_i and N_{In} (Fig. 1) and the fact that these defects are unlikely to influence the unintentional n-type conductivity in InN, Stampfl *et al.* did not investigate all the possible charge states, limiting themselves to the neutral charge state.

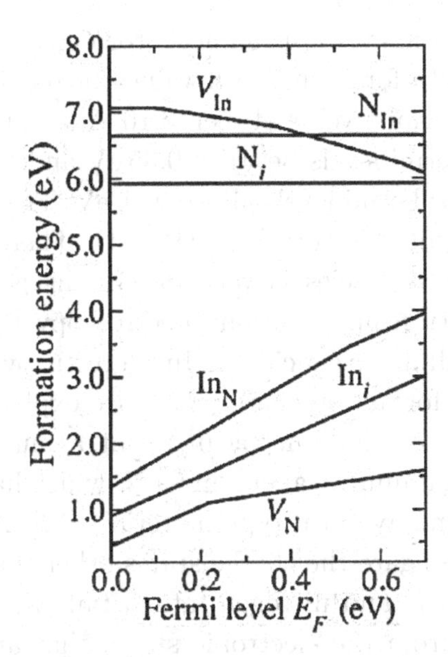

Figure 1. Calculated formation energies as a function of Fermi-level position for all native point defects in zinc-blende InN. The Fermi level E_F is referenced to the valence band maximum. Only results for In-rich conditions are shown Ref. [39].

The slopes in the formation energy versus Fermi-level curves in Fig. 1 indicate the charge state of the respective defects. For each defect, only the lowest-energy charge state is shown for a particular position of the Fermi level. The kinks indicate the transition levels, which are the Fermi-level values for which two charge states of the same defect have equal formation energies. For example, the nitrogen vacancy V_N has a transition level at 0.2 eV above the VBM. It is stable in the 3+ charge state for Fermi levels below 0.20 eV and in the + charge state for Fermi levels above 0.20 eV; hence, V_N is a shallow donor. Note that the 2+ charge state is not stable for any Fermi-level position in the bandgap; the reason for this so-called negative-U behavior will be discussed below. The indium interstitial In_i is stable exclusively in the 3+ charge state and is therefore also a shallow donor. The indium antisite In_N is a shallow donor, too. It is stable in the 4+ charge state for Fermi-level values below 0.55 eV and in the 3+ charge state for Fermi levels above 0.55 eV.

The indium vacancy V_{In} is an acceptor defect. It is stable in the neutral charge state for Fermi-level values below 0.10 eV, in the (1) charge state for Fermi levels between 0.10 eV and 0.35 eV, in the (2) charge state for Fermi levels between 0.35 eV and 0.65 eV, and in the (3) charge state for Fermi levels above 0.65 eV. The formation energy of the indium vacancy is very high (Fig. 1). Therefore, the concentration of indium vacancies is very low in the good-quality material. Indeed, positron annihilation spectroscopy [18] has confirmed an absence of indium vacancies in InN grown by molecular beam epitaxy. The high formation energy may also explain why irradiation with $^4He^+$ introduces indium vacancies at a much lower rate [19] than the rate for gallium vacancies in GaN, which have a significantly lower formation energy (in n-type material) [12]. Although Stampfl *et al.* did not investigate the charge states other than neutral for the nitrogen antisite N_{In} and nitrogen interstitial N_i, we expect N_{In} to be a deep donor from the electronic states that are induced in the band gap; similarly, we expect N_i to be a deep acceptor when occupying the octahedral interstitial site or electrically neutral when in the split interstitial form where two N atoms share the same N lattice site.

2.3. *Beyond-LDA results*

As the bandgap of InN is corrected, a more accurate calculation was performed by Stampfl *et al.* [20] By using self-interaction and relaxation-corrected (SIRC) pseudopotentials, the changes in the defect-induced states in InN are shown as follows. V_N induces a fully symmetric a_1 state near the VBM and a threefold degenerate t_2 state above the CBM. In the neutral vacancy V_N^0, the a_1 state is doubly occupied and the t_2 would be singly occupied. Since the t_2 state is above the CBM, the system can lower its energy by transferring the electron from the t_2 to the CBM. Therefore, the neutral charge state is never stable. In the SIRC pseudopotential calculations, it is found that the t_2 state is shifted with the CBM as the gap is corrected. The fact that defect states above the CBM shift upward with the conduction band is a very important finding. It confirms that V_N acts as a shallow donor, even when a band-gap correction is applied. The singlet state near the VBM is also shifted to higher energy, but by an amount that is much smaller than the band-gap correction. Furthermore, an inspection of the wave functions of the defect-induced states indicated that they are very similar in SIRC and LDA. Although the SIRC approach does not allow evaluation of total energies, it is possible to estimate how the calculated changes in the band structure affect the total energy. For V_N^0, the total energy should reflect the shift in the triplet state located above the CBM, that is, an increase in the formation energy of V_N^0. This keeps the transition state above the CBM and confirms that the nitrogen vacancy is still a shallow donor even when band-gap corrections are applied. This contribution to the energy is absent for V_N^+, for which the t_2 states are unoccupied. The formation energy of V_N^+ is mostly affected by the shift of the occupied a_1 state near the VBM. From the SIRC results, this shift is much smaller than that of the t_2 states and is expected to increase the formation energy of V_N^+ by $\sim$0.4 eV.

The indium vacancy creates a doubly occupied fully symmetric a_1 state that is resonant in the valence band and a threefold t_2 state close to the VBM. In the neutral charge state, $\mathrm{V}_{\mathrm{In}}^0$, t_2 is occupied by three electrons. The t_2 state can accept up to three more electrons, making the indium vacancy a triple acceptor ($\mathrm{V}_{\mathrm{In}}^{3-}$). The position of this state with respect to the VBM shifts by less than 0.01 eV in SIRC compared to LDA, and therefore, only a small increase in the formation energy of the indium vacancy is expected as the band gap is corrected. The nitrogen antisite N_{In} creates a doubly occupied singlet state in the band gap, as well as a higher lying triplet state that can accept up to six electrons. In the SIRC calculations, the singlet state is higher in the band gap by about 0.3 eV. This implies a higher formation energy of this defect than that obtained with the LDA calculations. Since the LDA energy was already so high as to imply very small concentrations, this correction would not change the conclusion that N_{In} is unlikely to affect the electrical properties of InN. The empty triplet state follows the CBM as the band gap is corrected and, because it is empty, it does not affect the formation energy of the neutral N_{In}.

Nitrogen interstitials (N_i) have high formation energies, as shown in Fig. 1. They introduce states in the band gap that have mostly N 2p character, and therefore, should follow the VBM as the band gap is corrected. In this case, their formation energy is well described by the LDA, and we expect the equilibrium concentration of N_i to be insignificant in as-grown InN.

According to results obtained above, the nitrogen vacancy V_N is the lowest energy defect in InN under any condition among six native point defects. Therefore, the properties of V_N have be investigated in more detail in wurtzite InN by using 96-atom supercells and based on both LDA and LDA+U [21]. Since LDA+U provides a partial correction to the band gap, transition levels and formation are consistently corrected according to the valence versus conduction band character of the states in the band gap. This method was previously successfully applied to the calculation of oxygen vacancies in ZnO [22].

Formation energies as a function of Fermi level for all charge states of V_N in wurtzite InN are shown in Fig. 2(a), and the local structural relaxations for the 3+ and + charge states are shown in Figs. 2(b) and 2(c). The results for V_N are consistent with the calculations obtained by Stampfl *et al.* [20] as shown in Fig. 1. V_N is stable in the 3+ and + charge states, with the $\varepsilon(3+/+)$ transition level at 0.1 eV above the VBM. The nitrogen vacancy V_N in InN is a negative-U system, in which the 2+ charge state is unstable for any position of the Fermi level in the band gap. This can be explained by the large difference in local structural relaxations around V_N for the different charge states. In the positive charge state (+), the a_1 state near the VBM is doubly occupied. Hence, V_N^+ is stabilized by an inward relaxation of the four surrounding In atoms with about 1.8% [23], maximizing their bonding, as shown in Fig. 2(b). As the electrons are removed from the a_1 state, the four In nearest neighbors strongly relax outward, strengthening their bonding with their remaining N neighbors, as shown in Fig. 2(b). These relaxations lower the formation energies of V_N^+ and V_N^{3+} with respect to that of V_N^{2+}, so that the latter is only metastable. Accordingly, the only relevant transition level is that between the 3+ and + charge states, $\varepsilon(3+/+)$.

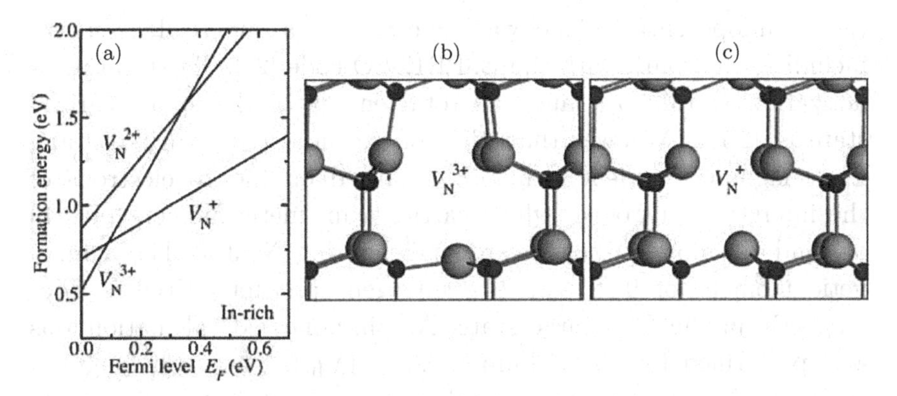

Figure 2. (a) Calculated formation energies as a function of Fermi level for the nitrogen vacancy in wurtzite InN, under In-rich conditions. Results for all three charge states, V_N^+, V_N^{2+}, and V_N^{3+}, are shown. E_F is referenced to the valence band maximum. The formation energies were corrected using LDA and LDA+U calculations as described in the text. Panels (b) and (c) show local structural relaxations around V_N^+ and V_N^{3+}. Indium atoms are represented by large spheres and N atoms by small spheres.

Finally, the possible role of nitrogen vacancies in the unintentional n type conductivity in as-grown InN is discussed. The formation energy of V_N in n-type material, for example, with the Fermi level E_F at the CBM, is 1.4 eV. This corresponds to an equilibrium concentration of V_N^+ of 2×10^{13} cm^{-3} at a typical growth temperature of $\sim$ 500°C [24]. The observed electron concentrations in as-grown InN are several orders of magnitude higher, and therefore cannot be explained by the presence of nitrogen vacancies. However, the formation energy of V_N for E_F at the valence band maximum is relatively low, and we expect V_N to be a possible source of compensation in p-type InN.

Although as Stampfl *et al.* concluded that nitrogen interstitials (N_i) and indium vacancies have high formation energies and play an insignificant role in carrier concentrations in as-grown InN, they found that the indium vacancies and nitrogen interstitials have some special properties. By performing spin-polarized calculations for the nitrogen split interstitial, they found that it is magnetic in the neutral charge state, where the value of the spin depends on the charge states. The calculations were performed using the LDA for the exchange correlation functional, as well as the pseudopotential plane wave method in a supercell geometry including 96 atoms, with the ESPRESSO code [25]. The total magnetization is 1.09 μ_B, and the total energy of the spin-polarized state is 125 meV lower than that of the non-spin-polarized state. The magnetic moment is mainly localized on the 2p electrons of the interstitial nitrogen, which carries a magnetic moment of 0.49 μ_B and the bonded host nearest-neighboring N atom has a magnetic moment of 0.39 μ_B. The nitrogen split interstitial is non-magnetic in the 3+ charge state. A spin-polarized calculation was also performed for the indium vacancy (V_{In}). The total magnetization is 3.0 μ_B for the neutral indium vacancy. For the singly negatively charged vacancy, V_{In}^{1-}, it has a total magnetic moment of 2.0 μ_B and the doubly negatively charged vacancy, V_{In}^{2-}, carries a magnetic moment of 1.06μ_B. The closed-shell V_{In}^{3-} defect does not carry a magnetic moment. The total energy of the spin-polarized state is 418 meV lower than that of non-spin-polarized state for the neutral indium vacancy. For V_{In}^{1-}, the preference of the spin-polarized state decreases to 206 meV, while there is a very small energy difference 49 meV between the spin-polarized state and the non-spin-polarized state for the 2− charge state. Analysis of the spin density shows that the magnetic moment is mainly localized on the unpaired 2p electrons of the four nearest-neighboring N atoms. For the neutral In vacancy, each N atom of the three equivalent neighbors carries a magnetic moment of 0.65 μ_B, and the remaining "apical" neighbor carries a magnetic moment of 0.67 μ_B. With increasing occupation of the defect levels, the surrounding nitrogen atoms have a smaller average magnetic moment, 0.455 μ_B/N for the 1− charge state and 0.25 μ_B/N for the 2− charge state.

3. Impurities in InN

We now discuss the behavior of impurities in InN. We focus on oxygen, silicon, and hydrogen as possible donors. In addition, we also discuss the effects of magnesium and carbon incorporation and their role as p-type dopants in InN.

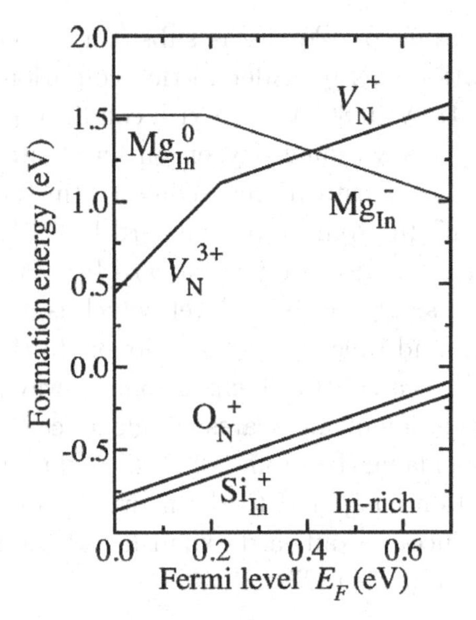

Figure 3. Calculated formation energies as a function of Fermi level for Si_{In}, O_N, and Mg_{In} in In-rich InN. Note that the Si_{In} and O_N donors have much lower formation energies than nitrogen vacancies (V_N).

3.1. *Donors (O, Si, and H)*

3.1.1. *O in InN*

Oxygen is a common unintentional impurity in nitrides and a shallow donor in GaN. Silicon is an n-type dopant in GaN and $Al_xGa_{1-x}N$. In InN, oxygen is expected to occupy the N site (O_N), while Si is expected to occupy the In site (Si_{In}). Stampfl *et al.* [20] calculated the formation energies of Si_{In} and O_N in In-rich InN. Their results are shown in Fig. 3. Both Si_{In} and O_N are shallow donors and have much lower formation energies than the nitrogen vacancy V_N. Based on the calculated formation energies of native defects shown in Fig. 1, it is reasonable that there will be no compensation of these impurities by acceptor-type native defects. Therefore, n-type doping of InN is easily attainable through extrinsic doping.

Because oxygen and silicon play an important role in affecting electronic properties of InN, these two impurities are studied in detail [26]. As shown above Stampfl *et al.* found that for the O impurity, the most favorable site is substitution on an N site O_N, where the formation energy under In-rich conditions in the neutral charge state is 0.47 eV (0.86 eV for N rich condition). Therefore, they only focused on substitutional oxygen on an N site for further calculations. In the O_N substitutional geometry (in the neutral charge state), the "apical" In atom moves outward by 1.67% and the three equivalent "planar" In atoms relax outward by 1.80%. The O impurity O_N induces a singly occupied level, which is a resonance in the conduction band, and thus, it acts as a donor. In the single positive charge state, there is an outward expansion — away from the oxygen atom — of the position of the nearest single apical In atom of 2.39% and of the three planar In atoms of 2.42%, thus a slightly larger breathing relaxation compared to the neutral state. For cubic GaN and AlN, O_N^+ induces an outward expansion of the nearest four Ga (Al) atoms of 4.2% (3.8%) [27, 28].

3.1.2. *Si in InN*

In the case of silicon, the situation is different from oxygen. Silicon has an atomic radius almost twice that of the nitrogen atom; thus, it causes a large strain if it replaces an N atom or occupies an interstitial site. Thus, Stampfl *et al.* [26] verified that for wz–InN, substitutional Si on an N site has a large formation energy of 5.04 eV under In-rich conditions and 7.75 eV under N-rich conditions. However, for Si on an In site, the formation energy is just 0.33 and 0.71 eV under In- and N-rich conditions, respectively. Thus, we only further consider silicon substituted on an In site. The Si_{In} defect induces a large inward contraction of the single apical neighboring N atom of 16.40% and of the nearest three planar N atoms by 15.90%. The huge contraction can be understood in that (i) the bond length of Si–N (~1.75 Å) in β-Si_3N_4 is much shorter than the apical N–In bond distance 2.15 Å and the planar N–In bond length 2.14 Å in InN, and (ii) the calculated heat of formation per atom of Si_3N_4 is greater by 1.36 eV than that of InN 0.58 eV, indicating the stronger Si–N bond. While Si_{Al}^+ in zb-AlN induces a relatively small inward relaxation of 5.0% of the neighboring N atoms [27], the bond length of Si–N is correspondingly only slightly shorter than that of Al–N in bulk. Like oxygen, Si_{In} induces a singly occupied level in the conduction band and thus acts as a donor. In the single positive charge state, Si_{In}^{+} induces an inward displacement of the apical O atom of 15.85% and for the three planar O atoms, the inward displacement is 15.36%, which is very similar to the neutral charge state.

3.1.3. *Hydrogen in InN*

We now turn to the effect of hydrogen on the properties of InN. Hydrogen is a common impurity in semiconductors, found in almost all growth and processing environments [29]. The presence of hydrogen is known to have a strong effect on the electronic properties of many materials. Hydrogen has been primarily thought of as an interstitial impurity that passivates intrinsic defects and other impurities, thereby significantly improving the electronic properties. In most semiconductors, interstitial hydrogen is amphoteric: it is stable as a donor in p-type and as an acceptor in n-type material, always counteracting the prevailing conductivity [29]. This behavior has been well established in GaN [30]. In contrast, interstitial hydrogen (H_i) in InN was predicted to be stable exclusively as a donor [29, 31] H_i strongly bonds to nitrogen, causing a breaking (or at least weakening) of an N–In chemical bond. Interstitial hydrogen can be found in the center bond of the or in the antibonding configurations, as shown in Figs. 4(a) and 4(b). Despite the strength of the N–H bond, calculations have shown that interstitial hydrogen is a fast diffuser in InN, with a migration barrier of 1.1 eV, causing it to be mobile at relatively modest temperatures [32]. Interstitial hydrogen incorporated during growth or processing will therefore still be mobile while the samples are cooled down.

In addition to being incorporated as an interstitial, hydrogen can also occupy a substitutional site in InN. This behavior is similar to what has been found in ZnO and MgO [33]. Substitutional hydrogen in InN was investigated using LDA and LDA+U calculations [32]. It was found that atomic hydrogen occupies a nitrogen site and forms a multicenter bond with its four In nearest neighbors, as discussed

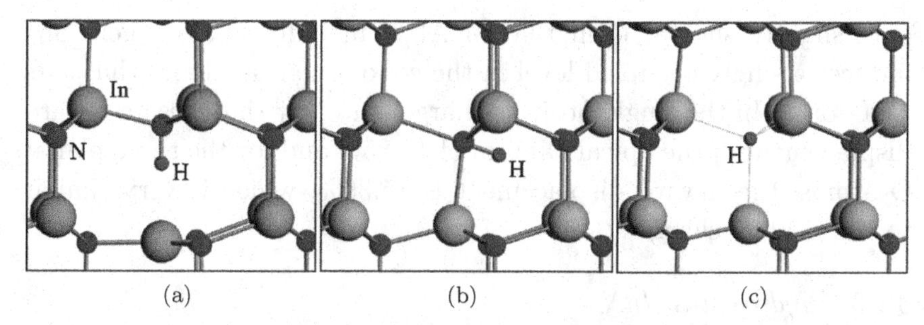

Figure 4. Local structure of interstitial hydrogen in (a) the bond-center configuration and (b) the antibonding configuration. (c) Substitutional hydrogen on a nitrogen site. Figure reproduced with permission from Ref. [32]. Copyright 2008, American Institute of Physics.

below. The calculated formation energies of interstitial and substitutional hydrogen are shown in Fig. 5. The formation energy of the nitrogen vacancy is also shown for comparison. Figure 5 shows that the formation energy of H_N is lower than that of V_N^+ for all Fermi-level positions within the band gap. Substitutional hydrogen is stable exclusively in the 2+ charge state V_N^{2+}. For E_F at the CBM, the formation energy of H_N^{2+} is 1.1 eV, that is, 0.3 eV lower than that of V_N. This formation energy corresponds to an equilibrium concentration of 2×10^{15} cm^{-3} at $T = 500°$C. Note that this estimate is based on equilibrium with H_2; in reality, equilibrium is more likely with an adsorbed species on the surface, which lowers the formation energy and raises the solubility. Substitutional hydrogen is therefore a plausible cause of unintentional n-type conductivity in InN. We note that even in InN grown by molecular beam epitaxy, hydrogen concentrations exceeding 10^{18} cm^{-3} have been found [34].

Frequencies of the local vibrational modes related to substitutional hydrogen in InN were calculated with the goal of facilitating future experimental detection and identification. H_N^{2+} gives rise to three almost degenerate local vibration modes, with the calculated frequencies close to 540 cm^{-1} [32]. Comparing with values of 570 cm^{-1} and 590 cm^{-1} for the highest longitudinal and transverse optical phonons in InN [35], strong coupling with bulk modes will make experimental observation of the H_N^{2+} local vibrational modes very challenging. The fact that hydrogen can act as a double donor

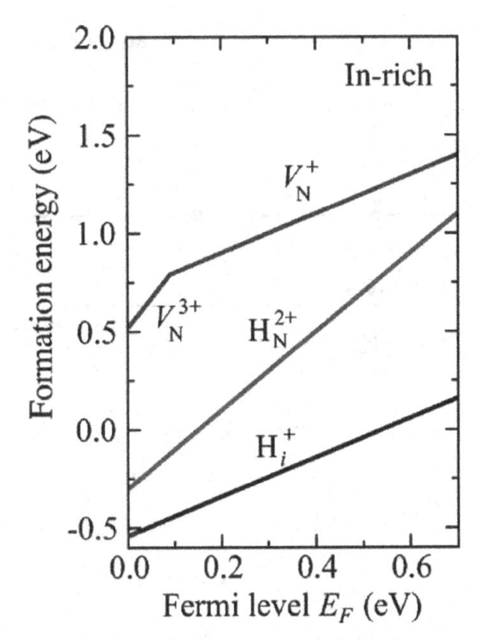

Figure 5. Formation energies of interstitial hydrogen (H_i^+), substitutional hydrogen (H_N^{2+}), and the nitrogen vacancy (V_N^+ and V_N^{3+}) as a function of Fermi level, under In-rich conditions. EF is referenced to the valence band maximum. Figure reproduced with permission from Ref. [32].

seems counterintuitive, since it has only one electron. Its electronic structure can be understood in a simple tight-binding or molecular orbital picture, as shown in Fig. 6. We start from the nitrogen vacancy. Removing a nitrogen atom from the InN lattice leaves four In dangling bonds (DBs), occupied with three electrons. In the near-tetrahedral environment of the wurtzite structure, these DBs combine into a doubly occupied fully symmetric a_1 state located in the band gap, plus three almost degenerate states located at ~2 eV above the CBM. The electron that would occupy the lowest of these three states is transferred to the CBM, resulting in the positive charge state of the nitrogen vacancy (V_N^+) as shown in Fig. 6. This is indeed consistent with the results of first-principles calculations (Figs. 1 and 2). When hydrogen is placed on the nitrogen site, the hydrogen 1s state combines with the a_1 state from V_N^+, resulting in a bonding state at 6.5 eV below the VBM and an antibonding state at 6 eV above

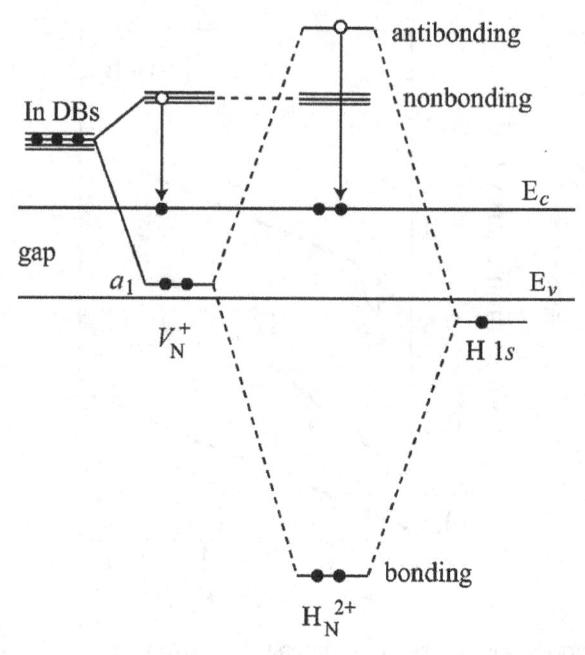

Figure 6. Schematic illustration of the electronic structure of substitutional hydrogen H_N^{2+} in InN. The a_1 state of V_N^+ and the hydrogen 1s state combine into bonding/antibonding states. The bonding state is identified as a multicenter bond involving H and the four In neighbors. The antibonding state located in the conduction band will be unoccupied. Figure reproduced with permission from Ref. [32].

the CBM. The electron that would occupy this antibonding state is transferred to the CBM, turning substitutional hydrogen into a double donor (H_N^{2+}). Note that the three almost degenerate V_N^+ states remain unaltered and constitute the non-bonding states in Fig. 6.

To complete the survey of potential donors, we return to interstitial hydrogen, which is stable exclusively in the positive charge state (Fig. 5). The calculated frequencies of the local vibrational modes are 3050 cm^{-1} for the stretching mode (including anharmonicity) and 626 cm^{-1} for the wagging modes [32]. The formation energy of H_i^+ is lower than that of either H_N^{2+} or V_N^+. In order to address the overall stability of interstitial and substitutional hydrogen, one also has to consider their migration barriers. The calculated migration barrier of interstitial hydrogen is 1.1 eV and corresponds to H breaking a bond with N and forming a bond with a nearest neighbor N. Using the calculated vibrational frequency as an attempt frequency, it is estimated that H_i^+ will become mobile at temperatures around 100°C. Although H_i is mobile at relatively modest temperatures, this does not necessarily imply that it easily diffuses out of the InN samples. Present InN films exhibit a high electron accumulation on the surface that is associated with a pinning of the Fermi level at ~0.8 eV above the CBM due to intrinsic surface states [36–38]. The associated band bending creates a potential barrier that impedes out diffusion of positively charged impurities. Therefore, a significant concentration of interstitial hydrogen can be trapped inside the material and contribute to the n-type conductivity.

It is also relevant to discuss the stability of substitutional hydrogen H_N^{2+} in InN. While interstitial hydrogen migrates by breaking and forming H–N bonds, H_N^{2+} can migrate via two distinct processes: (1) migration assisted by a nitrogen vacancy, where H jumps to a nearby V_N, leaving a vacancy behind; or (2) migration by a concerted exchange with a nearest-neighbor N atom. The activation energy for the vacancy-assisted process (1) is the sum of E_f (V_N^+) (1.4 eV for E_F at the CBM) and the barrier for the exchange process $H_N^{2+} \leftrightarrow V_N^+$ (estimated to be higher than 1 eV), resulting in a barrier of at least 2.4 eV [32]. A direct calculation of the energy barrier for the concerted exchange mechanism (2) gives 2.2 eV [32]. These barriers indicate that H_N^{2+} would be mobile at ~ 500°C. However, it is also important to consider the dissociation of substitutional hydrogen into a nitrogen vacancy and a hydrogen interstitial: $H_N^{2+} \rightarrow H_i^+ + V_N^+$. The activation energy for this dissociation process is 1.6 eV, based on the sum of the binding energy $E_b = E^f(H_i^+) + E^f(V_N^+) - E^f(H_N^{2+}) = 0.5$ eV and the migration barrier of H_i^+ (1.1 eV). Using the calculated vibrational frequency of 540 cm^{-1} for H_N^{2+} as an attempt frequency and an activation energy of 1.6 eV, substitutional hydrogen is expected to be stable up to at least 300°C [32]. Note that H_N^{2+} is subject to the same barriers to outdiffusion as discussed for H_i above. Therefore, both substitutional and interstitial H are the potential causes of unintentional n-type conductivity in InN.

3.2. *Acceptors (Mg and C)*

3.2.1. *Mg in InN*

Mg has been used to obtain p-type GaN and AlN and is thought to also be a p-type dopant for InN [39, 40]. However, early studies of Mg doping of InN did not show p-type conduction [41, 42]. Stampfl *et al.* [20] also investigated the effects of incorporation of Mg acceptors in InN (Mg_{In}). These results are also included in Fig. 3. The position of the acceptor level, that is, the transition level between neutral and negative charge states, is at about 0.2 eV above the VBM. This value is very similar to the Mg acceptor level in GaN [43]. The formation energy of Mg in InN is lower than that for in GaN, indicating higher solubility for Mg in InN than that for GaN. Just like in GaN, nitrogen vacancies are a potential source of compensation in Mg-doped InN. Jones *et al.* [44] have recently succeeded in probing a p-type Mg-doped layer underneath a surface electron accumulation layer in InN. Difficulties in overcoming the latter precluded a direct quantitative evaluation of the p-type conductivity in InN. Further research shows that Mg_{In} acts as a single acceptor, inducing a defect state with a hole below the VBM. The calculated ionization energy for the Mg acceptor is 0.12 eV, in good agreement with the experimentally reported value of $\sim$ 0.1 eV. In the neutral charge state, the Mg atom induces an inward movement of the neighboring apical N atom of 3.54% and the three planar N atoms of 3.14%. In the single negative charge state, Mg^- induces an inward displacement of the apical N atom and three planar N atoms of 3.26% and 3.40%, respectively, thus rather similar to the neutral charge state. The atomic relaxation of Mg in InN is in contrast to Mg in GaN and AlN, where the Mg atom induces a large outward movement of the surrounding N atoms of 10.1% for Mg^0_{Al} and 9.4% for Mg^-_{Al} in zb-AlN [45] and an outward movement of the surrounding N atoms of 6.15% for Mg^0_{Ga} in zb–GaN [46]. The different relaxation modes can be related to the bond length of Mg–N (2.13 Å in Mg_3N_2); in particular, it is shorter than that of N–In in InN by 0.7% but larger than the bond length of Ga–N in GaN by 9.2% and Al–N in AlN by 12.7%.

3.2.2. *C in InN*

The C acceptor has a higher formation energy than the Mg acceptor. Using LDA calculations, Ramos *et al.* [47] investigated the acceptor carbon impurities in InN (C_N). Their results indicate that C_N is a shallow acceptor with low formation energy and suggest that C incorporation in the N sublattice in InN is likely to occur even in the presence of a p-type background doping. Despite the prediction, to our knowledge, there are no experimental reports on the acceptor behavior of C impurities in InN.

4. Defect Complexes

Apart from single point defects, defect complexes will also be able to have an effect on electrical and optical properties of nitride compounds. For example, in earlier work, Neugebauer and Van de Walle [48] studied Ga vacancy-related complexes, which are responsible for the infamous yellow luminescence in GaN, and found that $V_{Ga}O_N$ is much more stable than $V_{Ga}Si_{Ga}$. Mattila and Nieminen [49] showed that the cation vacancy-related complexes ($V_{Ga}O_N$ and $V_{Al}O_N$) are energetically very favorable defects in zb–GaN and zb–AlN and induce gap states that are very likely to take part in the yellow luminescence or violet luminescence emission. To the best of our knowledge, there is, however, so far no detailed study of the incorporation of these impurities and, particularly the complexes that may form, in wz–InN. So, Stampfl *et al.* [50–52] investigated the structural and formation energies of all possible defect complexes including vacancy complexes, complexes with vacancies, and impurity complexes.

4.1. *Vacancy complexes*

4.1.1. *Nitrogen vacancy complexes*

Although nitrogen vacancies have the lowest formation energies in InN under any conditions, the nitrogen vacancies show different behaviors between p-type and more n-type materials. In a p-type material, the nitrogen vacancies are isolated acting as donors while in a more n-type material the nitrogen vacancies prefer to be situated close to one another on the nearest-neighbor sites forming "vacancy complexes or clusters." To study why the vacancies exhibit an attractive interaction, Stampfl *et al.* performed first-principles density functional theory calculations using the LDA [53] for the exchange correlation functional and the pseudopotential plane-wave method in a supercell geometry with the ESPRESSO code [54]. Up to six nitrogen vacancies are considered doping in the supercell by determining the lowest energy and charge state configurations. For two and three vacancies, a complete search is carried out in the 72-atom cell, resulting in 7 and 21 atomic configurations, respectively. All clearly favorable configurations are recalculated in a 96-atom cell. Since the most favorable two- and three-nitrogen vacancy configurations are clearly found to prefer clustering, i.e., to occupy the nearest-neighbor like species sites, for the larger vacancy complexes ($n>3$), they tested various configurations by adding the extra vacancy to the most favorable configuration with $n-1$ vacancies in the 96-atom cell. In all cases, the results show that the nitrogen vacancies prefer to be clustered together, occupying nearest-neighbor [either "in plane," i.e., the vacancies are located in the same (0001) plane, or "out of plane," i.e., the vacancies are located in different (0001) planes] sites. The most favorable structures are then refined in a 128-atom cell. The energies of the two most favorable structures as well as the corresponding values for well-separated vacancies (for $2V_N$ and $3V_N$) are listed in Table 1. Figure 7 shows the atomic geometries for the two lowest energy vacancy pair configurations, and Fig. 8 shows the lowest energy atomic geometries for the $3V_N$ to $6V_N$ complexes. The results are shown as in Fig. 7.

For the vacancy pair, the most favorable configuration is where the vacancies are located out of plane on the nearest-neighbor sites (Fig. 7(a)). The next most favorable structure consists of the two vacancies on the nearest-neighbor sites in the same (0001) plane (Fig. 7(b)), where the energy is 128 meV higher. For the favorable structure, four In–In distances have contracted, resulting in bond lengths very similar to those in bulk In (3.7% (−3.7%) larger (smaller) relative to the calculated (experimental) values).

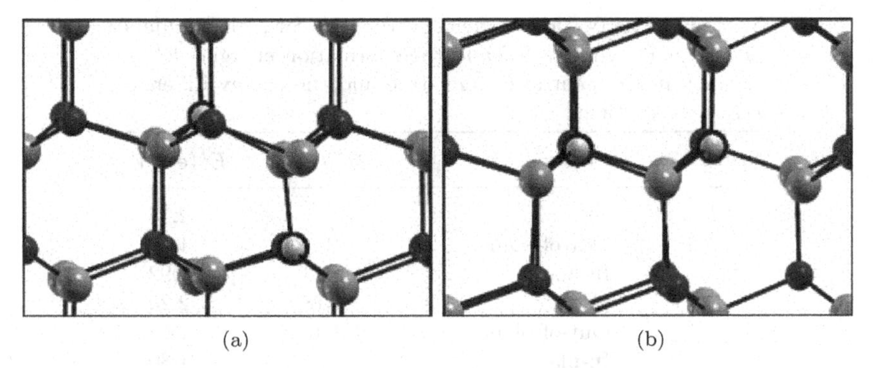

(a) (b)

Figure 7. Geometry of the local atomic relaxations around the nitrogen vacancy pair for the most (a) and the next-most (b) favorable configurations. The dark (red) and light gray spheres represent N and In atoms, respectively, and the open circles indicate the V_N sites. The vertical and horizontal directions of the figures are [0001] and $[2\bar{1}\bar{1}0]$ and the axis perpendicular to the paper is $[01\bar{1}0]$.

The favorable out-of-plane vacancy pair induces a number of single defect states in the conduction band region, where one band is doubly occupied, as seen in Fig. 9, which shows the band structure. The defect-induced states are indicated in Fig. 9 by the arrows, where all states at and between the arrows, are defect states. We assigned the defect states by plotting the spatial distribution of the square of the wave function at the point. For this vacancy pair, the 1+ and 2+ positive charge states as well as the 1− and 2− negative charge states are then considered and it is found that all are stable. With the exception of the single nitrogen vacancy (which has a singly occupied defect state above the conduction band maximum, which is labeled "B" in Fig. 9), all the "vacancy clusters," as for the vacancy pair, can either act as donors or acceptors, as will be seen below.

The calculated binding energy, however, for the vacancy pair in the 2+ positive charge state shows that the complex is unbound (see Table 2), so that it is energetically favorable for the vacancies to exist as singly charged isolated defects. Correspondingly, for the 2+ charge state, there is no contraction of the In–In bonds about the vacancies. Similar to the neutral charge state, there is a contraction of the In–In distances about the vacancies for the negative charge states,

Table 1. The two lowest energy structures for a given number of nitrogen vacancies V_N and their formation energies E^f per vacancy in the neutral charge state and the energy difference ΔE between them.

No. of V_N	Configuration	ΔE(eV)	E^f(eV/V_N)
1	—	—	2.15
2	Out-of-plane	0.0	1.86
2	In-plane	0.13	1.92
2	"Far"	0.78	2.25
3	Out-of-plane	0.0	1.75
3	In-plane	0.14	1.80
3	"Far"	1.50	2.25
4	Rhombus	0.0	1.56
4	Pyramid	0.35	1.65
5	Bi-linear	0.0	1.50
5	Bi-pyramid	0.18	1.54
6	Bi-linear	0.0	1.46
6	Rhombus-pyramid	0.39	1.53

Notes: For two and three vacancies, we also show the least favorable formation energies, corresponding to well-separated vacancies denoted as "far," and the energy differences with respect to the most favorable structures. Values in boldface indicate the most favorable configuration.

resulting in bond lengths that are very similar to those in bulk In (average of 2.0% (–5.3%) larger (smaller) relative to the calculated (experimental) values).

For doping of three vacancies in the neutral charge state, the most favorable configuration is the one where one vacancy is located out of plane and the other two are located in plane, again occupying nearest-neighbor sites (see Fig. 8(a)). The next most favorable structure, with an energy that is 140 meV higher is where all three vacancies are located at the nearest-neighbor sites in the same (0001) plane. For the favorable complex, there is a contraction of five of the In–In distances about the vacancies, again resulting in bond lengths very similar to those in bulk In [average of 3.9% (–3.5%) larger (smaller) relative to the calculated (experimental) values]. This constitutes an average effective inward displacement toward the vacancy of the surrounding In atoms by about 11%. The most favorable $3V_N$ complex induces a number of singlet defect states in the region of the conduction band, one of which is fully occupied by two electrons and another that is occupied by one electron, the rest being unoccupied (see Fig. 9). Therefore, the 1+, 2+, and 3+ charge states, as well as the 1−, 2−, and 3− negative charge states are taken into account, and we find that all are stable. Similar to the pair vacancy complex, the binding energy of the complex in the highest positive charge state (3+) is not bound, indicating that isolated singly charged vacancies are preferable (see Table 2). For all other charge states, the binding energy is positive, indicating that clustering is favorable. Similar to the neutral charge state, there is on average a contraction of five of the In–In distances about the vacancies for the negative charge states, resulting in bond lengths very similar to those in bulk In (average of 3.1% (−4.3%) larger (smaller) relative to the calculated (experimental) values).

For four nitrogen vacancies, clustering is still preferred. The most favorable structure out of a search of nine structures is that where the vacancies form a "rhombus structure," involving two in-plane and two out-of-plane sites (see Fig. 8(b)). The next most favorable is a "pyramid structure" (one out-of-plane site above three in-plane sites), which is less stable by 0.35 eV. For the favorable complex, there is a contraction of nine of the In–In distances about the vacancies, resulting in bond lengths very similar to those in bulk In [average of 4.1% (–3.3%) larger (smaller) relative to the calculated (experimental) values]. This constitutes an average effective inward displacement toward the vacancy of the surrounding In atoms by about 11%. The rhombus structure complex induces an increased number of defect states in the conduction band region compared to the $3V_N$ and $2V_N$ complexes, where there are two fully occupied singlet states, as well as several unoccupied levels, as seen in Fig. 9. Therefore, the positive charge states from 1+ to 4+, as well as the negative charge states from 1− to 4− are considered and it is found that all charge states except the 2± charge states are stable. As for the smaller complexes, the binding energy of the highest positive charge state (4+) is negative, showing that it is unbound (see Table 2). For all other charge states, the binding energy is positive, indicating that clustering is

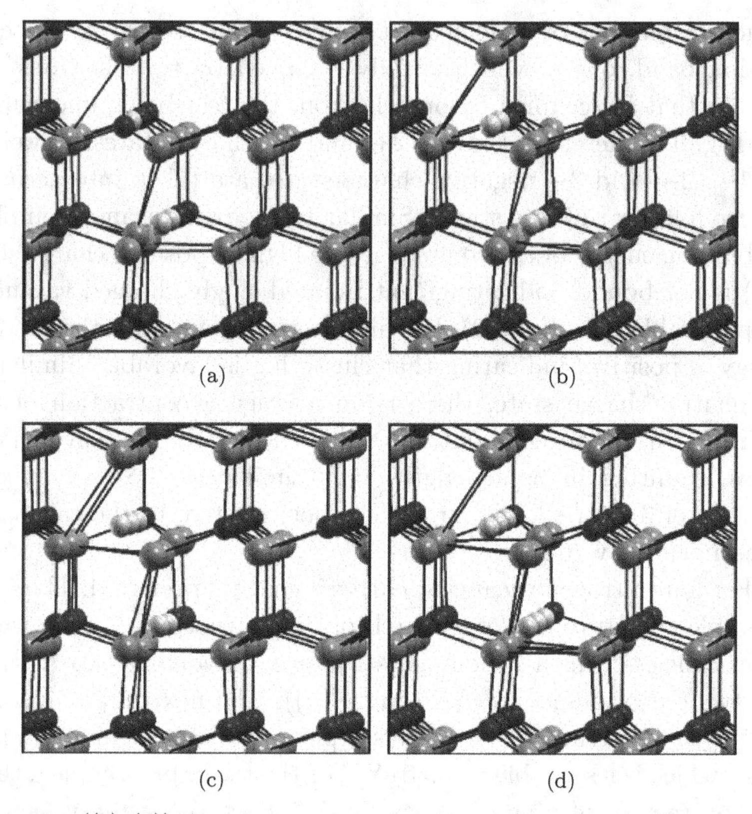

Figure 8. ((a)–(d)) Atomic geometry of the most favorable configurations for three to six nitrogen vacancies in the neutral charge state in InN. The dark (red) and light gray spheres represent N and In atoms, respectively. The vacancy sites are shown as the very pale spheres. Lines joining In atoms indicate that shorter bonds (by a minimum of 8% and a maximum of 14%) have formed compared to the bulk structure.

favorable. For these charge states, there is a contraction of the In–In distances about the vacancies, i.e., on average, 12 of the In–In distances are reduced, resulting in bond lengths very similar to those in bulk In (average of 2.6% (−4.7%) larger (smaller) relative to the calculated (experimental) values).

By adding another vacancy to the most favorable configuration with four vacancies in the 128-atom cell, we find that the most stable structure from a consideration of six configurations is the one where three vacancies are in a line at in-plane sites and the other two in

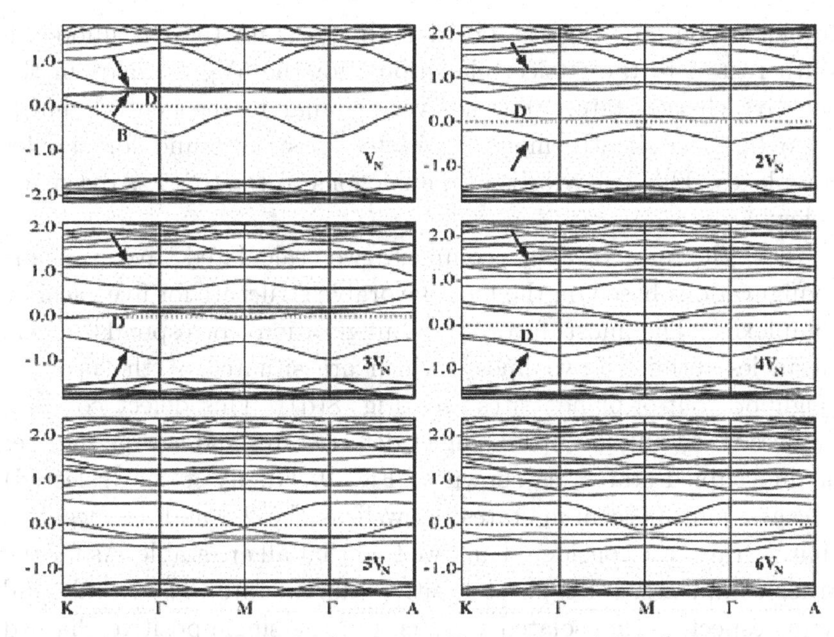

Figure 9. Band structures for (one to six, V_N to $6V_N$ as labeled) nitrogen vacancies in the neutral charge state in the 128-atom cell. The labels B and D represent bulk and defect states, respectively. The arrows indicate the regions of the defect states; that is, all states between and at the arrows are defect-induced states. The horizontal dotted lines indicate the Fermi level.

another line at nearest-neighbor out-of-plane sites (see Fig. 8(c)). The next favorable structure is a symmetric "bipyramid" structure, which is 0.183 eV higher in energy. For the favorable complex, there is a contraction of 11 of the In–In distances about the vacancies, again resulting in bond lengths very similar to those in bulk In (average of 4.8% (−2.9%) larger (smaller) relative to the calculated (experimental) values). This constitutes an average effective inward displacement of the In atoms by 10% relative to the In–N bulk distance. The most favorable structure induces two fully occupied defect levels and one singly occupied defect level, as well as several unoccupied defect states, as seen in Fig. 9. The number of induced states is again greater than that for the $4V_N$ and smaller complexes. Therefore, the (selected) 1+, 3+, and 5+ positive charge states as well as the 1− and 3− negative charge states are considered and find that all are stable. Again, the highest positive charge state (5+) is unbound. With regard to the lattice relaxations, for the $5V_N$ complex in the negative charge states, they are very similar to the neutral charge state but are slightly more contracted, just as found for smaller complexes. This behavior is also analogous to the $6V_N$, as discussed below.

For the six vacancy complex, we calculated five possible configurations based on the most favorable structure for five vacancy complexes. The most favorable configuration corresponds to six vacancies forming "two lines," which are situated at the nearest-neighbor "out-of-plane" sites (see Fig. 8(d)). This defect complex induces three fully occupied singlet defect states and a large number of unoccupied states, as shown in Fig. 9. For $6V_N$, the (selected) 2+, 4+, and 6+ charge states, as well as the 2− and 4− negative charge states are considered and we find that all are stable. As for the other complexes, the highest positive charge state (6+) is unbound with respect to the isolated vacancies in the single positive charged state.

The total density of states of the vacancy complexes in the neutral charge state is shown in Fig. 10(a), where the Fermi energy is indicated by the short black vertical line. In Fig. 10(b), the total density of states for the vacancy complexes in the highest positive charge state is shown, where, in this case, the Fermi energy coincides with the valence band maximum. For comparison, the total density of states of the isolated nitrogen vacancy is also listed in Fig. 10. The defect-induced states are localized at and between the In atoms nearest to the vacancies. As an example, Fig. 11 show the spatial distribution of the defect states, as indicated by the label "D" in Fig. 9 for the nV_N configurations (where $n = 1 - 4$) for the highest positive charge states $n+$.

In Table 2, the binding energies of the defect complexes are given per vacancy. All the calculations are performed in 128-atom supercells. As discussed earlier, it can be seen that for all charge states except the highest positive charge state for each complex, the binding energies are positive, which indicates that the interaction between vacancies is attractive. The magnitude of the binding energy

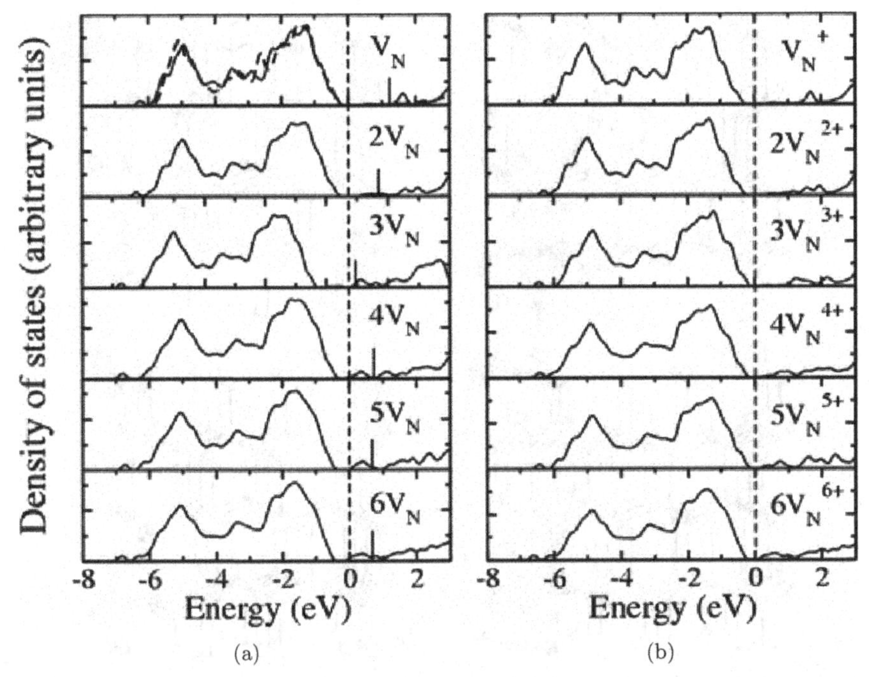

Figure 10. Total density of states for the most favorable configurations of up to six neutral (a) and positively charged (b) nitrogen vacancy complexes in wurtzite InN. The dashed lines indicate the valence band maximum (VBM). In the left panel, the short vertical lines indicate the Fermi level. In the right panel, the Fermi level coincides with the VBM. The density of states for bulk InN is included as a dashed line in the neutral isolated nitrogen vacancy plot.

per vacancy increases with the increasing number of vacancies in the complex for the neutral charge state, indicating that it is favorable to form larger clusters. For the highest positive charge states, which are present under more p-type conditions, all the binding energies are negative, although very small, showing that the defect clusters are unfavorable in comparison to isolated V_N^+. Note that, these same trends in all the binding energies are obtained irrespective of which reference defects that are used. For example, for the $5V_N^{3+}$ complex, the binding energy is positive when using any one of the following reference states: $(2 \times V_N^0 + 3 \times V_N^+), (2V_N^0 + 3V_N^{3+})$, or $(V_N^0{+}4V_N^{3+})$.

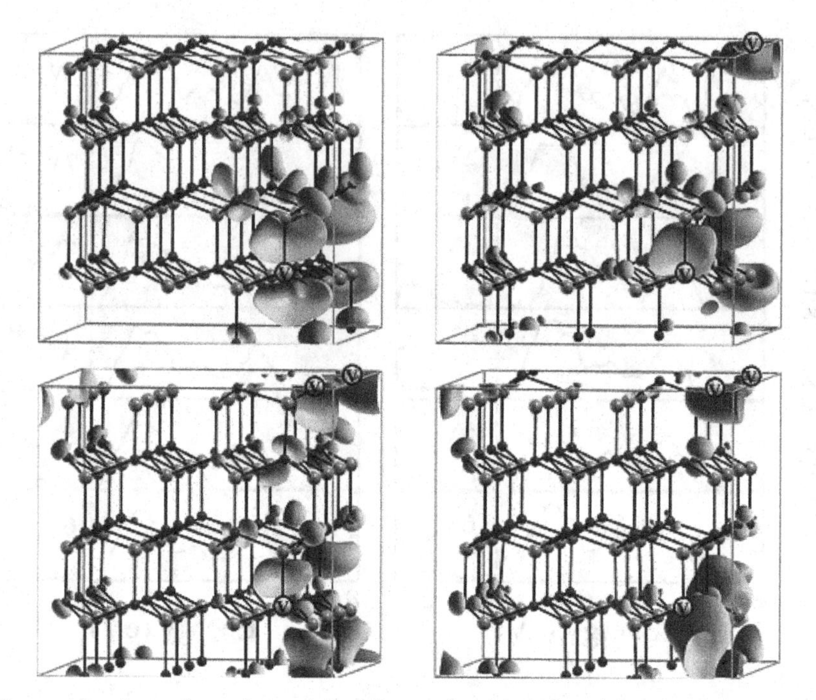

Figure 11. Isosurface plots of the charge density induced by the defect states (marked as D in Fig. 9) for V_N^+, $2V_N^+$(upper panel, left to right) and $3V_N^{3+}$ and $4V_N^{4+}$(lower panel, from left to right) in the 128-atom cell. The 0.001e/Bohr3 isovalues are shown. The dark (red) and light gray spheres represent N and In atoms, respectively, and the open circles with "V" indicate the V_N sites.

The fact that the N vacancies prefer to cluster together, i.e., giving local In-rich regions, could explain the reported metallic In inclusions in InN [55].

The formation energies of the defects as a function of the Fermi level E_F are also calculated, as shown in Fig. 12. The formation energies for the vacancy complexes are given per vacancy. It can be seen that for a p-type material, the single nitrogen vacancy V_N^+ has the lowest formation energy, but the formation energies of vacancy complexes and/or clusters per vacancy in the highest positive charge states are only slightly higher (practically degenerate). In more n-type materials, it can be seen that lower positive, neutral, and negatively charged vacancy complexes are predicted. The formation energy per vacancy of the vacancy complexes and/or clusters clearly

Table 2. The binding energies per nitrogen vacancy E_b/n (in eV) and the average relative displacements $\bar{d}(\%)$ of the In atoms neighboring the vacancy site, relative to the bulk In–N bond length, for the defect complexes in various charge states q.

Configuration	q	E_b/n	$\bar{d}$	q	E_b/n	$\bar{d}$	q	E_d/n	$\bar{d}$
V_N	0	—	−5.07		—		1+	—	−2.06
$2V_N$	0	0.30	−6.93	2−		−8.85	2+	−0.018	−1.56
$3V_N$	0	0.42	−6.43	3−		−6.33	3+	−0.024	−1.11
$3V_N$				2−	0.34	−7.97	2+	0.18	−2.78
$3V_N$				1−	0.25	−7.16	1+	0.35	−5.48
$4V_N$	0	0.59	−6.39	4−	0.43	−8.50	4+	−0.03	0.12
$4V_N$		0		3−	0.45	−7.89	3+	0.15	−2.09
$4V_N$		0		2−	0.40	−7.86	2+	0.32	−3.06
$4V_N$		0		1−	0.38	−6.88	1+	0.48	−4.87
$5V_N$	0	0.67	−6.12	3−	0.42	−7.31	5+	−0.06	−0.50
$5V_N$				1−	0.39	−6.67	3+	0.32	−5.15
$5V_N$							1+	0.57	−2.84
$6V_N$	0	0.70	−5.70	2−	0.52	−6.78	6+	−0.07	0.56
$6V_N$							4+	0.27	−2.29
$6V_N$							2+	0.52	−4.37

Notes: A negative number for $\bar{d}$ indicates relaxation toward the vacancy. The binding energies are given with respect to the isolated neutral and single positive charge states for the neutral and positive charge states of the complexes. Since the single nitrogen vacancy does not exist in the single negative charge state, the binding energy of negatively charged complexes is given with respect to the $2V_N^{0,q-}$ and $3V_N^{0,q-}$ complexes.

decreases with the increasing number of vacancies for all but the highest positive charge states.

4.1.2. *Indium vacancy complex*

Stampfl *et al.* [51] also considered the possibility of indium vacancy complex formation. The atomic and electronic structures, as well as energetics, of up to three indium vacancies were investigated by performing an exhaustive search for the optimal spatial distribution (7 and 21 atomic configuration for 2 and 3 In vacancies, respectively). The energetics show that the indium vacancies prefer to cluster together, occupying nearest-neighbor (either in the (0001) plane or out-of-plane) sites. For the neutral indium vacancy pair,

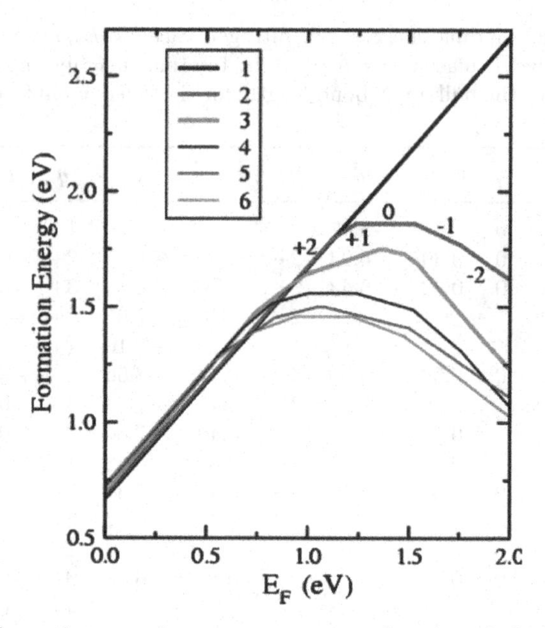

Figure 12. Formation energies per vacancy as a function of the Fermi level for the nitrogen vacancy and vacancy clusters in InN under In-rich conditions. The number of vacancies is listed in the legend, i.e., $1-6$. The zero of the Fermi level is at the top of the valence band maximum. Kinks in the curves indicate transitions between different charge states. For example, for $2V_N$, the $2\pm$, $1\pm$, and neutral charge states are shown in the plots.

the most favorable configuration is where the vacancies are located out of plane on the nearest-neighbor sites. Here, the V_{In}–V_{In} direction is not aligned parallel to the c axis but with an angle to it. The formation energy (per defect) for the neutral state is high, 5.95 eV under N-rich and 7.11 eV under In-rich conditions, and the binding energy indicates a strong attractive interaction (~ 1.8 eV per vacancy) as seen from Table 3.

After the relaxation (see Fig. 13(b)), an N–N bond forms with a distance of 1.23 Å, which is comparable to that of the N_2 dimer (1.10 Å). Due to the significant atomic relaxations, the $2V_{In}$ complex could be described as an N-spilt interstitial plus a mixed vacancy complex, $V_N + 2V_{In}$. Once formed, the N–N bond is so strong that the total energy of this configuration is lower than that of the others by about 3.6 eV.

Table 3. The binding energies, E_b, in eV, for the defect complexes in stable charge states. The binding energies of the defects are computed with respect to the defect combinations in the parentheses.

Defect	q		$E_b/V_{\rm In}$	q		$E_b/V_{\rm In}$	q		$E_b/V_{\rm In}$
$V_{\rm N} + V_{\rm In}$	0	$(V_N^+ + V_{\rm In}^-)$	2.16	1−	$(V_N^+ + V_{\rm In}^{2-})$	2.48	1+	$(V_N^+ + V_{\rm In}^0)$	2.24
	0	$(V_N^{3+} + V_{\rm In}^{3-})$	3.82	2−	$(V_N^+ + V_{\rm In}^{3-})$	2.77	1+	$(V_N^{3+} + V_{\rm In}^{2-})$	3.0
$V_{\rm N} + N_{\rm In}$	0	$(V_N^+ + N_{\rm In}^-)$	2.36	1+	$(V_N^+ + N_{\rm In}^0)$	2.09	2+	$(V_N^+ + N_{\rm In}^+)$	1.59
	0	$(V_N^{3+} + N_{\rm In}^{3-})$	5.33	3+	$(V_N^{3+} + N_{\rm In}^0)$	3.39	2+	$(V_N^{3+} + N_{\rm In}^-)$	4.29
				2−	$(V_N^+ + N_{\rm In}^{3-})$	2.35			
$2\,V_{\rm In}$	0	$(V_{\rm In}^0 + V_{\rm In}^0)$	3.66	3−	$(V_{\rm In}^- + V_{\rm In}^{2-})$	2.48	6−	$(V_{\rm In}^{3-} + V_{\rm In}^{3-})$	−0.76
				3−	$(V_{\rm In}^0 + V_{\rm In}^{3-})$	3.40			
$3\,V_{\rm In}$	0	$(V_{\rm In}^0 + V_{\rm In}^0 + V_{\rm In}^0)$	9.84	3−	$(V_{\rm In}^- + V_{\rm In}^- + V_{\rm In}^-)$	7.35	6−	$(V_{\rm In}^{2-} + V_{\rm In}^{2-} + V_{\rm In}^{2-})$	1.86
	5−	$(V_{\rm In}^0 + V_{\rm In}^{2-} + V_{\rm In}^{2-})$	3.78	3−	$(V_{\rm In}^0 + V_{\rm In}^- + V_{\rm In}^{2-})$	7.80	6−	$(V_{\rm In}^- + V_{\rm In}^{2-} + V_{\rm In}^{3-})$	2.28
	5−	$(V_{\rm In}^- + V_{\rm In}^- + V_{\rm In}^{3-})$	4.26	3−	$(V_{\rm In}^0 + V_{\rm In}^0 + V_{\rm In}^{3-})$	8.70	6−	$(V_{\rm In}^0 + V_{\rm In}^{3-} + V_{\rm In}^{3-})$	3.21
	5−	$(V_{\rm In}^0 + V_{\rm In}^{2-} + V_{\rm In}^{3-})$	4.68						

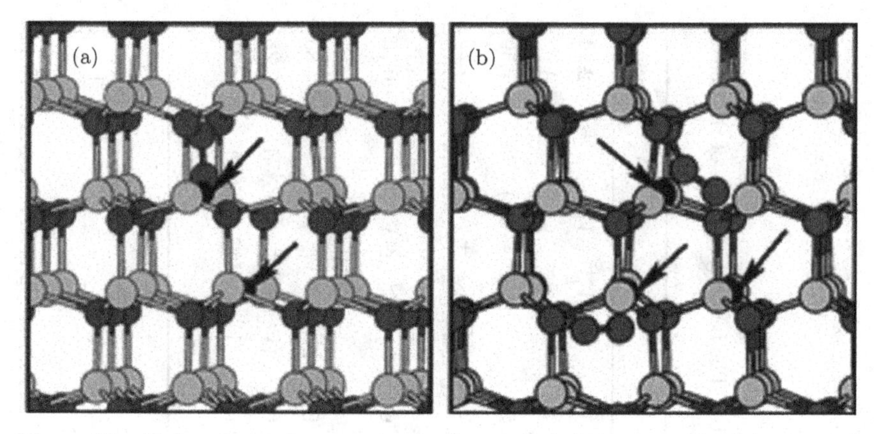

Figure 13. Relaxed atomic geometry in the neutral charge state for (a) two, and (b) three indium vacancies for the most favorable configurations. Dark and light spheres represent N and In atoms, respectively, and small black spheres indicate the V_{In} sites, which are also indicated by arrows.

The N–N bond can be clearly seen in the contour plot of the electron charge density in Fig. 14(a). The vacancy pair induces several defect states at the top of the valence band and in the conduction band, as indicated by the region between, and at, the arrows in Fig. 15(a). There are two doubly occupied singlet states and the rest are unoccupied. Therefore, charge states from 2+ to 6− are considered. The stable charge states are 2+, 0, and 3−, and the others are unstable. The binding energies of $2V_{In}$ in the neutral charge state are calculated with respect to the two neutral V_{In}, and for $2V_{In}^{3-}$, the binding energies (1.24 and 1.70 eV per vacancy) are computed with respect to V_{In}^{-} and V_{In}^{2-}, and V_{In}^{0} and V_{In}^{3-}, respectively. For $2V_{In}^{4-}$, the binding energies, calculated with respect to V_{In}^{2-} and V_{In}^{2-}, and V_{In}^{-} and V_{In}^{3-}, respectively, decrease to 0.41 and 0.65 eV per vacancy. The binding energies for the 5– and 6– charge states become negative, indicating a repulsive interaction between the two indium vacancies. In the 2+ charge state, the "nitrogen molecule" has a very similar N–N distance as for the neutral case. While in the 3– charge state, three N atoms around the vacancy pair form an independent N_3-like bonding arrangement with N–N bond lengths of 1.36 and 1.37 Å. For the higher negative charge states, the

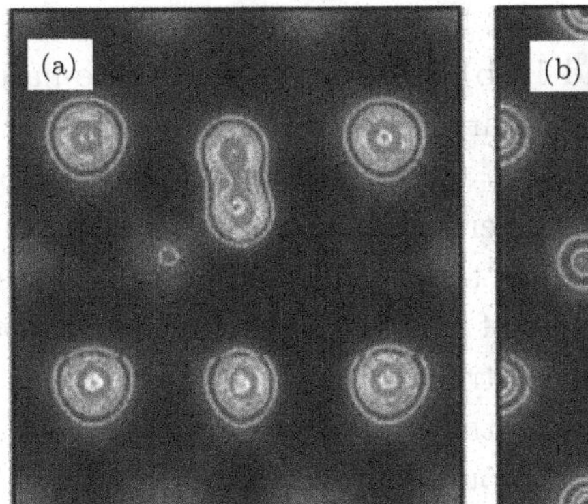

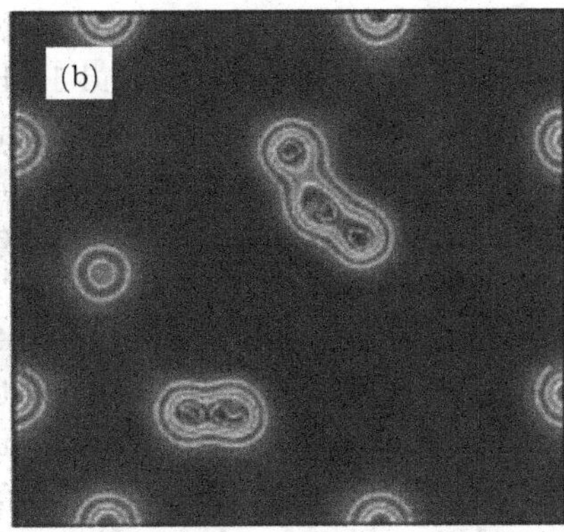

Figure 14. Contour plots of the electron charge density for (a) two and (b) three indium vacancies in InN in the neutral charge state. The lowest contour line value is at 0.0 e/Bohr^3, and the highest contour line has a value of 0.40 e/Bohr^3. Contour lines in between change successively by a factor of 0.08 e/Bohr^3.

N_3-like configuration disappears, and only the "N_2-like molecule" exists.

For three In vacancies, the most favorable configuration is the one where one vacancy is located out of plane and the other two are in plane, again occupying nearest-neighbor sites. This preference can be seen from the large values from Table 4, it's 3.82 eV/vacancy for the neutral state of the binding energies in Table 4. In the neutral charge state, the neighboring N atoms around the vacancies relax significantly to yield with an "N_2-like" geometry, and two other new N–N bonds are formed, resulting in an independent N_3-like configuration. The N–N bond length of "N_2" is 1.22 Å, and for the "N_3"-like configuration, the bond distances are 1.25 and 1.35 Å (see Fig. 13(b)), which are comparable to the N–N bond length (1.18 Å) of the gas phase azide (linear geometry) N_3 molecule [56]. The three In-vacancy complex, $3V_{\text{In}}$, induces many defect states, which are indicated in the region between, and at, the arrows in Fig. 15(b). In particular, the state at the Fermi level is occupied with one electron, and the three defect states below the Fermi level are each occupied with two electrons. We assign a state to be a defect state by plotting the square of the wave function and noting that it is weighted heavily on the defect, i.e., at, and between, the N atoms nearest to the vacancies. As an example, we show in Figs. 15(e) and 15(f) the defect states indicated by label "D" in Figs. 15(a) and 15(b) for the two- and three-indium vacancy clusters in the neutral charge state. They confirmed that all states lying between, and at, the arrows in Figs. 15(a) and 15(b) are indeed defect-related states. Therefore, charge states from 3+ to 6− are considered and we find that the 3+, 0, and 3− charge states are stable. The binding energies of all the negative charge states are larger than 0.6 eV, indicating attractive interactions between the vacancies. For instance, for $3V_{In}^{3-}$, the binding energies (2.45, 2.60, and 2.90 eV per vacancy) are computed with respect to three V_{In}^{-} , $V_{In}^{0} + V_{In}^{-} + V_{In}^{2-}$, and two V_{In}^{0} and V_{In}^{3-}, respectively. The N_2-like and "N_3-like" clusters remain for the various charge states, where the N–N bond lengths in both N_2 and N_3 increase slightly with increasing negative charge state. In the 3+ charge state, the N–N distance in N_2 is 1.19, and 1.23 and 1.33 Å in the N_3-like configuration. The respective values are 1.25, 1.30, and 1.39 Å in the 6-charge state. Figure 14(b) shows the contour plot of the electron charge density for $3V_{In}$ in the neutral charge state. The high density around the N_2- and N_3-like configurations can be clearly seen.

The density of states for the two- and three-indium vacancy clusters are shown in Figs. 15(c) and 15(d), respectively. It is evident that all the N atoms in the "molecules" induce additional peaks both in the valence band and in the conduction band, which are again quite different from the bulk (In-bonded) N atoms. Such In-vacancy complex-induced states may be a "fingerprint" for identifying the existence of these structures experimentally.

4.2. *Complexes with vacancies*

4.2.1. *Cation–anion vacancy pair*

It has been reported that the defect complex, $V_N + V_{Ga}$ in GaN and $V_N + V_{Al}$ in AlN, has a substantial binding energy and that its formation energy is lower than that of the isolated gallium vacancy under metal-rich conditions for p-type material [57]. It is interesting therefore to consider the interaction between the cation and anion vacancies ($V_N + V_{In}$) to see whether they prefer to bind together

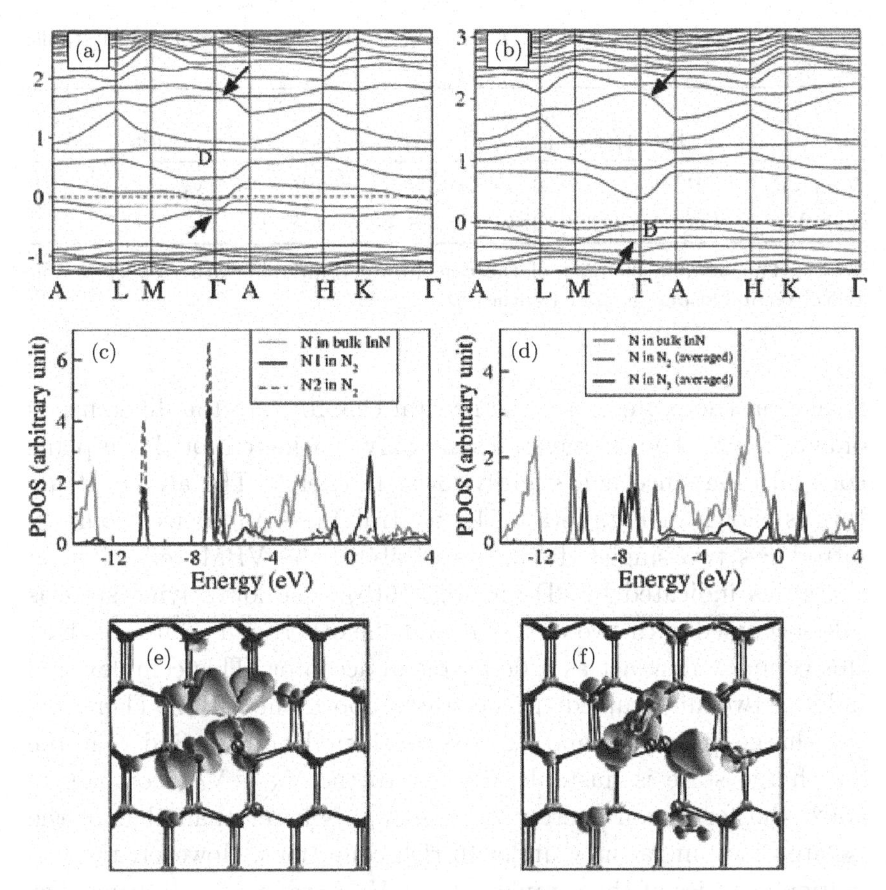

Figure 15. (a) and (b) Band structures; (c) and (d) partial density of states; (e) and (f) isosurface plots of the defect states (marked as D in (a) and (b)) of two and three indium vacancies in the neutral charge state, respectively. The arrows in (a) and (b) indicate the regions of the defect states; that is, all states between, and at, the arrows are defect-induced states. The horizontally dotted lines indicate the Fermi levels. The 0.001 e/Bohr^3 isovalues are shown. Dark and light spheres represent N and In atoms, respectively, and open circles with "V" indicate the V_{In} sites.

or be isolated in InN. It is found that the cation–anion vacancy pair is stable (as indicated by the large positive binding energies in Table 4). Configurations involving both in-plane (V_N and V_{In} both in the (0001) plane) and out-of-plane (V_N and V_{In} oriented in a direction parallel to the c axis) configuration have a very similar

Table 4. The binding energies, E_b, in eV, for the defect complexes in stable charge states.

Defect	q	Defect	E_b/ V_{In}	q	Defect	E_b/ V_{In}	q	Defect	E_b/ V_{In}
$V_N + V_{In}$	0	$(V_N^+ + V_{In}^-)$	2.16	1−	$(V_N^+ + V_{In}^{2-})$	2.48	1+	$(V_N^+ + V_{In}^0)$	2.24
$V_N + V_{In}$	0	$(V_N^{3+} + V_{In}^{3-})$	3.82	2−	$(V_N^+ + V_{In}^{3-})$	2.77	1+	$(V_N^{3+} + V_{In}^{2-})$	3.0

Notes: The binding energies of the defects are computed with respect to the defect combinations in the parentheses.

formation energy; e.g., for the neutral charge state the difference is only 3.5 meV. The charge states are only considered for the in-plane configuration since it is slightly lower in energy. The atomic structure is shown in Fig. 16(a). The neutral $V_N + V_{In}$ defect complex introduces two singlet defect states above the VBM very close in energy as indicated by "D" in Fig. 16(b). The lower lying state is fully occupied with two electrons and the other is unoccupied; thus, this complex may act as a donor or an acceptor. This complex also induces two unoccupied defect states above the CBM. Therefore, the charge states 1± and 2± are considered and we find that the 2+ charge state is unstable. Similar to the $V_N + V_{Ga}$ complex in GaN, the formation energy of $V_N + V_{In}$ is lower than that of the isolated indium vacancy under In-rich conditions. However, the formation energies of the complex $(V_N + V_{In})$ are high, indicating that in thermal equilibrium this defect will not be present in significant concentrations. The V_N in the complex induces an outward relaxation for the positive charge state (1+) and inward relaxations in the neutral and negative charge states. The breathing relaxation of the surrounding N atoms induced by V_{In} in the 1+ charge state is outward and becomes less so, or even slightly inward, with the increasing occupation.

4.2.2. $O_N V_{In}$ *complex and* $O_N V_{In}$

Rather than forming "vacancy complexes or clusters", the nitrogen vacancy V_N prefers to be bound to Mg_{In}, forming a neutral $Mg_{In}V_N$ complex, while the indium vacancy V_{In} prefers to bond to oxygen,

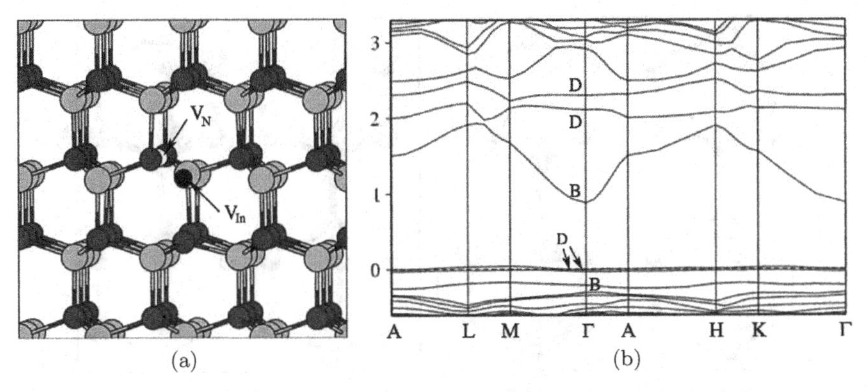

Figure 16. (a) Relaxed atomic geometry and (b) band structure of the vacancy complex $V_N + V_{In}$ in the neutral charge state. Dark and light spheres represent N and In atoms, respectively, and small white and black spheres indicate the V_N and V_{In} sites, respectively, which are also indicated by arrows. Labels B and D in (b) represent bulk and defect states, respectively. The horizontally dotted line in (b) indicates the Fermi level.

but the formation energy for the $O_N V_{In}$ complex (in all charge state) is high.

It is found that the defects Mg_{In} and V_N in wz–InN prefer to be located close together, where the associated binding energies for the neutral parallel and perpendicular structures are 0.87 and 0.81 eV, respectively (relative to Mg_{In}^- and V_N^+). The presence of nitrogen vacancies in GaN and InN therefore indicates p-type doping with Mg. The corresponding formation energies under In-rich conditions are 1.76 and 1.82 eV. In the neutral charge state, there are no defect-induced states in the band gap, but there are three unoccupied states at and above the CBM. In the neutral state, the neighboring atoms around the complex relax significantly as shown in Fig. 17(a). Specifically, the bond lengths of Mg and N for the parallel configuration are 2.01 Å, which are 5.6% shorter than the average bond length (2.13 Å) of crystalline Mg_3N_2, and 6.6% shorter than the N–In bond length. The distance between Mg and the "ideal" (i.e., bulk) position of the nitrogen vacancy is 15.3% larger than the N–In bond length. For the perpendicular case (Fig. 17(b), left panel), the bond lengths between Mg and N are 2.01 and 2.02 Å in the (0001) plane and 2.02 Å along the c axis, where the Mg atom moves away from the ideal position

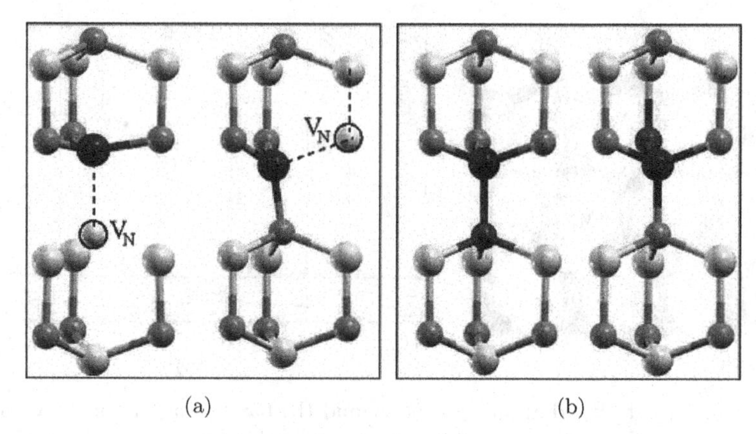

Figure 17. (a) "parallel" (left) and "perpendicular" (right) atomic geometries of the neutral ($Mg_{In}V_N$) complex. (b) the neutral $Mg_{In}O_N$ defect for the "parallel" (left) and "perpendicular" (right) configurations, in wz–InN. Large pale (light yellow) and smaller darker gray (pink) spheres indicate In and N atoms, respectively. Large and small black (blue and red) spheres indicate Mg and O atoms, respectively. The vacancies are indicated by the circled pale spheres. Only the neighboring atoms of the complexes are shown for clarity.

of the nitrogen vacancy by 12%. For both cases, the three In atoms next to V_N move toward the ideal vacancy position by 5% similar to the behavior for isolated V_N [58].

Motivated by the experimental studies which indicate that V_{In} has a tendency to couple with impurities, such as oxygen [59], and the *ab initio* investigation which found that the O_NV_{Ga} complex is stable with a large binding energy (~1.8 eV) [60], the O_NV_{In} complex is investigated in wz–InN. It is found that the defects O_N and V_{In} like to be located on neighboring sites and the formation of O_NV_{In} complex is favorable. In the neutral charge state, the binding energies calculated with respect to O_N^+ and V_{In}^- are 0.93 and 0.89 eV, and the formation energies are 7.26 and 7.30 eV under In-rich conditions, for the parallel and perpendicular configurations, respectively (see Table 5). Thus, further investigations are focused on the slightly favorable parallel configuration. For the defect in the parallel configuration, the average distance between the three planar nearest-neighboring N atoms and the ideal (i.e., bulk) In-vacancy position is 5% larger than the In–N bond length and the oxygen

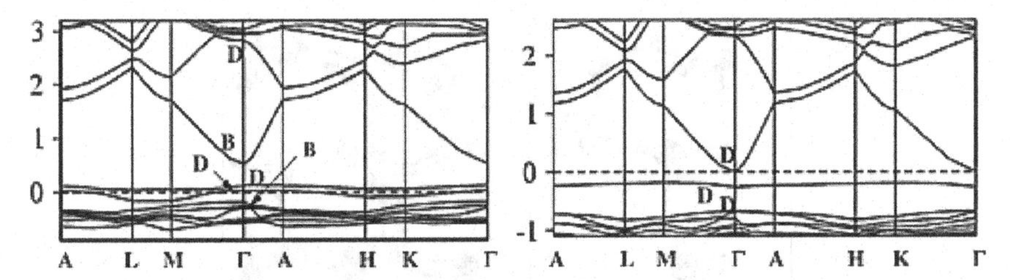

Figure 18. Band structures for the $V_{In}O_N$ (left) and $3O_NV_{In}$ (right) defect complexes in the neutral charge state in wz–InN. The labels "B" and "D" represent bulk and defect states, respectively. The horizontal dotted lines are the Fermi levels.

atom moves upward by 10% of the N–In bond length relative to the ideal In-vacancy position. Unlike single O_N, the O–In bond length is 1.1% smaller than the planar N–In distance. The complex induces a doubly occupied singlet defect state just above the VBM and an unoccupied singlet defect state in the gap, and thus, it can act as an acceptor (see Fig. 18). Therefore, the 2− and 1− charge states are considered. The atomic structure of this defect complex in the parallel configuration in the 2− charge state is shown in Fig. 19(a). The three planar N atoms nearest to the In vacancy relax inward by 1.9% and the oxygen atom moves away from the ideal N site by 14.3%. The average O–In bond length is 1.6% shorter than the planar N–In distance. The lattice relaxations of the complex O_NV_{Ga} in GaN were reported by Neugebauer and Van de Walle [60], where the oxygen atom moves away from the ideal Ga vacancy by 14.9%, and the three planar N atoms nearest to the Ga vacancy relax outward by 9.8% in the 2− charge state. The binding energy of the O_NV_{In} complex increases from $\sim$0.9 eV in the neutral charge state (with respect to O_N^+ and V_{In}^{3-}) to 1.56 eV in the $(O_NV_{In})^{2-}$ charge state relative to O_N^+ and V_{In}^{3-}. The formation energy of this complex is rather high (see Fig. 20) indicating that it will not exist in high concentrations in thermal equilibrium.

A configuration with one indium vacancy, V_{In}, and three planar oxygen atoms $3O_N$, as shown in Fig. 19(b), are also considered. The formation energy in the neutral charge state significantly decreases compared the O_NV_{In} complex under both In- and N-rich conditions

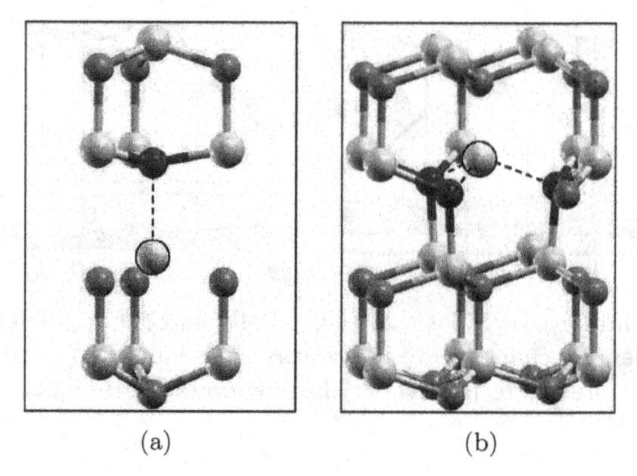

Figure 19. Atomic geometry of the (a) $(O_N V_{In})^{2-}$ and (b) $3O_N V_{In}$ defect complexes. Large pale (light yellow) and smaller darker gray (pink) spheres indicate In and N atoms, respectively. Small black (red) spheres and pale circled spheres indicate O atoms and vacancies, respectively. Only the neighboring atoms of the complexes are shown for clarity.

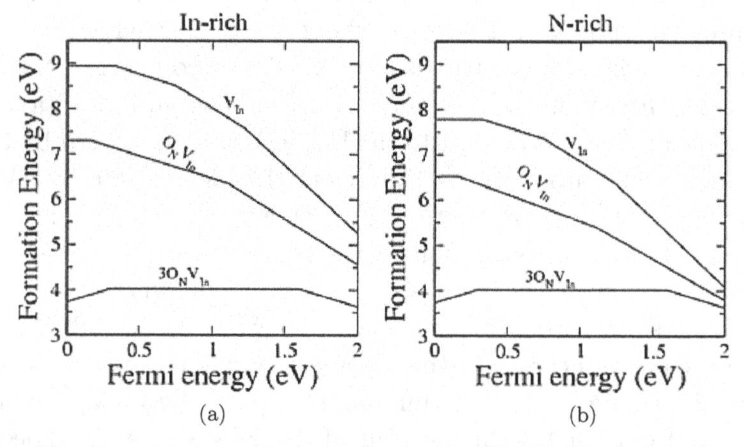

Figure 20. Formation energies as a function of the Fermi level for impurity complexes involving V_{In} and O_N in wz–InN under In-rich (a) and N-rich (b) conditions.

as shown in Fig. 20. The neutral defect induces three fully occupied singlet defect states: one in the bandgap and the other two at the top of the valence band, and an unoccupied defect state at the CBM (see Fig. 18(b)). Therefore, 3+, 2+, 1+, 1−, and 2− charge states are considered and we find that the 3+, 2+, and 2− charge

Table 5. Formation E^f and binding energies E_b (in eV) of defect complexes.

Defect	Configurations	E^f (In-rich)	E^f (N-rich)	E_b
$(\mathrm{Mg}V_\mathrm{N})^0$	$\perp$	1.82	2.60	0.81
	$\parallel$	1.76	2.54	0.87
$(\mathrm{O_N}V_\mathrm{In})^0$	$\perp$	7.30	6.52	0.89
$(\mathrm{O_N}V_\mathrm{In})^0$	$\parallel$	7.26	6.49	0.93
$(\mathrm{O_N}V_\mathrm{In})^-$	$\parallel$	$7.46 - E_F$	$6.68 - E_F$	1.48
$(\mathrm{O_N}V_\mathrm{In})^{2-}$	$\parallel$	$8.61 - 2E_F$	$7.84 - 2E_F$	1.56
$(3\mathrm{O_N}V_\mathrm{In})^0$	3O planar	4.027	4.027	3.99
$(3\mathrm{O_N}V_\mathrm{In})^+$	3O planar	$3.73 + E_F$	$3.73 + E_F$	3.06
$(3\mathrm{O_N}V_\mathrm{In})^-$	3O planar	$5.63 - E_F$	$5.63 + E_F$	

Notes: The binding energies of $(\mathrm{MgV_N})^0$ are computed with respect to $\mathrm{Mg_{In}^-}$ and $\mathrm{V_N^+}$ and the binding energies of $\mathrm{O_N V_{In}}$ in the 0, 1−, and 2− charge states are computed with respect to $\mathrm{O_N^+}$ and $\mathrm{V_{In}}$ in the charge states 1−, 2−, and 3−, respectively. The binding energies of $3\mathrm{O_N V_{In}}$ in the 0 and 1+ charged states are calculated with respect to $\mathrm{O_N^+}$ and $\mathrm{V_{In}}$ in charge states 3− and 2−, respectively. "$\parallel$" and "$\perp$" indicate the parallel and perpendicular configurations, respectively, as explained in the text. "Planar" indicates that the three O atoms are in the (0001) plane.

states are unstable. The 1+ charge state is stable in a small energy window ($E_F < 0.3\,\mathrm{eV}$). Therefore, further positive charge states are not considered. In the neutral charge state, the three oxygen atoms relax outward from the ideal N sites, away from the In vacancy by 10% of the In–N bond length, and the distance between each oxygen atom and its nearest-neighbor In atom is contracted by 1.2% relative to the ideal N–In bond length. In the 1− and 1+ charge states, the O–In distances are very similar to those in the neutral charge state, namely, the oxygen atoms move outward from the ideal N sites by 9.8% and 8% of the bulk N–In bond length, respectively.

4.3. *Other complexes*

Stampfl *et al.* [51] also predicted that the $\mathrm{Mg_m O_n}$ complexes could be important defects in InN under both In-rich and N-rich conditions for obtaining improved efficiency for n- and p-type conductivities through tuning of the relative concentrations of Mg and O, in particular by providing the maximum Mg solubility through the use of the chemical potential of Mg from bulk Mg. Complexes based on codoping of Si and C appear to offer no advantage in obtaining n- or p-type conductivity over isolated dopants. Details are as follows:

4.3.1. Mg_mO_n complexes

The Mg_{In} and O_N prefer to be located on nearest-neighbor sites, where the binding energy is 0.43 eV, relative to the single defects Mg_{In}^- and O_N^+. The total energy of the configuration with Mg_{In} and O_N far apart is 0.45 eV higher. For both perpendicular and parallel configurations, the total energies are identical as indicated in Table 6, and the relaxed Mg–O bond lengths are close to those of crystalline MgO, where the bond length is 2.10 Å. The atomic structures of the MgO complexes are shown in Fig. 17(b). As discussed above, Mg_{In} is an acceptor and O_N is a donor. In forming the $Mg_{In}O_N$ defect, we find that there are no longer any states in the band gap, but there is a doubly occupied singlet defect state at the VBM (see Fig. 23(a)). For the Mg_2O_2 complex in the parallel and perpendicular configurations, which form Mg–O armchair and zigzag structures, respectively (see Fig. 21), there is a very tiny energy difference (2 meV). The calculated binding energy of 1.12 eV shows that the defects still like to be located close to each other. In particular, per MgO unit, the binding energy is 0.56 eV, which is larger than the single MgO complex (0.43 eV), showing that it is favorable to form the larger cluster of Mg_2O_2. This defect induces a fully occupied state at the VBM and an unoccupied defect state above the CBM. When a larger complex, Mg_3O_3 is considered, it is found that the armchair structure is more stable than the zigzag structure by 0.23 eV. The binding energy of the armchair complex is 2.01 eV (see Table 6) or 0.67 eV per MgO pair, indicating that clustering is favorable. Correspondingly, the formation energies of the MgO, Mg_2O_2, and Mg_3O_3 complexes are 0.46, 0.65, and 0.65 eV, respectively, or 0.46, 0.32, and 0.21 eV per MgO pair. It can be concluded that these defects could play a role in the compensation of acceptor doping with Mg if oxygen is present as a

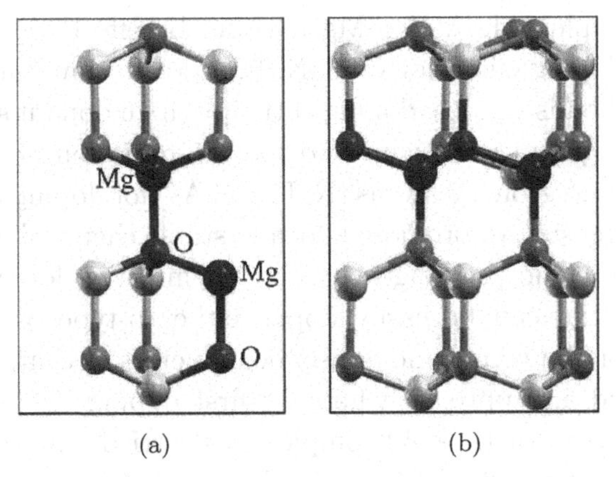

Figure 21. Atomic geometry of the Mg_2O_2 complexes in wz–InN with a (a) "parallel" ("armchair") configuration and a (b) "perpendicular" ("zig-zag") configuration. Large pale (light yellow) and smaller darker gray (pink) spheres indicate In and N atoms, respectively. Large (blue) and small dark gray (red) spheres indicate Mg and O atoms, respectively. Only the neighboring atoms of the complexes are shown for clarity.

contaminant. Note that the calculated binding energies are comparable to the values of 0.50 and 0.58 eV reported of for parallel and perpendicular MgO in wz–GaN [61] and 0.60 eV for MgO and 0.80 eV for Mg_2O_2 in zb–AlN [62].

The idea of "codoping" [61] is that both acceptors and donors are doped at the same time into a semiconductor and form metastable acceptor–donor–acceptor (A–D–A) complexes for obtaining p-type material or donor–acceptor–donor (D–A–D) complexes for obtaining n-type material. The formation of A–D–A or D–A–D complexes may be able to enhance the dopant solubility by lowering the formation energy through attractive interactions between the constituents of the defect complex and lower as well the defect transition energy levels through the coupling of donor–acceptor states.

First-principles electronic structure calculations and experiments have applied the codoping approach to address p-type GaN and AlN [61] where the results are promising. Codoping of GaN with magnesium and oxygen has been reported to result in high hole conductivities [63, 64]. Yamamoto and Katayama-Yoshida [65] proposed that the complex Mg_{Ga}–O_N–Mg_{Ga} could be effective in enhancing the p-type doping efficiency of GaN. In order to simultaneously produce stable bonds and low doping enthalpy (high dopant solubility), a cluster-doping approach, using two species from group III atoms (Al, Ga, and In) and group V atoms (N, P, and As) for doping p-type ZnO, has been suggested to promote a locally stable chemical environment [66]. To overcome the long-standing "asymmetry doping problem" (i.e., a material can be easily doped either p-type or n-type, but not both) for the wide-bandgap semiconductors, recently, Yan *et al.* [67] proposed an approach where by first doping the host by passive donor–acceptor (D + A) complexes is carried out, which creates fully occupied unoccupied impurity bands above (below) the VBM (CBM) of the host. Subsequently, further doping of excess dopant atoms will bind to the $(D + A)$ complexes and effectively dope the fully occupied (unoccupied) impurity bands, which leads to a lower ionization energy. With this approach, the experimental observations of B- and H-codoped n-type diamond [68] and Ga- and N-codoped p-type ZnO [69, 70] were explained.

In addition to the above-mentioned complexes, we also consider codoping and cluster doping involving Mg acceptors and O donors with regard to varying the relative concentrations of each species. For the complexes Mg_2O, Mg_3O, MgO_2, and MgO_3, both parallel and perpendicular configurations in the neutral charge state are considered. For the complexes Mg_4O and MgO_4, only the configurations shown in Figs. 22(c) and 22(f) are considered. From examination of the associated electronic structure and occupancies of the induced defect levels, relevant charge states of the (favorable) defect complexes are then considered.

Figure 22 shows the atomic geometries of the most stable Mg_mO_n complexes for $m = 1$ and $n = 2, 3$ (see Figs. 22(a) and 22(b))] and $n = 1$ and $m = 2, 3$ (see Figs. 22(d) and 22(e))]. In the neutral charge state, the Mg–O bond lengths, 2.10 and 2.09 Å in Mg_2O and 2.09 and 2.09 Å in MgO_2, are close to those of crystalline MgO (2.10 Å). The average Mg–O bond length is 2.07 and 2.06 Å in the neutral Mg_3O and MgO_3 complexes, respectively. With one more acceptor or donor added to the complexes, resulting in Mg_4O and MgO_4,

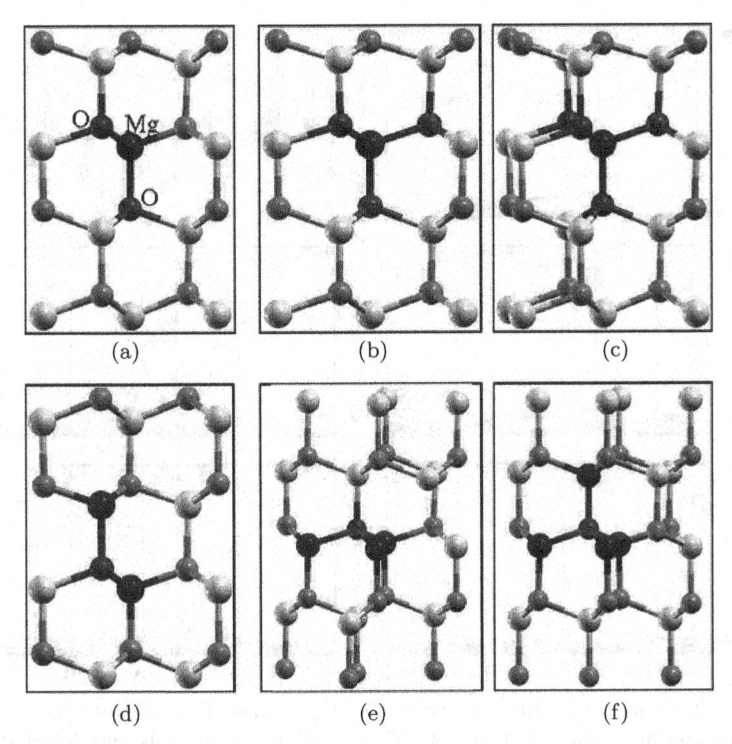

Figure 22. Atomic geometry of the most stable Mg_nO_m defect complexes in wz–InN: (a) "parallel" MgO_2, (b) "parallel" MgO_3, (c) MgO_4; (d) "parallel" Mg_2O, (e) "perpendicular" Mg_3O, and (f) Mg_4O. Large pale (light yellow) and smaller darker gray (pink) spheres indicate In and N atoms, respectively. Large (blue) and small dark gray (red) spheres indicate Mg and O atoms, respectively. Only the neighboring atoms of the impurities are shown for clarity.

the average distances between Mg–O become slightly contracted to 2.04 Å for both Mg_4O and MgO_4. The neutral MgO_2 and MgO_3 complexes give rise to unoccupied defect states above the CBM (see Figs. 23(b) and 23(c)). There are one (MgO_2) and two (MgO_3) electrons associated with these states, which due to the electron filling in the DFT calculations, occupy the CBM instead. Thus, these complexes are donors and we can consider the 1+ and 2+ charge states, respectively. The neutral MgO_4 complex induces two defect states in the conduction band with three associated electrons; thus, it acts as a donor. For this complex, we consider charge states 1+, 2+, and 3+. Neutral Mg_2O (Mg_3O) yields a hole (two holes) at (above) the

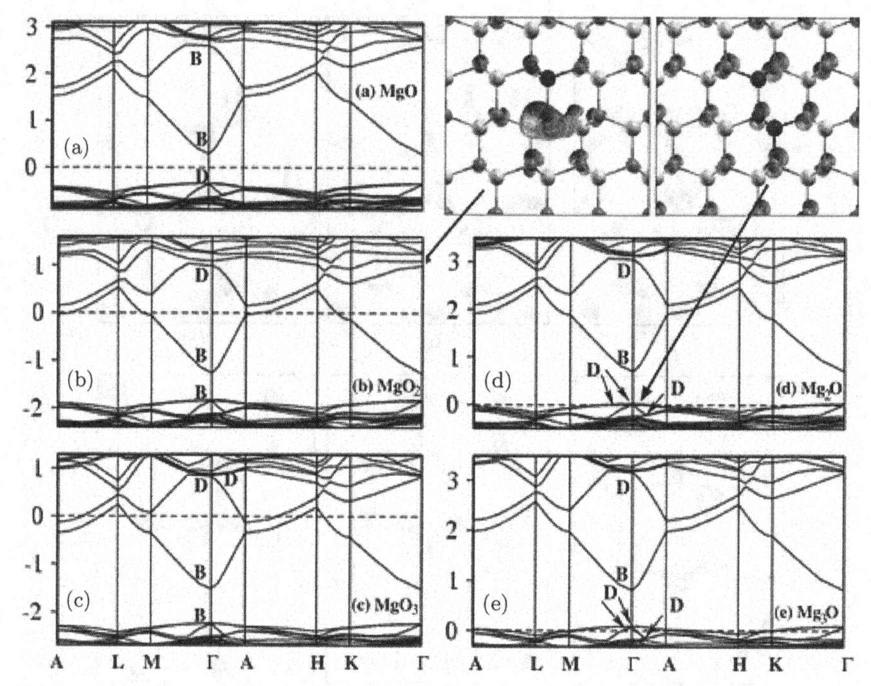

Figure 23. Band structures for the stable Mg_nO_m defect complexes in the neutral charge state in wz–InN. The labels "B" and "D" represent bulk and defect states, respectively. The horizontally dotted lines are the Fermi levels (except for (a) where it is placed in the center of the band-gap). For Mg_2O and MgO_2, isosurface plots are shown of the charge density induced by the defect states as indicated by the long arrows. The 0.001 e/Bohr^3 isovalues are shown. Large pale (light yellow) and smaller darker gray (pink) spheres indicate In and N atoms, respectively. Large (blue) and small black (red) spheres indicate Mg and O atoms, respectively.

VBM; thus, it can act as an acceptor (see Figs. 23(d) and 23(e)). The neutral Mg_4O complex induces three holes above the VBM. Thus, for these complexes, we consider 1–, 2–, and 3– charge states, respectively. The average Mg–O bond lengths are close to those of crystalline MgO (2.10 Å) in the charged states for all the complexes we considered. The formation (and binding) energies of the parallel and perpendicular configurations in the ionized states for the various complexes are close, as shown in Table 6. Complexes with more than one donor and acceptor that occur simultaneously, e.g., Mg_2O_3 and Mg_3O_2 in the 128-atom cell are also considered in parallel and perpendicular configurations. It is found that the parallel configuration is more favorable for both complexes (see Table 6). The neutral Mg_2O_3 complex induces a defect state above the CBM with an associated electron and thus acts as a single donor. The average Mg–O bond length of this complex is 2.08 Å in both the neutral charge state and in the 1+ charge state.

The neutral Mg_3O_2 complex induces a hole at the VBM and thus acts as an acceptor. The average Mg–O bond length is 2.09 Å in both the neutral charge state and in the 1− charge state. The ionization energy of the acceptor Mg_2O (Mg_3O_2) decreases to 0.05 eV (0.02 eV) compared to the value 0.12 eV of the isolated Mg acceptor.

Figure 24 shows the calculated formation energies of the Mg_mO_n complexes as a function of the Fermi level for both In-rich (left) and N-rich (right) conditions. Here the zero of the Fermi level is chosen at the VBM. The formation energies of O_N and Mg_{In} are also included for comparison (long-dashed lines in Fig. 24). It is interesting to note that for $n > m$ (of the Mg_mO_n complexes), the formation energies are low in p-type material where the formation energy decreases with the increasing oxygen content.

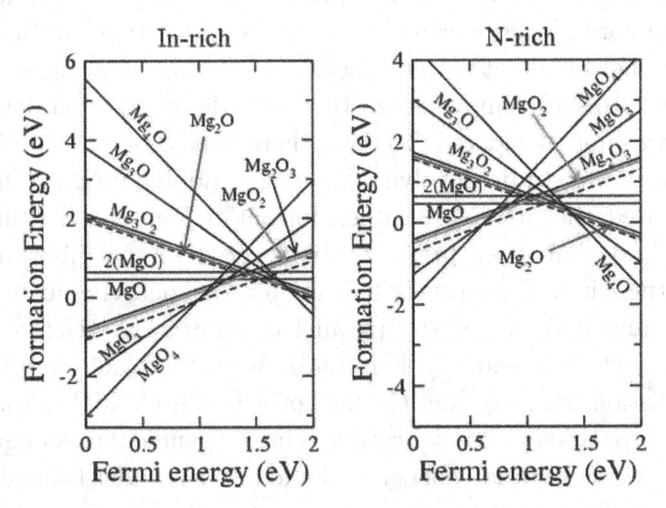

Figure 24. Formation energies as a function of the Fermi level for Mg_mO_n complexes in wz–InN under In-rich (left) and N-rich (right) conditions. The O_N^+ donor and Mg_{In}^- acceptor in dashed lines are included for comparison.

While for $n < m$ (of Mg_mO_n), the formation energies are higher in p-type but lower in n-type material, and the formation energy decreases with increasing Mg content in n-type InN and the binding energy slightly increases as well. Another feature worth noting is that the formation energies for MgO_2 and Mg_2O_3 (where $n-m=-1$) are comparable and similarly for Mg_2O and Mg_3O_2 (where $n-m=1$). Therefore, it can be predicted that the Mg_mO_n omplexes could be important defects in InN under both In-rich and N-rich conditions and their properties and behavior could be controlled by the Mg and O contents.

4.3.2. *Si_iC_j complexes*

Analogous to the Mg–O acceptor–donor system, the codoping of Si–C donors and acceptors is also considered and the behavior of this system is compared to that of Mg–O. Recent experimental investigations have studied Si and C in GaN [71–73]. The system is likely to be associated with the yellow luminescence in C-doped GaN films and is responsible for the semi-insulating behavior of GaN:C:Si films. It was found [71, 72], that when carbon concentrations were less than those of silicon, carbon should incorporate as C_N and compensate the Si donors. When C concentrations exceeded those of Si, carbon became the dominant active species and the material became insulating due to self-compensation. However, the interaction between C and Si was not considered. To date, there have been no experimental and theoretical studies involving Si and C codoping InN. We therefore investigated the possible complex formation between Si_{In} and C_N. It is found that they prefer to be located on neighboring sites, where the binding energy is 0.61 eV for the perpendicular configuration and 0.41 eV for the parallel configuration, relative to the single defects Si_{In}^+ and C_N^-. For comparison, the total energy of the configuration with Si_{In} and C_N far apart is 0.49 eV higher than that of the perpendicular configuration. The perpendicular configuration has the lower formation energy 1.73 and 3.28 eV under In- and N-rich rich conditions, respectively, and the parallel structure is 0.2 eV higher. We therefore only study the perpendicular configuration for other Si_iC_j configurations. The SiC complex induces two occupied singlet defect states close to the VBM and no defect state in the gap (see Fig. 26(a)). The Si–C bond length is 1.85 Å, close to that in bulk α-SiC (1.88 Å in the planar and 1.89 Å in the apical directions).

Similar to the Mg_mO_n complexes, both parallel and perpendicular configurations for the complexes Si_2C, Si_3C, SiC_2, and SiC_3 are considered. Figure 25 shows the relaxed atomic geometries of the most stable Si_iC_j complexes for $i = 1$ and $j = 2, 3$ (see Figs. 25(a) and 25(b)) and $j = 1$ and $i = 2, 3$ (see Figs. 25(d) and 25(e)). For the complexes Si_4C and SiC_4, the configurations shown in Figs. 25(c) and 25(f) are considered. The average Si–C

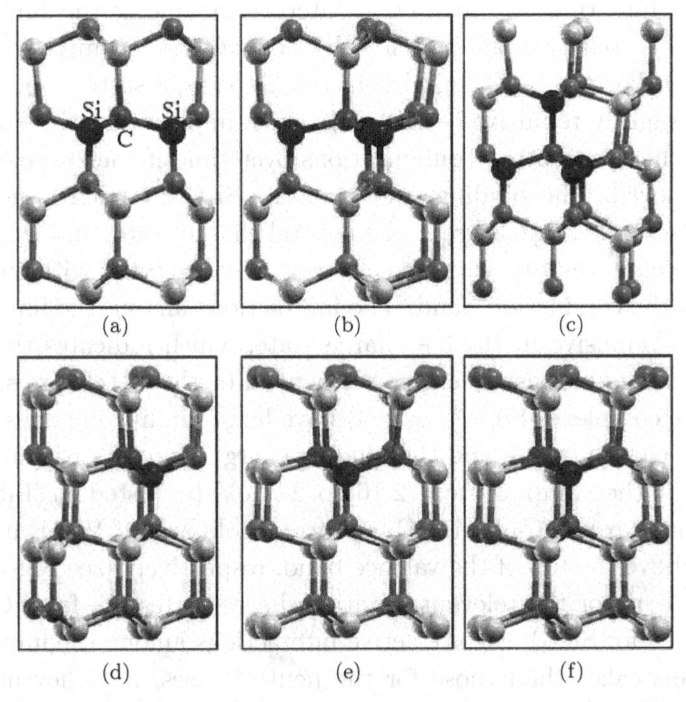

Figure 25. Atomic geometry of the most stable Si_iC_j defect complexes in wz-InN: (a) "zigzag" Si_2C, (b) Si_3C, (c) Si_4C; (d) "zigzag" SiC_2, (e) SiC_3, and (f) SiC_4. Large pale (light yellow) and smaller darker gray (pink) spheres indicate In and N atoms, respectively. Large (blue) and small dark gray (green) spheres indicate Si and C atoms, respectively. Only the neighboring atoms of the impurities are shown for clarity.

bond length is 1.86 Å for SiC_2, 1.88 Å for SiC_3, and 1.89 Å for SiC_4 complexes. The corresponding values are 1.94, 2.04, and 2.10 Å for the Si_2C, Si_3C, and Si_4C complexes, respectively. The complexes Si_2C and Si_3C have larger binding energies than the SiC complex in the neutral charge states (as listed in Table 7). The complex Si_2C induces a singlet defect state occupied with one electron in the conduction band and the complex Si_3C induces a fully occupied singlet defect state in the conduction band (see Figs. 26(b) and 26(c))]. As for the Mg_mO_n complexes ($m = 1$ and $n = 2$ and 3), these electrons are located at the CBM and not in the defect level due to the electron filling in the DFT calculations. These Si_2C and Si_3C complexes are therefore donors. The relevant positive charge states are investigated and we find that the attractive interaction among the impurities in Si_2C is weaker than that for the neutral case (compare 0.42 to 1.55 eV). For the Si_3C complex in the 2+ charge state, the interaction is slightly repulsive (−0.13 eV), which implies that this complex would change to other configurations over time if the two electrons are removed. The binding energy of the Si_4C complex is negative (−0.38 eV), i.e., repulsive, in the neutral charge state, and it induces two defect states, one with two electrons and the other with one electron in the conduction band. The interactions among the impurities are still repulsive in the 3+ charge state, which indicates that this complex cannot exist in either the neutral or the 3+ charge states.

The complexes SiC_j ($j = 2$–4) have large binding energies in the neutral charge states, and the binding energy increases with increasing C in the complex from 2.16 to 2.70 eV (as listed in Table 7). The complexes SiC_2 and SiC_3 induce a hole at the VBM and two holes above the top of the valence band, respectively (see Figs. 26(d) and 26(e)). For the relevant negative charge states (1− for SiC_2 and 1− and 2− for SiC_3), the attractive interactions among the impurities become weaker than those for the neutral cases, but they are still mutually attractive. The SiC_4 complex induces three holes close to the VBM and in the 3− charge state, the binding energy is ~1 eV similar to $(SiC_2)^-$ and $(SiC_3)^{2-}$.

Table 6. The formation E^f and binding energies E_b in eV of defect complexes.

Defect	Configuration	E^f (In-rich)	E^f (N-rich)	E^f (In-rich): $\mu_{Mg} = E_{Mg}$ (bulk)	E^f (N-rich): $\mu_{Mg} = E_{Mg}$ (bulk)	E_b
$(O_N)^0$		0.47	0.86	0.47	0.86	
$(O_N)^+$		$-1.08 + E_F$	$-0.69 + E_F$	$-1.08 + E_F$	$-0.69 + E_F$	
$(Mg_{In})^0$		1.84	1.45	1.64	0.48	
$(Mg_{In})^-$		$1.96 - E_F$	$1.57 - E_F$	$1.76 - E_F$	$0.6 - E_F$	
$(MgO)^0$	zz	0.46	0.46	0.26	−0.51	0.43
	ac	0.46	0.46	0.26	−0.51	0.43
$(Mg_2O_2)^0$	zz	0.65	0.65	0.25	−1.29	1.12
	ac	0.652	0.652	0.252	−1.29	1.12
$(Mg_3O_3)^0$	zz	0.88	0.87	0.28	−2.04	1.78
	ac	0.65	0.64	0.05	−2.27	2.01
$(MgO_2)^+$	zz	$-0.85 + E_F$	$-0.47 + E_F$	$-1.05 + E_F$	$-1.44 + E_F$	0.66
	ac	$-0.87 + E_F$	$-0.49 + E_F$	$-1.07 + E_F$	$-1.46 + E_F$	0.68
$(MgO_3)^{2+}$	zz	$-2.02 + 2E_F$	$-1.25 + 2E_F$	$-2.22 + 2E_F$	$-2.22 + 2E_F$	0.76
	ac	$-2.03 + 2E_F$	$-1.26 - 2E_F$	$-2.23 + 2E_F$	$-2.23 + 2E_F$	0.77
$(MgO_4)^{3+}$		$-3.04 + 3E_F$	$-1.88 + 3E_F$	$-3.24 + 3E_F$	$-2.85 + 3E_F$	0.70
$(Mg_2O)^-$	zz	$2.05 - E_F$	$1.66 - E_F$	$1.66 - E_F$	$-0.28 - E_F$	0.79
	ac	$2.03 - E_F$	$1.64 - E_F$	$1.64 - E_F$	$-0.3 - E_F$	0.81
$(Mg_3O)^{2-}$	zz	$3.81 - 2E_F$	$3.03 - 2E_F$	$3.22 - 2E_F$	$0.12 - 2E_F$	0.99
	ac	$3.94 - 2E_F$	$3.16 - 2E_F$	$3.35 - 2E_F$	$0.25 - 2E_F$	0.86
$(Mg_4O)^{3-}$		$5.55 - 3E_F$	$4.39 - 3E_F$	$4.76 - 3E_F$	$0.51 - 3E_F$	1.21
$(Mg_2O_3)^+$	zz	$-0.68 + E_F$	$-0.30 + E_F$	$-1.08 + E_F$	$-1.24 + E_F$	1.38
	ac	$-0.79 + E_F$	$-0.41 + E_F$	$-1.19 + E_F$	$-2.35 + E_F$	1.49
$(Mg_3O_2)^-$	zz	$2.22 - E_F$	$1.83 - E_F$	$1.62 - E_F$	$-1.08 - E_F$	1.51
	ac	$2.12 - E_F$	$1.73 - E_F$	$1.52 - E_F$	$-1.18 - E_F$	1.61

Notes: All the binding energies are computed with respect to Mg_{In}^- and O_N^+. "ac" and "zz" indicate the "armchair" (parallel) and "zigzag" (perpendicular) configurations, respectively (see text).

Table 7. The formation E^f and binding energies E_b in eV of defect complexes.

Defect	Config.	E^f (In-rich)	E^f (N-rich)	E^f (In-rich): $\mu_{Si} = E_{Si}$ (bulk)	E^f (N-rich): $\mu_{Si} = E_{Si}$ (bulk)	E_b
$(Si_{In})^0$		0.33	0.71	−1.30	−2.46	
$(Si_{In})^+$		$-1.10 + E_F$	$-0.72 + E_F$	$-2.73 + E_F$	$-3.89 + E_F$	
$(C_N)^0$		3.17	4.33	3.17	4.33	
$(C_N)^-$		$3.45 - E_F$	$4.61 - E_F$	$3.45 - E_F$	$4.61 - E_F$	
$(SiC)^0$	zz	1.73	3.28	0.10	0.10	0.61
$(SiC)^0$	ac	1.93	3.48	0.30	0.30	0.41
$(Si_2C)^0$	zz	2.27	4.21	−0.98	−2.14	1.55
$(Si_2C)^+$	zz	$0.82 + E_F$	$2.76 + E_F$	$-2.43 + E_F$	$-3.59 + E_F$	0.42
$(Si_3C)^0$	zz	3.42	5.74	−1.47	−3.79	0.73
$(Si_3C)^{2+}$	zz	$0.27 + 2E_F$	$2.59 + 2E_F$	$-4.62 + 2E_F$	$-6.94 + 2E_F$	−0.13
$(Si_4C)^0$		4.85	7.56	−1.66	−5.14	−0.38
$(Si_4C)^{3+}$		$-0.30 + 3E_F$	$2.41 + 3E_F$	$-6.81 + 3E_F$	$-10.29 + 3E_F$	−0.67
$(SiC_2)^0$	Zz	4.50	7.21	2.88	4.04	2.16
$(SiC_2)^-$	Zz	$4.83 - E_F$	$7.54 - E_F$	$3.20 - E_F$	$4.36 - E_F$	0.96
$(SiC_3)^0$	Zz	7.34	11.21	5.72	8.04	2.49
$(SiC_3)^{2-}$	Zz	$8.18 - 2E_F$	$12.05 - 2E_F$	$6.55 - 2E_F$	$8.87 - 2E_F$	1.06
$(SiC_4)^0$		10.30	15.33	8.67	12.15	2.70
$(SiC_4)^{3-}$		$11.67 - 3E_F$	$16.70 - 3E_F$	$10.04 - 3E_F$	$13.52 - 3E_F$	1.02

Notes: The binding energies of SiC are computed with respect to Si_{In}^+ and C_N^-. The binding energies of other complexes (Si_iC and SiC_j, i, and $j \neq 1$) in the neutral state are computed with respect to Si_{In}^0 and C_N^0; in charged states, they are computed with respect to Si_{In}^+ and C_N^-. "ac" and "zz" indicate the "armchair" (parallel) and "zigzag" (perpendicular) configurations, respectively (see text).

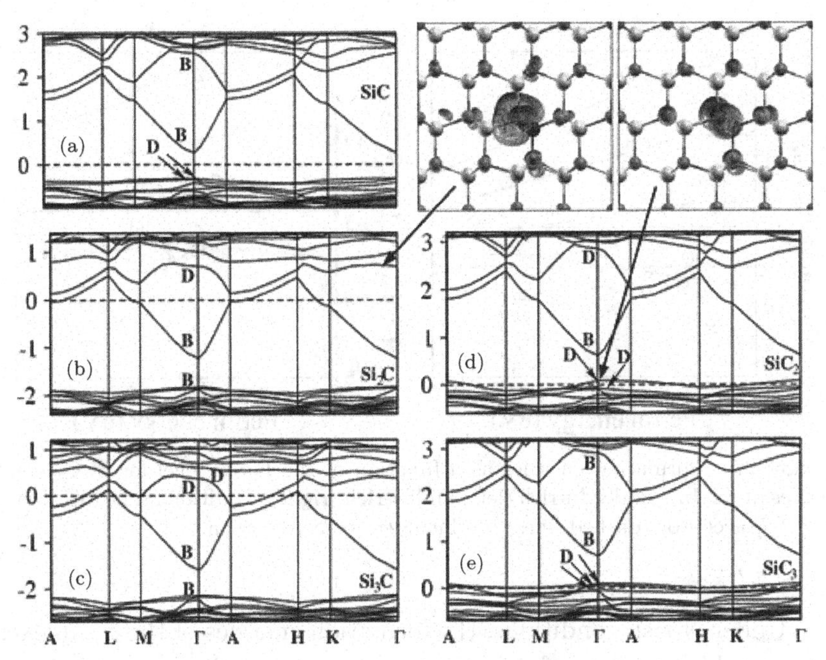

Figure 26. Band structures for Si_iC_j defect complexes in the neutral state in wz–InN. The labels "B" and "D" represent bulk and defect states, respectively. The horizontally dotted lines are the Fermi levels [except for (a) where it is placed in the center of the bandgap]. For Si_2C and SiC_2, isosurface plots of the charge density induced by the defect states are indicated by the long arrows. The 0.00 e/$Bohr^3$ isovalues are shown. Large pale (light yellow) and smaller darker gray (pink) spheres indicate In and N atoms, respectively. Large (blue) and small black (green) spheres indicate Si and C atoms, respectively.

The formation and binding energies are listed in Table 7, where the former are given under In- and N-rich conditions. Figure 27 shows the calculated formation energies of the Si_iC_j defect complexes as a function of Fermi level, E_F, for both In-rich (left) and N-rich (right) conditions. Here, the zero of Fermi level is chosen at the valence band maximum. For comparison, the result for the single Si and C impurities are included. Under In-rich conditions, for the complexes with $i < j$, the formation energies are all higher than 2.5 eV; thus, these defect complexes can hardly exist in significant concentration in thermal equilibrium, although the clusters have large binding energies.

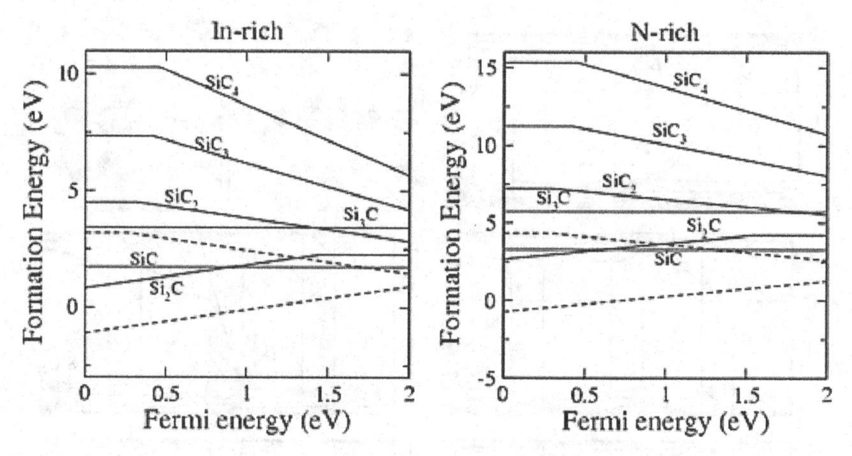

Figure 27. Formation energies as a function of the Fermi level for Si_iC_j complexes in wz–InN under In-rich (left) and N-rich (right) conditions. The Si_{In}^+ donor and C_N^- acceptor (dashed lines) are included for comparison.

Under N-rich conditions, the formation energies of the complexes are even higher. It can be furthermore seen that these complexes do not suggest any doping advantage over isolated Si atoms.

5. Summary

The current theory of native point defects, impurities and complexes in InN has been reviewed with an emphasis on first-principles calculations based on density functional theory. The results of these calculations indicate that native point defects cannot explain the universally observed n-type conductivity in as-grown InN. The nitrogen vacancy, which has been commonly invoked to explain the n-type conductivity in as-grown InN, is a shallow donor but has a high formation energy in n-type InN. The predicted equilibrium concentration is several orders of magnitude lower than the observed electron concentrations. The other native point defects have even higher formation energies and therefore will not occur in observable concentrations in n-type InN. Unintentional incorporation of impurities, such as oxygen and hydrogen, may well be responsible for the observed levels of n-type conductivity in InN. Hydrogen is a stable donor when present in an interstitial configuration, but it can also be stable as a substitutional impurity. It replaces nitrogen, forms a multicenter bond with its four In nearest neighbors, and, counterintuitively, acts as a double donor. Both interstitial and substitutional hydrogen have low formation energies, making them plausible candidates for unintentional n-type dopants. In addition, we also discuss the effects of Mg incorporation and its role as a p-type dopant in InN.

The theory of complexes and their effects on electrical properties are also reviewed in this chapter. It is known that nitrogen vacancy acts as a donor in a p-type material where there is very little interaction between these singly positive charged vacancies. However, in more n-type materials, the neutral charge state becomes favored and the vacancies then prefer to be situated close to one another on nearest-neighbor like species sites, forming "vacancy complexes or clusters." Such complex formation of nitrogen vacancy clusters results in a local In-rich region with a metallic-like bonding. It is possible that such structures are related to the experimentally reported In metal inclusions in InN. The energetics show that the indium vacancies also prefer to cluster together. Although the indium vacancy clusters up to three vacancies have rather higher formation energies, the formation of larger indium vacancy clusters cannot be excluded since the formation energy decreases per vacancy significantly with increasing number of vacancies in the cluster. For the cation and anion vacancies ($V_N + V_{In}$), the formation energy of $V_N + V_{In}$ is lower than that of the isolated indium vacancy under In-rich conditions. However, the formation energies of the complex ($V_N + V_{In}$) are high, indicating that in thermal equilibrium this defect will not be present in significant concentrations. The nitrogen vacancy, V_N, prefers to be bound to Mg_{In}, forming a $Mg_{In}V_N$ complex in thermal equilibrium; a similar behavior has been found for V_N and Mg_{Ga} in wz-GaN [74]. The O_NV_{In} complex has a high formation energy even though V_{In} has a tendency to bind with oxygen. The formation energy of the $3O_NV_{In}$ complex is significantly lower but still too high to be an important defect in InN. We predict the Mg_mO_n complexes could be important defects in InN under both In-rich and N-rich conditions for obtaining improved efficiency for n- and p-type conductivities through tuning of the relative concentrations of Mg and O; in particular, by providing the maximum Mg solubility through the use of the chemical potential of Mg from bulk Mg. Complexes based on codoping of Si and C appear to offer no advantage in obtaining n- or p-type conductivity over isolated dopants.

References

[1] A. G. Bhuiyan, A. Hashimoto, A. Yamamoto, Indium nitride (InN): A review on growth, characterization, and properties, *J. Appl. Phys.* **94**, 2779 (2003).
[2] K. S. A. Butcher, T. L. Tansley, InN, Latest development and a review of the band-gap controversy, *Superlattices Microstruct.* **38**, 1 (2005).
[3] W. Walukiewicz, J. W. Ager III, K. M. Yu, Z. Liliental-Weber, J. Wu, S. X. Li, R. E. Jones, J. D. Denlinger, Structure and electronic properties of InN and In-rich group III-nitride alloys, *J. Phys. D* **39**, R83 (2006).
[4] S. N. Mohammad, H. Morkoc, Progress and prospects of group-III nitride semiconductors, *Prog. Quantum Electron.* **20**, 361 (1996).
[5] B. E. Foutz, S. K. O'Leary, M. S. Shur, L. F. Eastman, Transient electron transport in wurtzite GaN, InN, and AlN, *J. Appl. Phys.* **85**, 7727 (1999).
[6] H. Lu, W. J. Schaff, L. F. Eastman, J. Wu, W. Walukiewicz, D. C. Look, R. J. Molnar, Growth of thick InN by molecular beam epitaxy, *Mat. Res. Soc. Symp. Proc.* **743**, L4.10.1–L4.10.6 (2003).
[7] M. Higashiwaki, T. Matsui, High-quality InN film grown on a lowtemperature-grown GaN intermediate layer by plasma-assisted molecularbeam epitaxy, *Japn. J. Appl. Phys.* **41**, 542 (2002).
[8] M. Higashiwaki, T. Matsui, Effect of low-temperature-grown GaN intermediate layer on InN growth by plasma-assisted MBE, *Physica Status Solidi* (c), **360** (2002).
[9] P. Hohenberg, W. Kohn, Inhomogeneous Electron Gas, Physical Review, 136 (1964) B864; W. Kohn and L. J. Sham, Self-consistent equations including exchange and correlation effects, *Phys Rev* **140**, 1138 (1965).
[10] M. C. Payne, M. P. Teter, D. C. Allan, T. A. Arias, J. D. Joannopoulos, Iterative minimization techniques for ab initio total-energy calculations molecular dynamics and conjugate gradients, *Rev. Modern Phys.* **64**, 1045 (1992).
[11] R. M. Martin, *Electronic Structure*: *Basic Theory and Methods* (Cambridge University Press, Cambridge, 2004).
[12] C. G. Van de Walle, J. Neugebauer, First-principles calculations for defects and impurities: Applications to III-nitrides, *J. Appl. Phys.* **95**, 3851 (2004).

[13] D. B. Laks, C. G. Van de Walle, G. F. Neumark, S. T. Pantelides, Role of native point defects in wide-band-gap semiconductors, *Phys. Rev. Lett.* **66**, 648 (1991).
[14] D. W. Jenkins, J. D. Dow, Electronic structures and doping of InN, InxGa1−xN, and InxAl1−xN, *Phys. Rev. B* **39**, 3317 (1989).
[15] T. L. Tansley, R. J. Egan, Point-defects in the nitrides of aluminum, gallium, and indium, *Phys. Rev. B* **45**, 10942 (1992).
[16] S. Strite, H. Morkoç, GaN, AlN, and InN: A review, *J. Vacu. Sci. Technol. B* **10**, 1237 (1992).
[17] C. Stampfl, C. G. Van de Walle, D. Vogel, P. Kruger, J. Pollmann, Native defects and impurities in InN: First-principles studies using the localdensity approximation and self-interaction and relaxation-corrected pseudopotentials, *Phys. Rev. B* **61**, 7846 (2000).
[18] J. Oila, A. Kemppinen, A. Laakso, K. Saarinen, W. Egger, L. Liszkay, P. Sperr, H. Lu, W. J. Schaff, Influence of layer thickness on the formation of In vacancies in InN grown by molecular beam epitaxy, *Appl. Phys. Lett.* **84**, 1486 (2004).
[19] F. Tuomisto, A. Pelli, K. M. Yu, W. Walukiewicz, W. J. Schaff, Compensating point defects in 4He+-irradiated InN, *Phys. Rev. B* **75**, 193201 (2007).
[20] C. Stampfl, C. G. Van de Walle, D. Vogel, P. Kruger, J. Pollmann, Native defects and impurities in InN: First-principles studies using the local density approximation and self-interaction and relaxation-corrected pseudopotentials, *Phys. Rev. B* **61**, 7846 (2000).
[21] A. Janotti, C. G. Van de Walle, Sources of unintentional conductivity in InN, *Appl. Phys. Lett.* **92**, 032104 (2008)
[22] A. Janotti, C. G. Van de Walle, Oxygen vacancies in ZnO, *Appl. Phys. Lett.* **87**, 122102 (2005).
[23] X. M. Duan, C. Stampfl, Nitrogen vacancies in InN: Vacancy clustering and metallic bonding from first principles, *Phys. Rev. B* **77**, 115207 (2008).
[24] C. S. Gallinat, G. Koblmuller, J. S. Brown, S. Bernardis, J. S. Speck, G. D. Chern, E. D. Readinger, H. Shen, M. Wraback, In-polar InN grown by plasma-assisted molecular beam epitaxy, *Appl. Phys. Lett.* **89**, 032109 (2006).
[25] S. Baroni, A. Dal Corso, S. de Gironcoli, P. Giannozzi, http://www.pwscf.org
[26] X. M. Duan, C. Stampfl, Defect complexes and cluster doping of InN: First-principles investigations, *Phys. Rev. B* **79**, 035207 (2009).
[27] C. Stampfl, C. G. Van de Walle, Theoretical investigation of native defects, impurities, and complexes in aluminum nitride, *Phys. Rev. B* **65**, 155212 (2002).
[28] C. H. Park, D. J. Chadi, Stability of deep donor and acceptor centers in GaN, AlN, and BN, *Phys. Rev. B* **55**, 12995 (1997).
[29] C. G. Van de Walle, J. Neugebauer, Universal alignment of hydrogen levels in semiconductors, insulators, and solutions, *Nature* **423**, 626 (2003).

[30] J. Neugebauer, C. G. Van de Walle, Theory of hydrogen in GaN, In Hydrogen in semiconductors II, edited by N. H. Nickel, *Semiconductors and Semimetals*, Vol. 61, edited by R. K. Willardson and E. R. Weber (Academic Press, Boston, 1999).

[31] E. A. Davis, S. F. J. Cox, R. L. Lichti, C. G. Van de Walle, Shallow donor state of hydrogen in indium nitride, *Appl. Phys. Lett.* **82**, 592 (2003).

[32] A. Janotti, C. G. Van de Walle, Sources of unintentional conductivity in InN, *Appl. Phys. Lett.* **92**, 032104 (2008).

[33] A. Janotti, C. G. Van de Walle, Hydrogen multicentre bonds, *Nature Mater.* **6**, 44 (2007).

[34] C. S. Gallinat, G. Koblmuller, J. S. Brown, S. Bernardis, J. S. Speck, G D. Chern, E. D. Readinger, H. Shen, M. Wraback, In-polar InN grown by plasma-assisted molecular beam epitaxy, *Appl. Phys. Lett.* **89**, 032109 (2006).

[35] O. Madelung (ed.) *Semiconductors — Data Handbook*, 3rd (edn.), (Springer, Berlin, 2004).

[36] I. Mahboob, T. D. Veal, C. F. McConville, H. Lu, W. J. Schaff, Intrinsic electron accumulation at clean InN surfaces, *Phys. Rev. Lett.* **92**, 036804 (2004).

[37] T. D. Veal, L. F. J. Piper, I. Mahboob, H. Lu, W. J. Schaff, C. F. McConville, Electron accumulation at InN/AIN and InN/GaN interfaces, *Physica Status Solidi* (c), **2**, 2246 (2005).

[38] D. Segev, C. G. Van de Walle, Origins of Fermi-level pinning on GaN and InN polar and nonpolar surfaces, *Europhys. Lett.* **76**, 305 (2006).

[39] C. Stampfl, C. G. Van de Walle, D. Vogel, P. Krüger, J. Pollmann, Native defects and impurities in InN: First-principles studies using the local-density approximation and self-interaction and relaxation-corrected pseudopotentials, *Phys. Rev. B* **61**, R7846 (2000).

[40] D. W. Jenkins, J. D. Dow, Electronic-structures and doping of InN, InxGa1-xN, and InxAl1-xN, *Phys. Rev. B* **39**, 3317 (1989).

[41] P. Specht, R. Armitage, J. Ho, E. Gunawan, Q. Yang, X. Xu, C. Kisielowski, E. R. Weber, The influence of structural properties on conductivity and luminescence of MBE grown InN, *J. Cryst. Growth* **269**, 111 (2004).

[42] X. Wang, S.-B. Che, Y. Ishitani, A. Yoshikawa, Polarity inversion in high Mg-doped in-polar InN epitaxial layers, *Appl. Phys. Lett.* **91**, 081912 (2007).

[43] J. Neugebauer, C. G. Van de Walle, Role of hydrogen in doping of GaN, *Appl. Phys. Lett.* **68**, 1829 (1996).

[44] R. E. Jones, K. M. Yu, S. X. Li, W. Walukiewicz, J. W. Ager, E. E. Haller, H. Lu, W. J. Schaff, Evidence for p-type doping of InN, *Phys. Rev. Lett.* **96**, 125505 (2006).

[45] C. Stampfl, C. G. Van de Walle, Cohesive properties of group-III nitrides: A comparative study of all-electron and pseudopotential calculations using the generalized gradient approximation, *Phys. Rev. B* **65**, 155212 (2002).

[46] J. Neugebauer, C. G. Van de Walle, Chemical trends for acceptor impurities in GaN, *J. Appl. Phys.* **85**, 3003 (1999).
[47] L. E. Ramos, J. Furthmuller, L. M. R. Scolfaro, J. R. Leite, F. Bechstedt, Substitutional carbon in group-III nitrides: Ab initio description of shallow and deep levels, *Phys. Rev. B* **66**, 075209 (2002).
[48] J. Neugebauer, C. G. Van de Walle, Gallium vacancies and the yellow luminescence in GaN, *Appl. Phys. Lett.* **69**, 503 (1996).
[49] T. Mattila, R. M. Nieminen, Point-defect complexes and broadband luminescence in GaN and AlN, *Phys. Rev. B* **55**, 9571 (1997).
[50] X. M. Duan, C. Stampfl, Nitrogen vacancies in InN: Vacancy clustering and metallic bonding from first principles, *Phys. Rev. B* **77**, 115207 (2008).
[51] X. M. Duan, C. Stampfl, Vacancies and interstitials in indium nitride: Vacancy clustering and molecular bondlike formation from first principles, *Phys. Rev. B* **79**, 174202 (2009).
[52] X. M. Duan, C. Stampfl, Defect complexes and cluster doping of InN: First-principles investigations, *Phys. Rev. B* **79**, 035207 (2009).
[53] J. P. Perdew, A. Zunger, self-interaction correction to density-density-functional approximations for many-election systems, *Phys. Rev. B* **23**, 5048 (1981).
[54] S. Baroni, A. Dal Corso, S. de Gironcoli, P. Giannozzi, http://www.pwscf.org
[55] T. V. Shubina, S. V. Ivanov, V. N. Jmerik, D. D. Solnyshkov, V. A. Vekshin, P. S. Kop'ev, A. Vasson, J. Leymarie, A. Kavokin, H. Amano, K. Shimono, A. Kasic, B. Monemar, Mie resonances, infrared emission, and the band gap of InN, *Phys. Rev. Lett.* **92**, 117407 (2004).
[56] C. R. Brazier, P. F. Bernath, J. B. Burkholder, C. J. Howard, fourier-transform spectroscopy of the nu-3 band of the n3 radical, *J. Chem. Phys.* **89**, 1762 (1988).
[57] M. G. Ganchenkova, R. M. Nieminen, Nitrogen vacancies as major point defects in gallium nitride, *Phys. Rev. Lett.* **96**, 196402 (2006).
[58] X. M. Duan, C. Stampfl, Nitrogen vacancies in InN: Vacancy clustering and metallic bonding from first principles, *Phys. Rev. B* **77**, 115207 (2008).
[59] A. Uedono, S. F. Chichibu, M. Higashiwaki, T. Matsui, T. Ohdaira, R. Suzuki, Vacancy-type defects in Si-doped InN grown by plasma-assisted molecular-beam epitaxy probed using monoenergetic positron beams, *J. Appl. Phys.* **97**, 043514 (2005).
[60] J. Neugebauer, C. G. Van de Walle, Gallium vacancies and the yellow luminescence in GaN, *Appl. Phys. Lett.* **69**, 503 (1996).
[61] H. Katayama-Yoshida, T. Nishimatsu, T. Yamamoto, N. Orita, Codoping method for the fabrication of low-resistivity wide band-gap semiconductors in p-type GaN, p-type AlN and n-type diamond: Prediction versus experiment, *J. Phys.: Condens. Matter* **13**, 8901 (2001)
[62] C. Stampfl, C. G. Van de Walle, Theoretical investigation of native defects, impurities, and complexes in aluminum nitride, *Phys. Rev. B* **65**, 155212 (2002).

[63] R. Y. Korotkov, J. M. Gregie, B. W. Wessels, Electrical properties of p-type GaN : Mg codoped with oxygen, *Appl. Phys. Lett.* **78**, 222 (2001).
[64] G. Kipshidze, V. Kuryatkov, B. Borisov, Y. Kudryavtsev, R. Asomoza, S. Nikishin, H. Temkin, Mg and O codoping in p-type GaN and AlxGa1-xN, *Appl. Phys. Lett.* **80**, 2910 (2002).
[65] T. Yamamoto, H. Katayama-Yoshida, *Jpn. J. Appl. Phys. Part 2* **36**, L180 (1997).
[66] L. G. Wang, A. Zunger, Cluster-doping approach for wide-gap semiconductors: The case of p-type ZnO, *Phys. Rev. Lett.* **90**, 256401 (2003).
[67] Y. Yan, J. Li, S. H. Wei, M. M. Al-Jassim, Possible approach to overcome the doping asymmetry in wideband gap semiconductors, *Phys. Rev. Lett.* **98**, 135506 (2007).
[68] R. Kalish, C. Saguy, C. Cytermann, J. Chevallier, Z. Teukam, F. Jomard, T. Kociniewski, D. Ballutaud, J. E. Butler, C. Baron, A. Deneuville, Conversion of p-type to n-type diamond by exposure to a deuterium plasma, *J. Appl. Phys.* **96**, 7060 (2004).
[69] M. Joseph, H. Tabata, T. Kawai, p-type electrical conduction in ZnO thin films by Ga and N codoping, *Jpn. J. Appl. Phys., Part 2* **38**, L1205 (1999).
[70] D. C. Look, B. Claflin, Y. I. Alivov, S. J. Park, The future of ZnO light emitters, *Phys. Status Solidi A* **201**, 2203 (2004).
[71] C. H. Seager, A. F. Wright, J. Yu, W. Götz, Role of carbon in GaN, *J. Appl. Phys.* **92**, 6553 (2002).
[72] C. H. Seager, D. R. Tallant, J. Yu, W. Götz, *J. Lumin.* **106**, 115 (2004).
[73] A. Armstrong, A. R. Arehart, D. Green, U. K. Mishra, J. S. Speck, S. A. Ringel, Impact of deep levels on the electrical conductivity and luminescence of gallium nitride codoped with carbon and silicon, *J. Appl. Phys.* **98**, 053704 (2005).
[74] A. F. Wright, N vacancy diffusion and trapping in Mg-doped wurtzite GaN, *J. Appl. Phys.* **96**, 2015 (2004).

CHAPTER 6

Dopants and Impurity-Induced Defects in ZnO

M. AZIZAR RAHMAN, MATTHEW R. PHILLIPS and CUONG TON-THAT*

School of Mathematical and Physical Sciences,
University of Technology Sydney, Ultimo, NSW 2007, Australia
*cuong.ton-that@uts.edu.au

In recent years, there has been resurgent interest in ZnO due to its efficient UV emission and potential application in opto-electronic devices. The problem of self-compensating defects, especially those related to acceptor-like dopants, remains a major challenge to date. In this chapter, we provide an overview of the fundamental properties of point defects and dopants as well as their complexes in bulk crystals and nanostructures. Nominally undoped ZnO is typically n-type, which has been widely ascribed to O vacancies or Zn interstitials; however, these defect assignments are controversial as O vacancies are deep donors while Zn interstitials are highly mobile at room temperature and considered to be unstable. Accordingly, H and group III impurities have also been suggested to be responsible for the observed n-type conductivity. Although there are still many unanswered questions concerning defect-related luminescence bands in ZnO, great progress has been made in recent years to identify and characterize them using spatially resolved luminescence spectroscopies and first-principles calculations. The ubiquitous green emission has several possible origins, including O and Zn vacancies as well as Cu impurities. The properties of group I (Li, Na, and Cu), group III (Ga, Al, and In), and group V (N, P, and Sb) impurities as well as their complexes with native point defects and H are discussed, along with a concluding outlook for future research into the optical properties of ZnO.

1. Introduction

Zinc oxide is a wide-bandgap semiconductor with great potential for numerous applications such as solid-state lighting [1, 2], solar cells [3], lasing [4], and sensors [5]. With a large exciton binding energy of 60 meV, ZnO emits light efficiently in the UV region of the spectrum, making it a strong candidate for optical devices. In contrast to GaN, large crystals of ZnO can be grown easily and the low cost of Zn makes ZnO economically competitive for large-scale device fabrication. While ZnO has inherent advantages, the lack of control over native defects and impurities presents a major obstacle to the realization of practical devices. The presence of such defects and impurities can inhibit the performance of devices due to detrimental effects such as nonradiative recombination, carrier trapping, as well as parasitic luminescence and absorption. The literature in this area is vast, and only reasonably well-established results are discussed in this chapter. For readers interested in a complete critical discussion on the structural and physical properties of ZnO materials and devices, excellent comprehensive review articles have been published by other workers [6–8].

2. Point Defects in ZnO

2.1. *Intrinsic point defects*

Several theoretical and experimental studies of defect formation in ZnO have been reported [6, 9, 10], and the calculated energy levels of different defects as a function of Fermi level position are displayed in Fig. 1. These data show that defects such as O and Zn vacancies have lowest formation energies in Zn- and O-rich ZnO, respectively. Accordingly, Zn vacancies play a dominant role as compensating acceptors in n-type ZnO. O vacancies are present in significant

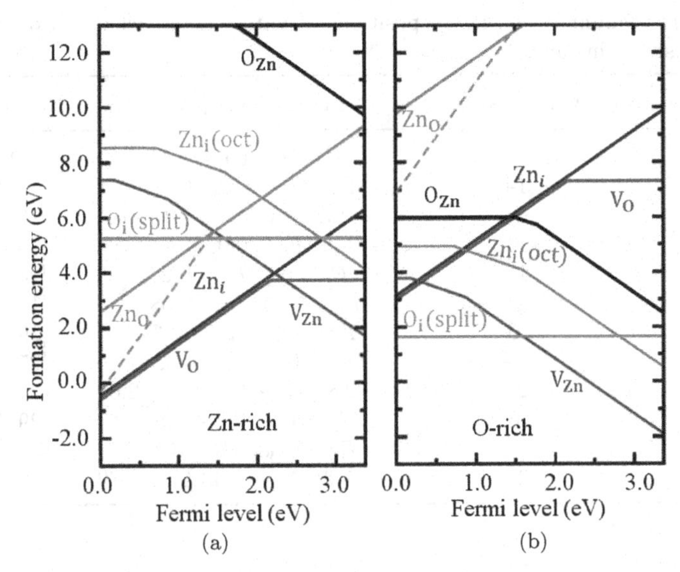

Figure 1. Calculated formation energies of native defects as a function of Fermi level for (a) Zn-rich and (b) O-rich ZnO. The zero of Fermi-level indicates the valence band maximum and the slope corresponding to the charge states. Positive slopes correspond to positively charged defects, indicating their donor-like behavior. Negative slopes indicate negatively charged defects and acceptor-type behavior. Reprinted with permission from [12].

concentration but unlikely to be the dominant cause of n-type conductivity in nominally undoped ZnO due to their deep donor nature, while Zn interstitials are unstable at room temperature due to their high mobility [11, 12]. Other native point defects including oxygen antisites (O_{Zn}) and zinc antisites (Zn_O) are typically not abundant in ZnO under equilibrium conditions due to their high formation energies; the effect of these defects on the properties of Zn_O is not discussed further in this chapter. Table 1 provides a summary of the reported energy-level positions of the four main point defects in ZnO: zinc vacancy (V_{Zn}), oxygen vacancy (V_O), zinc interstitial (Zn_i), and oxygen interstitial (O_i).

2.1.1. *Oxygen vacancies*

O vacancies have the lowest formation energy among the native donor-like defects (Fig. 1) and have frequently been cited as the cause

Table 1. Summary of energy positions of native point defects in different charge states in ZnO.

Native defect	Charge state	Energy level (eV)	References
V_O	0	$E_c - 0.05$	[25]
	1+	$E_c - 2.0$, $E_c - 1.92$, $E_c - 2.24$	[25–27]
	2+	$E_c - 1.0$, $E_c - 1.1$	[13, 19]
Zn_i	0	$E_c - 0.05$	[25, 26]
	1+	$E_c - 0.5$, $E_c - 0.2$	[25, 26]
	2+	$E_c - 0.15$, $E_c - 0.08$	[14, 28]
V_{Zn}	0	$E_v + 0.3$, $E_v + 0.31$	[29, 30]
	1−	$E_v + 0.7$, $E_v + 0.82$	[25, 26]
	2−	$E_v + 2.8$, $E_v + 2.91$, $E_v + 2.67$	[25, 27]
O_i	0	$E_v + 1.08$, $E_v + 0.9$	[29, 31]
	1−	$E_v + 0.38$, $E_v + 0.4$	[31, 32]
	2−	$E_v + 0.99$, $E_v + 1.43$ $E_v + 0.79$	[30, 32]

of n-type conductivity in ZnO. However, first-principles calculations show that the V_O is a deep, negative U donor, where the 1+ charge state is thermodynamically unstable [6, 12, 13], converting spontaneously to either its 2+ or 0 charge state. O vacancies become neutral when the Fermi level is close to the conduction band and above the V_O 0/2+ charge transition level. Conversely, when the Fermi level is below the 0/2+ level, V_O have a charge of $+2e$ and can act as a source of compensation in p-type ZnO. The position of the 0/2+ charge transition level for O vacancies was calculated to be 1–2 eV below the conduction band maximum [14, 15].

A green luminescence band is commonly observed in both bulk and nanostructured ZnO materials and has been controversially attributed to O vacancies [16–18]. The difficulty in identifying the origin of this defect band is the fact that it is broad and often overlaps with other broad visible emission peaks, which leads to inaccuracy in the measurement of the green emission peak's position, width, and intensity without the use of careful curve fitting. The reported peak energy of V_O-related green luminescence varies from 2.42 to 2.54 eV (Table 2). Ye *et al.* [19] reported that the radiative recombination of an electron from the V_O^+ state to the valence band is responsible for the 2.5 eV green emission band, whereas Hofmann *et al.* [20] reported

Table 2. Summary of energy positions and the proposed defects responsible for deep-level emissions in ZnO.

Peak position (eV)	Emission color	Proposed defect	References
2.2–2.5	Green	V_o	[16–18, 20, 43, 47]
2.3–2.5	Green	V_{Zn}	[17, 18, 47]
2.4	Structured green		[48, 49]
2.7–3.1	Blue		[27, 35, 41]
1.7	Red		[50]
2.3	Green	O_i	[47]
2.0–2.1	Yellow/Orange		[43, 44, 51]
1.8–2.0	Red		[35, 41, 45]
2.5–2.7	Blue	Zn_i	[40, 52]
2.85	Violet		[41]

that the optical excitation can convert only a fraction of V_O to V_O^+, which is thermodynamically unstable. It is noted that the different energies of the "green" band in different studies could be associated with different defects; we observed two distinct emission bands at 2.30 and 2.52 eV, attributable to Zn and O vacancies, respectively (see Fig. 2) [18].

2.1.2. *Zinc vacancies*

Zn vacancies in ZnO are double acceptors and have a negligible concentration in p-type ZnO because of their high formation energy. On the contrary, V_{Zn} has the lowest formation energy among native point defects and is present in a moderate concentration in donor-doped ZnO [12]. These acceptors are more favorable to form in the O-rich environment (Fig. 1). Positron annihilation experiments have confirmed that Zn vacancies are dominant compensating acceptors in n-type ZnO [21]. Theoretical and experimental results have been reported for the energy level of V_{Zn} in different charge states (Table 1). In semiconductor terminology, the charge transition level of a defect is defined as the position of Fermi level at which the two charge states of this defect have equal formation energies. First-principles calculations reported the 0/1− and 1−/2− acceptor charge transition levels of V_{Zn} at 0.18–0.2 eV and 0.87–1.2 eV above the

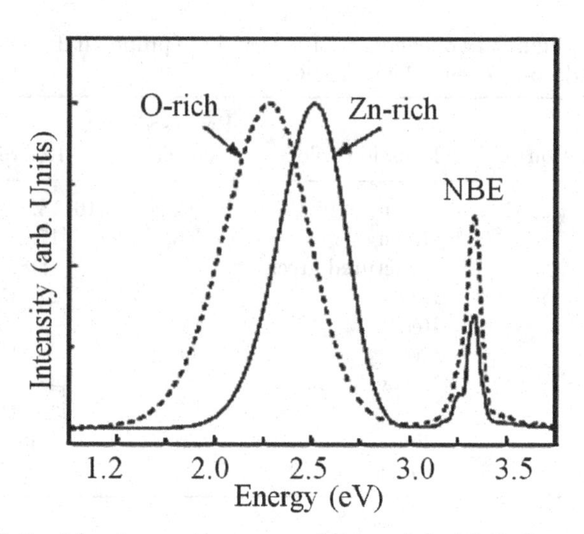

Figure 2. Cathodoluminescence spectra of Zn- and O-rich ZnO particles at 80 K (beam energy = 15 keV, beam current = 0.25 nA) normalized to the green peak height. Both Zn- and O-rich ZnO exhibit near-band-edge emission at 3.35 eV but different green emission peak energies of 2.52 and 2.30 eV, respectively. Reprinted with permission from [18].

valence band maximum, respectively [11, 12], whereas these levels were calculated to be at 0.9 and 1.5 eV, respectively, by generalized gradient approximation (GGA) calculations [22]. When the Fermi level lies above the 1−/2− level, the equilibrium charge state is a negatively charged double acceptor V_{Zn}^{2-}, which is formed by accepting two electrons from the valence band. Electron paramagnetic resonance (EPR) experiments have shown that the 1−/2− level of V_{Zn} lies at ~1.0 eV above the valence band maximum [23, 24]. The slight variation in these data can be attributed to different calculation methods, or in the case of experiment, to diverse sample types, fabrication techniques, and growth conditions.

Zn vacancies are also considered to be responsible for the green emission band in ZnO [17, 18]. Sekiguchi *et al.* provided a strong argument in favor of Zn vacancies being the origin of green emission [33, 34]. The authors reported that the hydrogen plasma treatment strongly passivates the green emission band in ZnO as Zn vacancies can readily interact with and be passivated by hydrogen. The peak positions of the reported V_{Zn}-related green luminescence are found in the 2.2–2.5 eV energy range (Table 2). Different research groups suggested different types of electronic transitions involving V_{Zn} defects to describe the green emission band, such as from a shallow donor to a deep V_{Zn} acceptor [35], from the conduction band to V_{Zn} acceptor [27], from a Zn_i donor to V_{Zn} acceptor level [36], and from a hole transfer from divalent zinc vacancy (V_{Zn}^{2-}) and a monovalent (V_{Zn}^{-}) defect [37]. The defect center responsible for the green luminescence in ZnO remains controversial.

2.1.3. *Zinc interstitials*

In the ZnO wurtzite lattice, Zn interstitials (Zn_i) act as donors and can occupy either octahedral or tetrahedral sites. However, Zn_i is more stable at the octahedral site because the tetrahedral site exhibits a higher formation energy [48]. Among the three charge states of zinc interstitials (Zn_i^o, Zn_i^+, and Zn_i^{2+}), the most stable state is Zn_i^{2+}, which is formed by donating two electrons to the conduction band. Under equilibrium conditions, Zn_i has a high formation energy in n-type ZnO (as shown in Fig. 1) and is likely to be present in a negligible concentration [48]. Accordingly, Zn_i defects cannot be the source of the background n-type conductivity in ZnO even under Zn-rich conditions [48]. Additionally, there is general consensus that Zn_i is unstable at room temperature due to their high mobility. Nevertheless, it has been suggested that Zn_i defects can be a source of n-type conductivity under non-equilibrium conditions or by forming complexes involving N impurities [10]. For example, Hutson *et al.* reported the presence of Zn_i shallow donors with an activation energy of 51 meV in Hall experiments [38] and Look *et al.* observed the presence of Zn_i shallow donors when ZnO samples were irradiated by a high-energy electron beam [39]. The formation energy of Zn_i falls with decreasing Fermi energy, indicating that they act as a compensating defect in p-type ZnO [48].

Zn_i is not considered to be an optically active luminescence center under equilibrium conditions, but this defect has been reported to be a source of blue and violet emissions in ZnO that was produced in highly non-equilibrium processes such as laser ablation and Zn-rich annealing [40, 41]. Halliburton *et al.* attributed an increased concentration of free carriers in the ZnO crystal annealed in Zn vapor [42] to Zn_i and proposed that the non-equilibrium conditions promote the formation of this defect. The peak positions of the reported Zn_i-related blue/violet luminescence are found in the 2.54–2.85 eV energy range as summarized in Table 2. Zeng *et al.* assigned the violet and blue emissions to the transitions from Zn_i and extended Zn_i states to the valance band, respectively [40].

2.1.4. *Oxygen interstitials*

Oxygen interstitials (O_i) can occupy either the octahedral interstitial site or tetrahedral interstitial site and are deep acceptors. First-principles calculations suggest that the tetrahedral interstitials are unstable, transforming into a split-interstitial configuration, also known as dumbbell configuration [14]. Conversely, O_i in the octahedral interstitial site are more stable and electrically active [49]. The octahedral O_i introduce –/2– and 0/– acceptor charge transition levels at 1.59 and 0.72 eV above the valence band maximum, respectively [6]. The octahedral O_i configuration has high formation energy and its concentration is negligible in ZnO under equilibrium conditions. Janotti *et al.* reported that oxygen interstitials are electrically inactive in *p*-type ZnO and act as a deep acceptor when the Fermi level is greater than 2.8 eV [6].

A broad yellow emission peaking in the energy range of 2.0–2.1 eV has been reported in ZnO (Table 2) [43, 44]. This yellow emission band in ZnO has been attributed to O_i defects [43, 44]. Moreover, several groups reported that the red emission centered at $\sim$ 1.80 eV in ZnO nanostructures is related to O_i defects [35, 45]. Figure 3 shows the deep level emission of the ZnO film that was annealed in Ar gas at 700°C and 900°C and subsequently treated by remote hydrogen plasma. The annealed ZnO films exhibit green (at 2.35 eV) and yellow (at 2.0 eV) emissions, which are both completely passivated by a hydrogen plasma, where the yellow emission is attributed to O_i defects [46]. The hydrogen plasma passivation of the O_i-related yellow emission band in ZnO is consistent with its assignment to O_i. In contrast, a similar yellow emission attributed to Li impurities

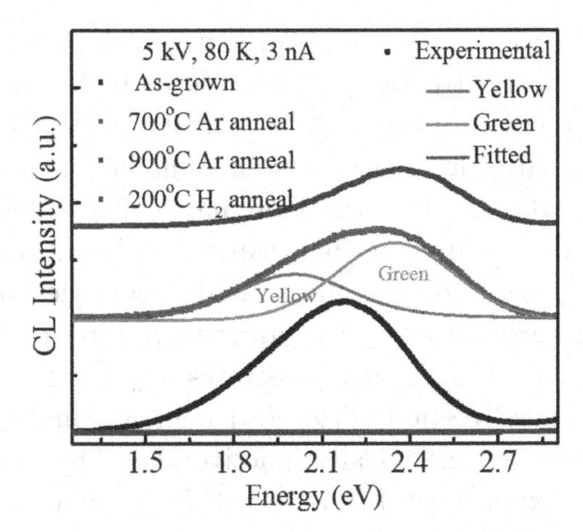

Figure 3. Cathodoluminescence spectra of the as-grown, 700°C and 900°C annealed undoped ZnO films (beam energy = 5 keV, beam current = 3 nA). The modeled CL spectra fitted with the yellow and green luminescence components. The emission spectrum from the film after being treated with remote hydrogen plasma (bottom spectrum) is also included and shows that the plasma treatment completely passivates both the green and yellow emission bands.

in ZnO is not passivated by an equivalent hydrogen plasma treatment [46].

3. Impurities and Defect Complexes

The control of impurities and their associated charge states is of paramount importance in applications that exploit the wide range of properties of doped ZnO. Intense research efforts have been focused on the fabrication and characterization of acceptor-doped ZnO; however, reports of p-type samples are highly controversial. Incorporation of dopants into ZnO can also change the growth direction of 1D nanostructures; for example, ZnO nanorings and nanobelts can be produced using In as the dopant element [53, 54].

3.1. *Group I elements and hydrogen*

Group I (Li, Na, and K) elements are useful potential dopants for p-type ZnO and have been theoretically predicted to be shallow acceptors when substituting Zn atoms, whereas H (also a group I element) is a shallow donor. First-principles calculations and experiments have drawn attention to the role of hydrogen on the electronic and optical properties ZnO [55, 56]. It has been reported that that interstitial hydrogen acts as a shallow donor and its positive charge state (H_i^+) is thermodynamically stable [6]. Hydrogen is a common impurity that can be unintentionally incorporated during growth of the most used techniques and can contribute to background n-type conductivity. First-principles calculations showed that hydrogen passivates V_{Zn} by the formation of a thermodynamically stable (V_{Zn}–H_2) neutral complex [34]. These defect complexes are electrically inactive [34]. This result is consistent with the experimental study that H plasma treatment completely passivates the V_{Zn}-related green emission band [18]. More recent theoretical work based on density function calculations suggests that hydrogen can form a H–V_O defect complex with the O vacancy, which has a low formation energy of about 1.7 eV lower than that of H_i [55]. Unlike H_i, this shallow donor complex is considered to be immobile at elevated temperatures and is responsible for the stability of high conductivity in H-doped ZnO films up to 250°C [55].

Group I impurities (Li, Na, and K) at the interstitial sites are donors but act as acceptors at the Zn substitutional site. First-principles calculations have shown that Li, Na, and K elements are found to be deep acceptors located at 0.74, 0.85, and 1.26 meV above the valence band maximum, respectively [32, 57]. Based on the optical measurements, Meyer *et al.* and Kushnirenko *et al.* reported that Li_{Zn}, Na_{Zn}, and K_{Zn} are optically active acceptor centers, which give rise to shallow donor to deep acceptor recombination in the visible spectrum [58, 59]. Meyer *et al.* also reported that Li and Na, incorporated either during growth or by diffusion, can introduce acceptors with a hole-binding energy of around 0.3 eV, which shows a donor–acceptor pair recombination [58]. Park *et al.* calculated the ionization energy of 0.09, 0.17, and 0.32 eV for substitutional Li, Na, and K, respectively [60].

A Li-related pair complex (Li_{Zn} + Li_i) can also be formed in ZnO under O-rich conditions [61]. Figures 4(a), 4(b) shows the local

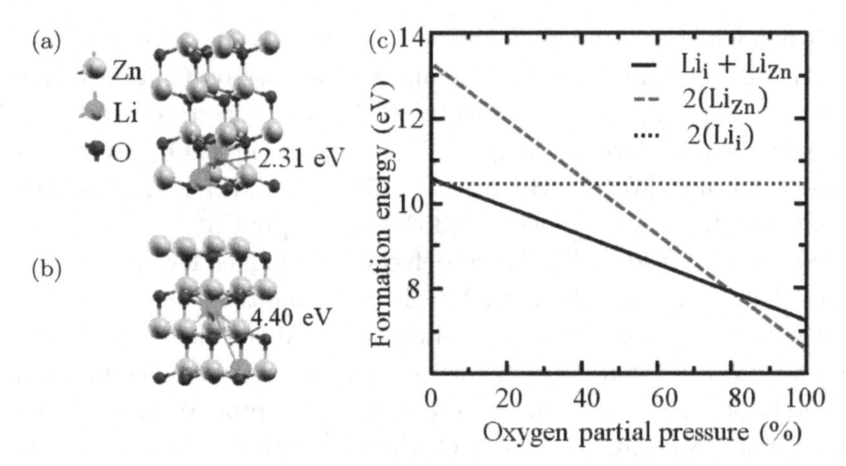

Figure 4. The local atomic geometry of $Li_{Zn} + Li_i$ pair complexes for (a) nearest and (b) well-separated Li species (Reprinted with permission from [62]). (c) Formation energy of the $Li_{Zn} + Li_i$ pair complex as a function of oxygen partial pressure. (Reprinted with permission from [57]).

atomic geometry of the $Li_{Zn} + Li_i$ complex [62]. For both the O-rich and Zn-rich conditions, the formation energy decreases for Li_{Zn} and increases for Li_i with are increase in the Fermi level. Under O-rich conditions, the $Li_{Zn} + Li_i$ pair complex has lower formation energy than either isolated Li_{Zn} or Li_i defect (Fig. 4) [57]. Based on the first-principles calculations and photoluminescence measurements, Sahu *et al.* proposed that this acceptor-like complex is optically active and exhibits a visible emission band at 3.0 eV. EPR experiments showed that Li has two acceptor states at 0.85 and 0.15 eV above the valence band maximum [63].

Group IB elements (Cu, Au, and Ag) act as acceptors when incorporated at Zn sites. Under O-rich conditions, the formation energies of these acceptors are very low, but are considerably higher when located at interstitial sites [64]. The charge transition energies (0/−) for Cu_{Zn}, Ag_{Zn}, and Au_{Zn} are relatively high at 0.7, 0.4, and 0.5 eV above the valence band maximum, respectively [64]. Group IB elements are better p-type dopants than IA elements due to their high ionization energy and less self-compensation by native donor defects. Cu favorably occupies the Zn site, (Cu_{Zn}), which acts as an acceptor leading to the p-type conductivity and ferromagnetism

[65, 66]. Cu is a commonly incorporated impurity up to the ppm level in II–VI semiconductors; a large amount of work on Cu-doped ZnO exists concerning various aspects of Cu acceptor states and their possible role in structured green luminescence in ZnO crystals and nanostructures [67, 68]. Huang *et al.* [69] reported that Cu has three charge states, i.e. Cu_{Zn}^{+} or $Cu^{3+}(3d^8 4s^0)$, Cu_{Zn}^{0} or Cu^{2+} $(3d^9 4s^0)$, and Cu_{Zn}^{-} or $Cu^{1+}(3d^{10} 4s^0)$. These defects have higher formation energies than native defects in ZnO in Zn-rich conditions and have lower formation energies in O-rich conditions [69]. When the Fermi level is above the Cu^{2+}/Cu^{1+} (0/–) charge transition level, the equilibrium state is Cu^{1+}, which is a negatively charged acceptor. With increasing Cu doping concentration in ZnO, the Cu^{1+} state causes the *p*-type conductivity and the Fermi level shifts toward the valence band until the Cu^{2+}/Cu^{1+} level is reached. First-principles calculations showed that the Cu-doped ZnO is ferromagnetic when Cu atoms are present in ZnO in their Cu^{3+} and Cu^{2+} charge states, which exist only in *p*-type ZnO [69]. Conversely, Cu-doped ZnO was found to be non-magnetic when Cu atoms are in the Cu^{1+} state. The charge states of Cu can be also mediated by doping ZnO with Ga, which pushes the Fermi level above the Cu^{2+}/Cu^{1+} transition level [68]. The position of this Cu^{2+}/Cu^{1+} level is still a matter of great controversy: Wardle *et al.* [70] estimated this transition level to be at 1.0 eV above the valence band maximum, while Yan *et al.* [64] gave a value of 0.7 eV. Electrical measurements on Cu-doped ZnO revealed that the 0/– level lies at 0.2 eV below the conduction band minimum [71]. Admittance spectroscopy and photoluminescence experiments have shown that Cu has two acceptor levels at $E_C-0.17$ and $E_V+0.4$ eV [72, 73].

3.2. *Group-II element (Mg)*

Mg energetically prefers to occupy the Zn site and induces substitutional defects acting as acceptors in ZnO[74]. Alloying ZnO films with MgO is a commonly used method for tuning the bandgap of ZnO [75]. Mg can also occupy either octahedral interstitial site

(Mg_i^O) or tetrahedral interstitial site (Mg_i^T). The formation energy of Mg_i is higher than that of Mg_{Zn} for all Fermi energies except under Zn-rich conditions where the formation energy of Mg_i is lower when the Fermi level is near the valence band maximum (see Fig. 5) [74]. Since the Fermi level is always close to the conduction band in ZnO due to its inherent n-type conductivity, Mg_{Zn} is the most thermodynamically stable defect type under both O- and Zn-rich conditions. The substitutional Mg_{Zn} in ZnO lowers the formation energy of Zn vacancies (V_{Zn}) by ~2.1 eV and thereby promotes the formation of V_{Zn} centers [76]. This prediction is consistent with the photoluminescence results, revealing that doping ZnO with Mg significantly enhances the V_{Zn}-related green luminescence band [77]. First-principles calculations showed that Mg_{Zn} can interact with the donor-like defects (V_O and Zn_i) of ZnO, forming the Mg_{Zn}–Zn_i and Mg_{Zn}–V_O defect complexes [76]. This is in agreement with the experimental result that Mg doping quenches the V_O-related visible emission band in ZnO [78].

3.3. *Group III elements*

The group III elements (Ga, Al, and In) are well-known n-type dopants in ZnO. Al and Ga are favored donors for transparent conductor applications; high doping levels of $n \sim 10^{20}$ cm^{-3} can make ZnO electrically conductive and optically transparent [79]. Group III elements are preferably incorporated at Zn sites and induce substitutional defects in ZnO [80]. They can also be accommodated at either the octahedral interstitial site (Ga_i^O) or the tetrahedral interstitial site (Ga_i^T). The formation energies of these dopants at interstitial sites are much higher than those on substitutional sites and their concentrations are negligible when grown under equilibrium conditions [80, 81]. Theoretical and experimental studies showed that the group III elements can interact with native acceptor-like defects (V_{Zn} and O_i), generating high concentrations of Ga_{Zn}–V_{Zn} and Ga_{Zn}–O_i defect complexes due to the Coulomb interaction between donor and acceptor centers [82, 83]. The local atomic geometry of Ga_{Zn}–V_{Zn}

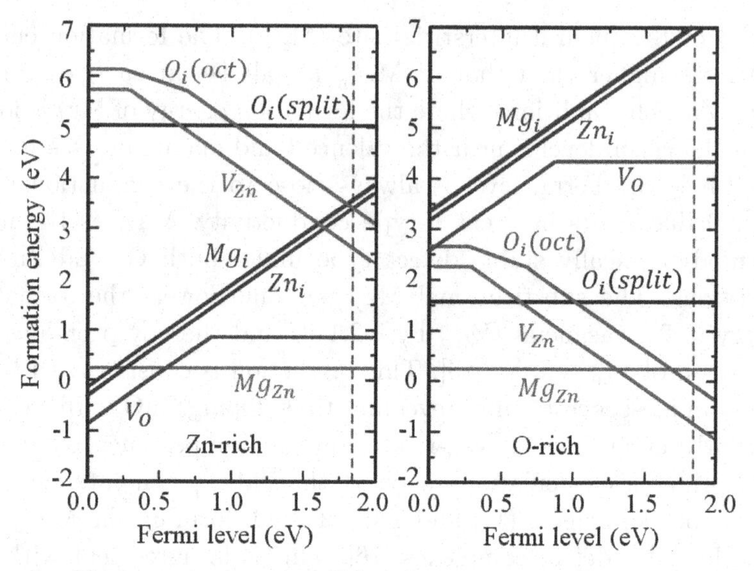

Figure 5. The formation energies of Mg defects in comparison with native defects in ZnO under Zn-rich and O-rich conditions. The Fermi level is referenced to the valence band maximum. Reprinted with permission from [74].

and Ga_{Zn}–O_i defect complexes are shown in Fig. 6. These defects act as acceptors and are electrically active in ZnO.

Under O-rich conditions, the Ga_{Zn}–V_{Zn} acceptor complex has significantly lower formation energy than the isolated substitutional Ga_{Zn} donors when the Fermi energy is close to the conduction band minimum (Fig. 6) and is responsible for appreciable defect compensation in highly Ga-doped ZnO. These theoretical predictions were verified by experimental results, which reported that both the electrical conductivity and carrier concentration significantly decrease at high Ga doping concentrations when samples were grown under O-rich conditions [84, 85]. The formation energy of Ga_{Zn}–O_i is higher than Ga_{Zn}–V_{Zn}, indicating that the compensation effect due to the Ga_{Zn}–O_i complex in Ga-doped ZnO is weaker. In O-poor conditions, the formation energies of these defect complexes are substantially higher (Fig. 7), suggesting that the compensation mechanism is less pronounced. Several experimental and theoretical studies showed that the Ga_{Zn}–V_{Zn} and Ga_{Zn}–O_i complexes have ionization

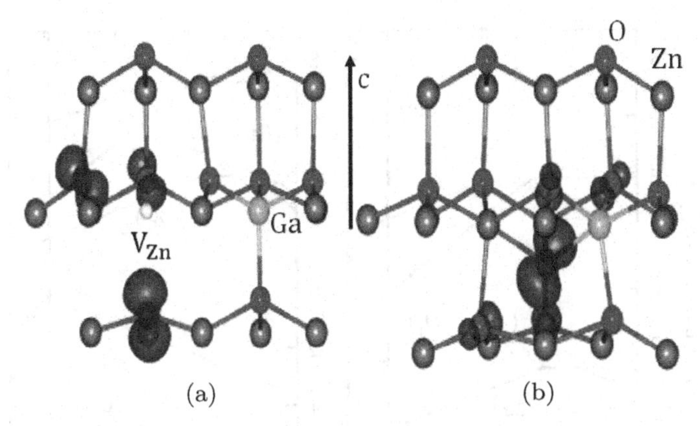

Figure 6. Local atomic geometry for (a) Ga_{Zn}–V_{Zn} and (b) Ga_{Zn}–O_i complexes in Ga-doped ZnO. Both defect complexes act as deep acceptors and are electrically active in ZnO. Reprinted with permission from [82].

energies of about 0.75 and 0.66 eV above the valence band maximum, respectively [82, 86].

3.4. *Group IV elements*

Theoretical investigations showed that group IV elements (C, Si, and Sn) act as donors when occupying the Zn site in ZnO and that ZnO can be made highly conductive and transparent in the visible region by doping with Si [87]. However, there has been little evidence that these elements are incorporated at an appreciable level into ZnO crystals. Carbon is a common impurity in chemical vapor transport growth with graphite and MOCVD; it was theoretically predicted that C could either occupy in the substitutional Zn- or O-sites, forming C_{Zn} donor or C_O acceptor [88]. Based on first-principles calculations, Lu *et al.* [89] predicted that C_O interacts with the donor-like defects (V_O and Zn_i), forming C_O–V_O, C_O–Zn_i, and $2C_O$–V_O–Zn_i defect complexes. The authors claimed that the C_O–Zn_i and $2C_O$–V_O–Zn_i complexes have a low formation energy as shown in Fig. 8 and are the possible source of green and red emissions in ZnO, respectively. This prediction is consistent with photoluminescence studies of C-doped ZnO [90]. Combination of density function calculations and optical measurements showed that Sn doping of

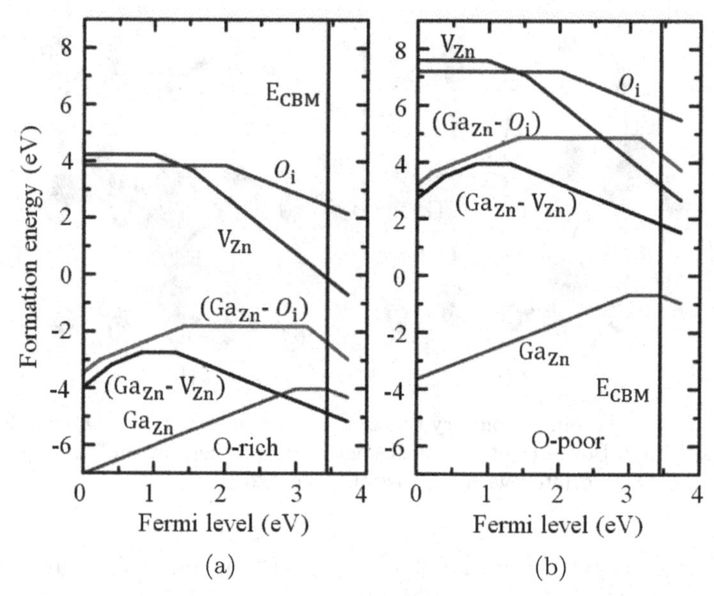

Figure 7. Formation energies of Ga-related defects as a function of Fermi level in (a) O-rich and (b) O-poor conditions. The zero value of the Fermi level corresponds to the valence band maximum and the "kinks" in the curves indicate transitions between the defect charge states. Under the O-rich condition, both Ga_{Zn}–V_{Zn} and Ga_{Zn}–O_i complexes can efficiently form due to their low formation energies and play an important role as compensating defects in Ga-doped ZnO. Reprinted with permission from [82].

ZnO quantum dots eliminates V_O while creating O_{Zn} and O_i defects as a result of conversion from Sn^{4+} to Sn^{2+} state with increasing Sn doping concentration [91].

3.5. *Group V elements*

The greatest challenge in research on ZnO is the achievement of reliable and reproducible p-type doping. Group V (N, P, As, and Sb) elements are the most promising candidates for p-type doping, but unfortunately, most of these dopants form deep acceptors. The substitutional X_O (where $X = N, P, As, and Sb$) at the O site is a deep acceptor and is unlikely to contribute for the p-type conduction in ZnO, while the substitutional X_{Zn} at the Zn site exhibits donor-like behavior [7, 92]. First-principles calculations show that N substituting an O is a deep acceptor with a hole binding energy

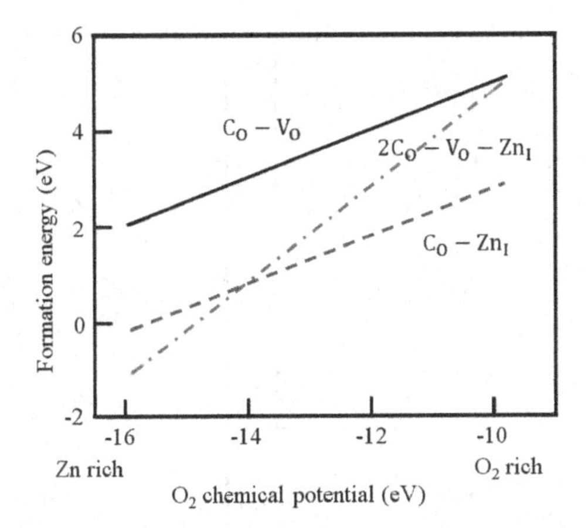

Figure 8. Calculated formation energies of carbon defect complexes as a function of O_2 chemical potential. The complexes are C_O–V_O (black line), C_O–Zn_i(red) and $2C_O$–V_O–Zn_i (green). Reprinted with permission from [89].

of 0.4 eV [60], while photoluminescence experiments revealed a hole binding energy of ~0.2 eV [93]. X-ray electron and X-ray absorption spectroscopies on N-doped ZnO showed that N exists in three chemical states, attributed to N_O, N_{Zn}, and N_2 molecules, and a direct correlation was observed between the N-related donor–acceptor pair emission and the concentration of N_2 [94, 95]. The experimental observations are in agreement with first-principles calculations which predict that N_2 at the Zn site is a shallow acceptor [96].

P, As, and Sb have significantly larger ionic radii than O, and according to the first-principles calculations, they act as deep acceptors when accommodated at the O substitutional site [60]. Nevertheless, p-type conductivity in P-doped ZnO has been reported by several groups [97, 98]. Modeling also predicted that the substitutional X_{Zn} interacts with Zn vacancies, forming an acceptor complex $X_{Zn}-2V_{Zn}$ [92, 99]. The formation energies of P-induced defects and complexes in ZnO are displayed in Fig. 8. Under O-rich conditions, the $P_{Zn}-2V_{Zn}$ acceptor complex has a lower formation energy than the other defects, the calculated transition level (+/−) for this complex is 150 meV above the valence band maximum [99]. This calculated value seems to be in qualitative agreement with the experimental investigations that report an acceptor level in the range

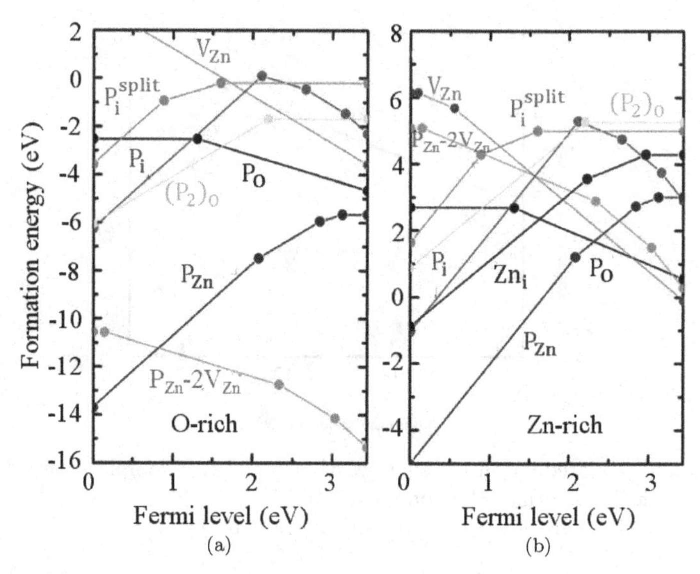

Figure 9. Formation energy of native and P-induced defects in ZnO under O-rich and Zn-rich conditions. Reprinted with permission from [99].

of 120 – 180 meV for P-doped ZnO [98, 100]. Photoluminescence and Hall measurements of Sb-doped ZnO films also showed that the incorporation of Sb induces the formation of $Sb_{Zn}-2V_{Zn}$ shallow acceptor complex with a binding energy 161 meV, responsible for their p-type conduction [101]. Further experiments and modeling need to be performed to confirm the microscopic structure of these acceptor complexes.

4. Conclusions

Recent theoretical and experimental investigations have provided further insight into the nature of point defects, impurities, and dopants in ZnO and their effect on the electrical and optical properties. Significant progress has been made in recent years in fabricating of high-quality ZnO samples suitable for detailed and systematic investigations of defect characteristics. Yet, several significant questions concerning the fundamental properties of defects and their roles in the performance of ZnO-based devices remain largely unresolved. The ultimate goal of being able to reproducibly fabricate, high-quality p-type ZnO has not been met. This review emphasizes that comprehensive studies aimed at understanding and controlling native and impurity-induced defects are essential for the development of practical ZnO devices with bespoke properties for specific applications. Elements in several periodic table groups have been considered as potential dopants in ZnO. The electronic and optical properties of the host ZnO were found to be influenced in a similar way upon doping with elements from the same group. Group I elements act as deep acceptors when residing on a substitutional Zn site and are responsible for the yellow emission in ZnO. Group I atoms at interstitial sites behave as shallow acceptors and have been reported to yield p-type conductivity in ZnO, but the exact nature of these acceptors is not yet clear. Mg from group II has been proved effective in engineering the bandgap of ZnO. Group III elements, specifically Al and Ga, as well as H, account for the n-type behavior of ZnO in most circumstances. Al and Ga are the preferred donors for transparent conductor applications, but these donors could be compensated via the formation of acceptor complexes with either Zn vacancies or O interstitials. Both of the acceptor complex types are known to be electrically and optically active. Little is known about the incorporation of C and Si, but many studies claim that these group IV dopants can make ZnO films highly conductive for transparent electrical contacts. Group V dopants, in particular N and P, are considered to be the natural choice for p-type doping; however, the origin of p-type response in ZnO doped with N or P remains controversial due to the formation of compensating defects and their low solubility. ZnO holds great promise for a range of photonic and optoelectronic applications, it is hoped that new fundamental insights from reliable characterization of defects will lead to transformative breakthroughs in practical ZnO devices.

Acknowledgments

Financial support from the Australian Research Council (ARC) under the Discovery Project funding scheme (project number DP150103317) is acknowledged.

References

[1] X. Huang, *Nat. Photon.* **8**, 748 (2014).
[2] M. A. Rahman, J. A. Scott, A. Gentle, M. R. Phillips, C. Ton-That, *Nanotechnology* **29**, 425707 (2018).
[3] Y. Tang, Z. Chen, H. Song, C. Lee, H. Cong, H. Cheng, W. Zhang, I. Bello, S. Lee, *Nano Lett.* **8**, 4191 (2008).
[4] S. Chu, G. Wang, W. Zhou, Y. Lin, L. Chernyak, J. Zhao, J. Kong, L. Li, J. Ren, J. Liu, *Nat. Nanotechnol.* **6**, 506 (2011).
[5] S. N. Das, J. P. Kar, J.-H. Choi, T. I. Lee, K.-J. Moon, J.-M. Myoung, *J. Phys. Chem. C* **114**, 1689 (2010).
[6] A. Janotti and C. G. Van de Walle, Rep. Prog. Phys. **72**, 126501 (2009).
[7] Ü. Özgür, Y. I. Alivov, C. Liu, A. Teke, M. Reshchikov, S. Doğan, V. Avrutin, S.-J. Cho, H. Morkoc, *J. Appl. Phys.* **98**, 11 (2005).
[8] A. B. Djurisic, A. M. C. Ng, X. Y. Chen, *Prog. Quant. Electron.* **34**, 191 (2010).
[9] F. Selim, M. Weber, D. Solodovnikov, K. Lynn, *Phys. Rev. Lett.* **99**, 085502 (2007).
[10] D. C. Look, G. C. Farlow, P. Reunchan, S. Limpijumnong, S. Zhang, K. Nordlund, *Phys. Rev. Lett.* **95**, 225502 (2005).
[11] A. Kohan, G. Ceder, D. Morgan, C. G. Van de Walle, *Phys. Rev. B* **61**, 15019 (2000).
[12] A. Janotti, C. G. Van de Walle, *Phys. Rev. B* **76**, 165202 (2007).
[13] A. Janotti, C. G. Van de Walle, *Appl. Phys. Lett.* **87**, 122102 (2005).
[14] P. Erhart, K. Albe, A. Klein, *Phys. Rev. B* **73**, 205203 (2006).
[15] A. Janotti, C. G. Van de Walle, *Nat. Mater.* **6**, 44 (2007).
[16] F. Leiter, H. Alves, A. Hofstaetter, D. Hofmann, B. Meyer, *Phys. Status Solidi B* **226**, R4 (2001).
[17] T. M. Børseth, B. Svensson, A. Y. Kuznetsov, P. Klason, Q. Zhao, M. Willander, *Appl. Phys. Lett.* **89**, 262112 (2006).
[18] C. Ton-That, L. Weston, M. Phillips, *Phys. Rev. B* **86**, 115205 (2012).
[19] J. Ye, S. Gu, F. Qin, S. Zhu, S. Liu, X. Zhou, W. Liu, L. Hu, R. Zhang, Y. Shi, *Appl. Phys. A* **81**, 759 (2005).
[20] D. Hofmann, D. Pfisterer, J. Sann, B. Meyer, R. Tena-Zaera, V. Munoz-Sanjose, T. Frank, G. Pensl, *Appl. Phys. A* **88**, 147 (2007).
[21] F. Tuomisto, V. Ranki, K. Saarinen, D. C. Look, *Phys. Rev. Lett.* **91**, 205502 (2003).
[22] S. Lany, A. Zunger, *Phys. Rev. Lett.* **98**, 045501 (2007).
[23] X. Wang, L. Vlasenko, S. Pearton, W. Chen, I. A. Buyanova, *J. Phys. D: Appl. Phys.* **42**, 175411 (2009).
[24] S. Evans, N. Giles, L. Halliburton, L. Kappers, *J. Appl. Phys.* **103**, 043710 (2008).
[25] S. Lima, F. Sigoli, M. Jafelicci Jr, M. R. Davolos, *Int. J. Inorg. Mater.* **3**, 749 (2001).

[26] V. Nikitenko, *Zinc Oxide — A Material for Micro-and Optoelectronic Applications* (Springer, 2005), p. 69.
[27] P. Thiyagarajan, M. Kottaisamy, N. Rama, M. R. Rao, *Scripta Mater.* **59**, 722 (2008).
[28] A. Djurišić, A. Ng, X. Chen, *Prog. Quant. Electron.* **34**, 191 (2010).
[29] B. Lin, Z. Fu, Y. Jia, *Appl. Phys. Lett.* **79**, 943 (2001).
[30] S. Choi, M. R. Phillips, I. Aharonovich, S. Pornsuwan, B. C. Cowie, C. Ton-That, *Adv. Opt. Mater.* **3**, 821 (2015).
[31] P. Xu, Y. Sun, C. Shi, F. Xu, H. Pan, *Nucl. Instr. Meth. Phys. Res. B* **199**, 286 (2003).
[32] J. Hu, B. Pan, *J. Chem. Phys.* **129**, 154706 (2008).
[33] C. G. Van de Walle, *Phys. Rev. Lett.* **85**, 1012 (2000).
[34] E. Lavrov, J. Weber, F. Börrnert, C. G. Van de Walle, R. Helbig, *Phys. Rev. B* **66**, 165205 (2002).
[35] N. Alvi, K. Ul Hasan, O. Nur, M. Willander, *Nanoscale Res. Lett.* **6**, 130 (2011).
[36] A. Escobedo-Morales, U. Pal, *Appl. Phys. Lett.* **93**, 193120 (2008).
[37] C. Xu, X. Sun, X. Zhang, L. Ke, S. Chua, *Nanotechnol.* **15**, 856 (2004).
[38] D. Thomas, *J. Phys. Chem. Solids* **3**, 229 (1957).
[39] D. C. Look, J. W. Hemsky, J. Sizelove, *Phys. Rev. Lett.* **82**, 2552 (1999).
[40] H. Zeng, G. Duan, Y. Li, S. Yang, X. Xu, W. Cai, *Adv. Funct. Mater.* **20**, 561 (2010).
[41] C. H. Ahn, Y. Y. Kim, D. C. Kim, S. K. Mohanta, H. K. Cho, *J. Appl. Phys.* **105**, 013502 (2009).
[42] L. Halliburton, N. Giles, N. Garces, M. Luo, C. Xu, L. Bai, L. A. Boatner, *Appl. Phys. Lett.* **87**, 172108 (2005).
[43] X. Wu, G. Siu, C. Fu, H. Ong, *Appl. Phys. Lett.* **78**, 2285 (2001).
[44] A. R. Gheisi, C. Neygandhi, A. K. Sternig, E. Carrasco, H. Marbach, D. Thomele, O. Diwald, *Phys. Chem. Chem. Phys.* **16**, 23922 (2014).
[45] X. Liu, X. Wu, H. Cao, R. Chang, *J. Appl. Phys.* **95**, 3141 (2004).
[46] Z. Wang, C. Luo, W. Anwand, A. Wagner, M. Butterling, M. A. Rahman, M. R. Phillips, C. Ton-That, M. Younas, S. Su, *Sci. Rep.* **9**, 3534 (2019).
[47] J. Lv, C. Li, *Appl. Phys. Lett.* **103**, 232114 (2013).
[48] D. Reynolds, D. C. Look, B. Jogai, *J. Appl. Phys.* **89**, 6189 (2001).
[49] S. Anantachaisilp, S. M. Smith, C. Ton-That, S. Pornsuwan, A. R. Moon, C. Nenstiel, A. Hoffmann, M. R. Phillips, *J. Lumin.* **168**, 20 (2015).
[50] S. Anantachaisilp, S. M. Smith, C. Ton-That, T. Osotchan, A. R. Moon, M. R. Phillips, *J. Phys. Chem. C* **118**, 27150 (2014).
[51] V. Kumar, H. Swart, O. Ntwaeaborwa, R. Kroon, J. Terblans, S. Shaat, A. Yousif, M. Duvenhage, *Mater. Lett.* **101**, 57 (2013).
[52] M. Patra, K. Manzoor, M. Manoth, S. Vadera, N. Kumar, *J. Lumin.* **128**, 267 (2008).
[53] X. Y. Kong, Y. Ding, R. Yang, Z. L. Wang, *Science* **303**, 1348 (2004).
[54] X. Y. Kong, Z. L. Wang, *Nano Lett.* **3**, 1625 (2003).

[55] W. Chen, L. Zhu, Y. Li, L. Hu, Y. Guo, H. Xu, Z. Ye, *Phys. Chem. Chem. Phys.* **15**, 17763 (2013).
[56] R. C. Wang, C. F. Cheng, *Plasma Process. Polym.* **12**, 51 (2015).
[57] R. Vidya, P. Ravindran, H. Fjellvåg, *J. Appl. Phys.* **111**, 123713 (2012).
[58] B. K. Meyer, J. Stehr, A. Hofstaetter, N. Volbers, A. Zeuner, J. Sann, *Appl. Phys. A* **88**, 119 (2007).
[59] V. Kushnirenko, I. Markevich, T. Zashivailo, *J. Lumin.* **132**, 1953 (2012).
[60] C. Park, S. Zhang, S.-H. Wei, *Phys. Rev. B* **66**, 073202 (2002).
[61] A. Carvalho, A. Alkauskas, A. Pasquarello, A. Tagantsev, N. Setter, *Phys. Rev. B* **80**, 195205 (2009).
[62] R. Sahu, K. Dileep, B. Loukya, R. Datta, *Appl. Phys. Lett.* **104**, 051908 (2014).
[63] C. Rauch, W. Gehlhoff, M. Wagner, E. Malguth, G. Callsen, R. Kirste, B. Salameh, A. Hoffmann, S. Polarz, Y. Aksu, *J. Appl. Phys.* **107**, 024311 (2010).
[64] Y. Yan, M. Al-Jassim, S.-H. Wei, *Appl. Phys. Lett.* **89**, 181912 (2006).
[65] D. Buchholz, R. Chang, J.-Y. Song, J. Ketterson, *Appl. Phys. Lett.* **87**, 082504 (2005).
[66] C.-L. Hsu, Y.-D. Gao, Y.-S. Chen, T.-J. Hsueh, *ACS Appl. Mater. Interfaces* **6**, 4277 (2014).
[67] N. Garces, L. Wang, L. Bai, N. Giles, L. Halliburton, G. Cantwell, *Appl. Phys. Lett.* **81**, 622 (2002).
[68] M. A. Rahman, M. T. Westerhausen, C. Nenstiel, S. Choi, A. Hoffmann, A. Gentle, M. R. Phillips, C. Ton-That, *Appl. Phys. Lett.* **110**, 121907 (2017).
[69] D. Huang, Y.-J. Zhao, D.-H. Chen, Y.-Z. Shao, *Appl. Phys. Lett.* **92**, 182509 (2008).
[70] M. Wardle, J. Goss, P. Briddon, *Phys. Rev. B* **72**, 155108 (2005).
[71] E. Mollwo, G. Müller, P. Wagner, *Solid State Commun.* **13**, 1283 (1973).
[72] M. A. Reshchikov, V. Avrutin, N. Izyumskaya, R. Shimada, H. Morkoc, S. Novak, *J. Vac. Sci. Technol.* **27**, 1749 (2009).
[73] Y. Kanai, *Jpn. J. Appl. Phys.* **30**, 703 (1991).
[74] N. Palakawong, J. Jutimoosik, J. T-Thienprasert, S. Rujirawat, S. Limpijumnong, *Integrated Ferroelectrics* **156**, 72 (2014).
[75] A. K. Sharma, J. Narayan, J. F. Muth, C. W. Teng, C. Jin, A. Kvit, R. M. Kolbas, O. W. Holland, *Appl. Phys. Lett.* **75**, 3327 (1999).
[76] R. Dutta, N. Mandal, *Appl. Phys. Lett.* **101**, 042106 (2012).
[77] E. B. Manaia, R. C. K. Kaminski, B. L. Caetano, V. Briois, L. A. Chiavacci, C. Bourgaux, *Eur. J. Nanomed.* **7**, 109 (2015).
[78] J. Perkins, G. Foster, M. Myer, S. Mehra, J. Chauveau, A. Hierro, A. Redondo-Cubero, W. Windl, L. Brillson, *APL Mater.* **3**, 062801 (2015).
[79] H. J. Ko, Y. F. Chen, S. K. Hong, H. Wenisch, T. Yao, D. C. Look, *Appl. Phys. Lett.* **77**, 3761 (2000).
[80] R. Saniz, Y. Xu, M. Matsubara, M. Amini, H. Dixit, D. Lamoen, B. Partoens, *J. Phys. Chem. Solids* **74**, 45 (2013).

[81] Y.-S. Lee, Y.-C. Peng, J.-H. Lu, Y.-R. Zhu, H.-C. Wu, *Thin Solid Films* **570**, 464 (2014).
[82] D. Demchenko, B. Earles, H. Liu, V. Avrutin, N. Izyumskaya, Ü. Özgür, H. Morkoç, *Phys. Rev. B* **84**, 075201 (2011).
[83] M.-H. Lee, Y.-C. Peng, H.-C. Wu, *J. Alloys Compd.* **616**, 122 (2014).
[84] T. Yamada, K. Ikeda, S. Kishimoto, H. Makino, T. Yamamoto, *Surf. Coat. Technol.* **201**, 4004 (2006).
[85] C.-Y. Tsay, K.-S. Fan, C.-M. Lei, *J. Alloys Compd.* **512**, 216 (2012).
[86] A. Tang, Z. Mei, Y. Hou, L. Liu, V. Venkatachalapathy, A. Azarov, A. Kuznetsov, X. Du, *Sci. China Phys., Mech. & Astron.* **61**, 077311 (2018).
[87] J. Liu, X. Fan, C. Sun, W. Zhu, *Chem. Phys. Lett.* **649**, 78 (2016).
[88] S. Panpan, S. Xiyu, H. Qinying, L. Yadong, C. Wei, *J. Semicond.* **30**, 052001 (2009).
[89] Y. Lu, Z. Hong, Y. Feng, S. Russo, *Appl. Phys. Lett.* **96**, 091914 (2010).
[90] L. Tseng, J. Yi, X. Zhang, G. Xing, H. Fan, T. Herng, X. Luo, M. Ionescu, J. Ding, S. Li, *AIP Adv.* **4**, 067117 (2014).
[91] W. Yang, B. Zhang, Q. Zhang, L. Wang, B. Song, Y. Ding, C. Wong, *RSC Adv.* **7**, 11345 (2017).
[92] G. Petretto, F. Bruneval, *Phys. Rev. Appl.* **1**, 024005 (2014).
[93] B. Meyer, Alves, H, D. Hofmann, W. Kriegseis, D. Forster, F. Bertram, J. Christen, A. Hoffmann, M. Straßburg, M. Dworzak, *Phys. Status Solidi B* **241**, 231 (2004).
[94] C. Ton-That, L. Zhu, M. N. Lockrey, M. R. Phillips, B. C. C. Cowie, A. Tadich, L. Thomsen, S. Khachadorian, S. Schlichting, N. Jankowski, A. Hoffmann, *Phys. Rev. B* **92**, 7 (2015).
[95] C. L. Perkins, S. H. Lee, X. N. Li, S. E. Asher, T. J. Coutts, *J. Appl. Phys.* **97**, 7 (2005).
[96] W. R. L. Lambrecht, A. Boonchun, *Phys. Rev. B* **87**, 195207 (2013).
[97] K. K. Kim, H. S. Kim, D. K. Hwang, J. H. Lim, S. J. Park, *Appl. Phys. Lett.* **83**, 63 (2003).
[98] A. Allenic, W. Guo, Y. Chen, M. B. Katz, G. Zhao, Y. Che, Z. Hu, B. Liu, S. B. Zhang, X. Pan, *Adv. Mater.* **19**, 3333 (2007).
[99] P. Li, S.-H. Deng, J. Huang, *Appl. Phys. Lett.* **99**, 111902 (2011).
[100] A. Allenic, W. Guo, Y. Chen, Y. Che, Z. Hu, B. Liu, X. Pan, *J. Phys. D: Appl. Phys.* **41**, 025103 (2007).
[101] X. Pan, W. Guo, Z. Ye, B. Liu, Y. Che, H. He, X. Pan, *J. Appl. Phys.* **105**, 113516 (2009).

CHAPTER 7

Ferromagnetism in B2-Ordered Alloys Induced *via* Lattice Defects

RANTEJ BALI

Helmholtz-Zentrum Dresden-Rossendorf, Institute of Ion Beam Physics & Materials Research
Bautzner Landstr. 400, Dresden 01328, Germany
r.bali@hzdr.de

This chapter considers the case of B2-ordered alloys that are initially non-ferromagnetic and where the introduction of lattice defects can cause the onset of ferromagnetism. This disorder-induced ferromagnetism is confined to the regions where the defects are concentrated. In general, the lattice can be thermally re-ordered, removing the defects and erasing the magnetized regions. Using B2 $Fe_{60}Al_{40}$ thin films as a prototype, the use of ion irradiation as well as pulsed laser irradiation for inducing antisite defects in the crystalline lattice is demonstrated. Ion beams can be applied as broad beams in combination with shadow masks for printing magnetic patterns over large areas, or focused down to approximately nanometer diameters for stylus-like writing of nanomagnets of desired geometries. The patterning resolution is limited by the lateral scattering of ions and can be estimated by semi-empirical modelling, described in this chapter. In the case of laser pulsing, disordering can be induced at thin film surfaces for pulse fluences above the melting threshold. Pulsing below the threshold can lead to surface re-ordering, erasing the magnetic regions and achieving all-laser re-writeable patterning. Localized disordering of B2 ordered systems thus enables a versatile path to embedding highly resolved non-volatile magnets at room temperature, with potential in magnetic device applications.

1. Introduction

Materials exhibiting phase transitions can enable advances in information technology, not only in terms of the raw data storage capacity but also by providing pathways for realizing novel optical and transport devices. Understanding magnetic phase transitions, in particular those where ferromagnetic onsets lead to an increase of the intrinsic saturation magnetization (M_s) can be of huge relevance to data storage and sensing technologies.

Ferromagnetic onsets from an initially non-ferromagnetic precursor can be realized in a variety of alloy systems via the subtle introduction of lattice defects, of which antisite defects are a well-known type. The alloys in question are typically composed of a 3d magnetic metal and a non-ferromagnetic metal, forming stable B2 structures such as $Fe_{60}Al_{40}$ [1–5], $Fe_{50}Rh_{50}$ [6–10], $Fe_{60}V_{40}$ [11], $Fe_{60}Ru_{40}$ [12], $Co_{50}Ga_{50}$ [13], etc.

In an ordered alloy, atoms of component species occupy specific crystallographic sites on the unit cell. Violating the order in the site occupancy, results in point defects. For instance, starting with an ordered lattice, swapping atoms between two sites occupied by non-identical atoms generates a pair of antisites. In a B2 ordered lattice, antisite defects can be induced by swapping atoms between the 100 and 200 planes, thereby forming a fully randomized bcc, also known as the A2 structure. This phase transition can be realized *via* a number of paths, including ion irradiation [4, 5, 14], mechanical strain [15–18], nanoindentation [19] as well as laser pulses irradiation [20, 21]. This chapter will focus on the implementation of ions as well as lasers for locally inducing antisites and other lattice defects, such as open-volume defects and static disorder that can influence ferromagnetic onsets in alloys.

Ion beams have proven to be a highly versatile tool to modify material properties, including the magnetic behavior of thin films [22, 23]. Ion beams are highly compatible with industrial processes, where they are widely used in semiconductor foundries for *p*- and *n*-type doping. A broad range of ion species, including noble gases and molecules are available for irradiation, offering an effectively limitless scope for material modification. In addition, advanced lithographic processes for generating doping patterns already exist and can be readily applied to irradiation-based magnetic patterning, for instance to generate magnetic arrays for patterned magnetic media [24]. A variety of extrinsic properties such as the coercive field necessary for magnetization reversal as well as the magnetic anisotropy axes that determine the preferential direction of the magnetization can be strongly modified *via* intermixing and atomic re-arrangements. However, here we confine the discussion largely to the modification of intrinsic magnetic properties, *viz.* the M_s and crystal structure.

Disordering can cause a large variation in nearest-neighbor interactions between the magnetic moment-carrying atoms. Furthermore, changes of the lattice parameter can be observed as the B2 structure transforms to the A2 structure. From the viewpoint of fundamental observations of magneto-structural correlations in binary alloy systems, ion beams can enable a sensitive and gradual disordering of an ordered alloy precursor. Investigations aimed at understanding nearest-neighbor magnetic interactions in alloy systems benefit from the systematic and clean ion irradiation process. Disordering can be performed using noble gas ions, thereby avoiding any chemical reactions and observing property changes purely due to atomic re-arrangements. Thin films that have been systematically disordered with noble gas ions can act as near-ideal specimens for direct spectroscopic probing, such as in the use of extended X-ray absorption fluorescence spectroscopy (EXAFS) for observing the local environment and X-ray magnetic circular dichroism (XMCD) studies for magnetic observations [16, 25, 26]. The above approach has been applied to B2 $Fe_{60}Al_{40}$ as well as B2 $Fe_{50}Rh_{50}$ thin films to understand the influence of varying Fe–Fe nearest-neighbor interactions on the magnetic and structural properties.

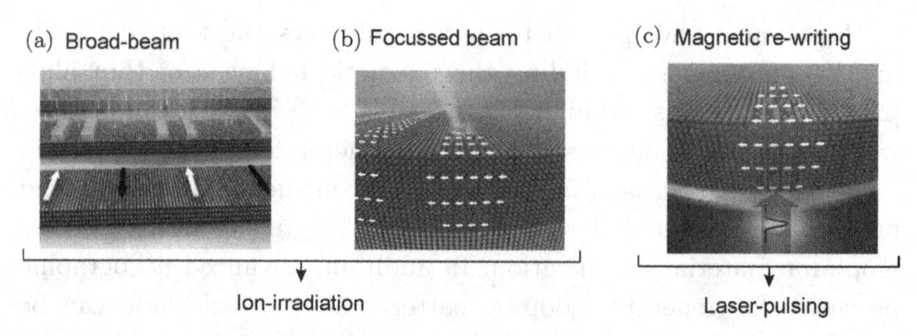

Figure 1. Pathways to generate disordering in B2 ordered alloys. (a) Broad-beam irradiation of light noble gas ions, in combination with a shadow mask. (b) The use of a focused ion beam to generate the desired magnetic patterns without the necessity of lithographic masks. (c) Using laser pulsing to write and erase magnetic regions at the film surface. (a) & (c) Reproduced with permission from Refs. [5, 20]. Copyrights 2013 & 2018 American Chemical Society. (b) Reproduced with permission from Ref. [20].

Lattice defects can be exploited for the manipulation of magnetic properties, at the nanoscale, in thin film alloys. Figure 1 depicts the disordering of a precursor alloy thin film of the B2 structure, achieved by ions as well as laser beams. A broad beam of ions is used to induce disorder, as shown in Fig. 1(a), where lattice defects are induced in the desired regions using a patterned shadow mask of a thickness that is sufficient to fully stop the ions over pre-determined regions. State-of-the-art tools such as gas field ion sources, available in ion microscopes can realize noble gas ion beams of approximately nanometer diameters and can be used for direct magnetic writing as shown in Fig. 1(b). Highly focused beams can be used to generate localized disorder and produce prototype magnetic devices [27–29].

In a recent approach, instead of ions, laser pulses were used not only for generating ferromagnetism but also for erasing it [20]. As depicted in Fig. 1(c), a laser pulse of fluence above a certain "writing" threshold induces ferromagnetism at the film surface. Above the threshold, the laser pulses melt the entire film cross-section, from the point of incidence at the substrate–film interface to the top surface. A lack of nucleation sites in the fully melted film can delay re-solidification, despite the temperature dropping well below the melting point. A long-lived (a few nanoseconds) supercooled state suppresses the time–temperature window for solid-state re-ordering, trapping atoms at the disordered sites and rendering a magnetic region at the point of laser incidence. Conversely, for laser pulses that only partially melt the film cross-section, the unmelted fraction acts as a nucleation template and re-solidification commences once the temperature falls below the melting point, providing sufficient time ($\sim$ns) for the solid-state diffusion at elevated temperature.

Here, B2 $Fe_{60}Al_{40}$ has been used as a prototype to demonstrate the above ion- and laser-induced effects. The main reasons for this are the sensitivity of the alloy to antisite defects in particular, as well as the ease with which $Fe_{60}Al_{40}$ can be processed. Magnetron sputtering from a single target onto substrates held at room temperature followed by *ex situ* annealing is sufficient to achieve the B2 ordered structure. Thin films can be prepared on a variety of substrates including SiO_2 and MgO. B2 $Fe_{60}Al_{40}$ can be easily disordered to the A2 structure using the above-mentioned methods, causing a transition of the magnetic behavior from a paramagnetic (PM) to the ferromagnetic (FM) phase. Post-disordering, the alloy can be re-annealed to form the B2 structure, with a large number of switching cycles.

Although one of the simplest cases of a magnetic phase change, viz. the PM $\rightarrow$ FM transition, understanding its mechanism in B2 $Fe_{60}Al_{40}$ is nevertheless non-trivial, since antisite defects also vary the electron density of states (DOS) manifesting observable changes to the inter-atomic distances. Antisite defects and the lattice parameter may therefore be inter-dependent and strongly related to the onset of ferromagnetism. The gradual and systematic introduction of antisite defects, as achieved using ion irradiation is therefore an invaluable tool to understanding the mechanism of phase transitions in these materials.

Another fascinating case is that of B2 $Fe_{50}Rh_{50}$. This alloy is well known for its temperature-driven meta-magnetic phase transition, whereby at a $T = 380\,\text{K}$, the initially antiferromagnetic (AFM) ordering transforms to being FM [30]. Disorder in $Fe_{50}Rh_{50}$ leads to two phase transitions; slight disordering of B2 $Fe_{50}Rh_{50}$ causes an AFM $\rightarrow$ FM transition. Further disordering leads to the formation of the A1 (fcc structure), which is PM. Vacuum annealing of A1 $Fe_{50}Rh_{50}$ can return the alloy to the well-ordered B2 structure [8]. Recently, the variation of lattice parameter and disorder under ion irradiation has been investigated [31], which is a step toward understanding the mechanism of the disorder-induced AFM $\rightarrow$ FM transition, whereas the FM $\rightarrow$ PM transition is still relatively unexplored. In this chapter we will discuss the AFM $\rightarrow$ FM transition in B2 $Fe_{50}Rh_{50}$ further.

We begin with a detailed discussion on B2 $Fe_{60}Al_{40}$. In particular, the semi-empirical modelling of the disorder-induced magnetization, used to predict the necessary irradiation parameters for magnetic writing, can be generalized to a variety of other systems.

2. Antisite Disorder in B2 $Fe_{60}Al_{40}$

B2 $Fe_{60}Al_{40}$ is perhaps the most common example of an alloy where irradiation of light noble gas ions is sufficient to induce ferromagnetic behavior. Figure 2(a) schematically shows the arrangement of Fe and Al atoms in the B2 ordered structure. Antisite defects are formed by exchanging Fe and Al site occupancies, as shown by the arrows in Fig. 2(b). The changes to the structure can be tracked using X-ray diffraction. Figure 2(c) shows the 100 superstructure reflection that originates from the alternating pure-Fe and Al-rich atomic planes of the B2 ordering. Antisite defects gradually destroy this structure, to form the A2 structure, and this can be observed as a suppression of the superstructure reflection, as seen in Fig. 2(d).

Corresponding to the above structural changes, there is a drastic change in the magnetic behavior. The B2 structure is paramagnetic, or in the case of thin films, weakly ferromagnetic (Fig. 2(e)). With the increase in antisite defects, a large increase of the magnetization is observed, as shown in Fig. 2(f). In this example, disordering has been induced in a thin film of B2 $Fe_{60}Al_{40}$ via irradiation of Ne^+ ions. In Fig. 2(f), a 40-nm-thick $Fe_{60}Al_{40}$ film has been irradiated with 10 keV Ne^+ ions, at a fluence of 6×10^{14} ions cm^{-2}. Here the film has been partially magnetized; a maximum M_s of 780 kAm^{-1} can be reached with appropriate choice of irradiation conditions [5].

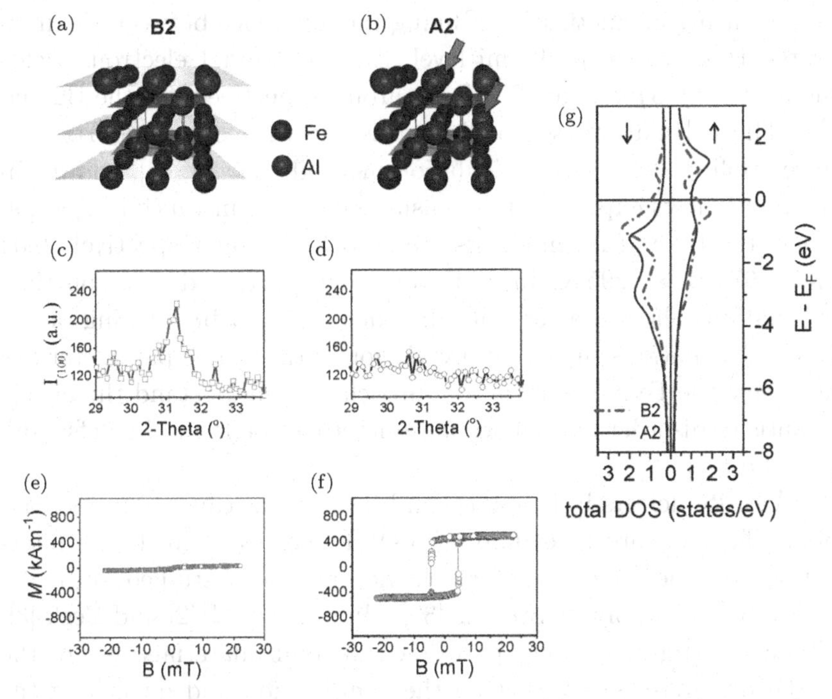

Figure 2. The effect of disordering in B2 $Fe_{60}Al_{40}$. (a) Schematic diagram showing a part of an ordered B2 $Fe_{60}Al_{40}$ lattice. (b) Ion irradiation leads to antisite defects, indicated by arrows. The disordered lattice is of the A2 structure. Disordering can be characterized using the 100 superstructure reflection, which is observed for (c) B2 $Fe_{60}Al_{40}$ and (d) vanishes in A2 $Fe_{60}Al_{40}$. Corresponding changes are observed in the plots of magntization, M vs. magnetic field, B, with (e) the B2 structure possessing very low saturation magnetization (M_s), which increases with the formation of (f) the A2 structure. DFT calculations of the total densities of states, DOS, for spin-up (↑) and spin-down (↓) electrons are shown in (g). (a)–(f) Reproduced with permission from Ref. [5]. Copyright 2013 American Chemical Society. (g) Reproduced with permission from Ref. [35].

The spin-resolved densities of states, DOS, for the B2 as well as A2 $Fe_{60}Al_{40}$ structures can be calculated (see, Ref. [35]) and are shown in Fig. 2(g). To understand the effect of antisite defects on magnetism, the total DOS as a function of energy E, for spin-up (↑) and spin-down (↓) electrons, each for B2 and A2 $Fe_{60}Al_{40}$ has been considered [35]. The calculation shows a distinct increase of the spin-splitting between the ↓ and ↑ DOS, which is expected for an increasing magnetization. Taking the difference between the integrated DOS up to the Fermi level, E_F, for $\downarrow$ and $\uparrow$ electrons yields moments of $0.5\,\mu B$ and $1.25\,\mu B$/Fe-atom respectively for the B2 and A2. The calculation is broadly consistent with other reports that have applied a variety of calculation methods [17, 25, 44], and qualitatively follows experimental measurements, for instance in Ref. [5], where the B2 and A2 moments are reported to be, respectively 0.04 and $1.67\,\mu B$, at 295 K. In addition to the increased spin-splitting, calculations show that antisite disorder causes a broadening of the DOS. Calculations show that a variation of the lattice parameter can also vary the Fe-moment, and approaches to understand the effects of antisite disorder and lattice parameter can be found in Refs. [16], [34] and [35].

Antisite defects in $Fe_{60}Al_{40}$ can be viewed as causing an increase of the Fe–Fe nearest neighbors (nns). The effect of nn Fe–Fe interactions on the ferromagnetic behavior has been studied since the 1960s, most notably by Stearns [32], Huffman *et al.* [2] and Beck [3]. Those investigations were performed on bulk materials, where the Fe–Fe nns were set by varying the composition and quenching the alloys to achieve a random atomic distribution. It was possible to show that the Fe–Al alloy becomes ferromagnetic as the Fe–Fe nns increases from 3 to 4; however, due to the inhomogeneity present in bulk samples, for instance at the sample surfaces and at grain boundaries, as well as the varying composition, any variation of the lattice parameter associated purely with disorder could not be identified.

In more recent experiments, the disorder has been induced without varying the alloy composition, by ball-milling $Fe_{60}Al_{40}$ powders [16–18]. This made it possible to observe an $\sim$1% increase of the lattice parameter occurring as the B2 $Fe_{60}Al_{40}$ transforms to the A2 structure [33]. These observations opened a debate on the respective contributions of the Fe–Fe nn interactions and the increasing lattice parameter on the induced magnetization [33, 34]. The intraparticle defect distribution from ball-milling can however be inhomogeneous, and the method is not ideal for surface-sensitive probing of the disorder and magnetic properties.

Thin films are ideal samples for surface-sensitive probing as well as for device applications. Disorder generated by broad-beam ion irradiation tends to be homogeneously distributed along the film plane. Along the film depth, however, disorder follows a quasi-Gaussian distribution. Ion irradiation has been applied to disorder B2 $Fe_{60}Al_{40}$ foils by Menendez *et al.*, thereby generating a magnetized surface layer [4]. Fassbender *et al.* showed that a variety of ion species can be used to modify surface disorder [14]. Bali *et al.* applied ion irradiation to $Fe_{60}Al_{40}$ films, of <50 nm thickness, achieving depth-homogeneous M_s and limited effects of lateral scattering of ions, thereby realizing high-resolution magnetic patterns [5].

Studies on ion-irradiated thin film can provide insights into the mechanism of disorder-induced ferromagnetism. Figure 3 shows the variation of saturation magnetization (M_s) and lattice parameter (a_0) with the disorder (1-S), for 250-nm-thick $Fe_{60}Al_{40}$ films. The order parameter, S, is proportional to the square root of the intensities of the 100 superstructure reflection and the 110 fundamental peak. A perfectly ordered lattice should possess an S of 1 or conversely, a 1-S = 0. In the present case, the initial, fully ordered films possess a 1-S = 0.2, because of the excess of Fe atoms present in the Al-rich planes. Ion-irradiation leads to an increase in the antisite defects and a corresponding increase in 1-S, reaching a value of 1 for fully disordered films.

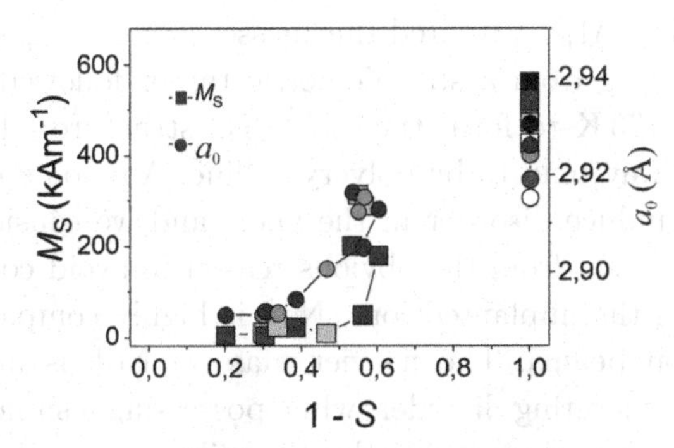

Figure 3. Relationship between saturation magnetization (M_s), lattice parameter (a_0) and the degree of disorder, 1-S, in 250-nm thick $Fe_{60}Al_{40}$ films. For given 1-S, the greyscale identifies the M_s and a_0 measured on the same film [35].

The variation in disordering shown in Fig. 3 has been achieved using H^+, He^+ as well as Ne^+ ions, with the films held at temperatures varying between 100 and 523 K. Thus process-independent relationships between M_s, a_0 and 1-S are obtained [35]. It is seen that during the initial stages of disordering (1-S <0.6), the a_0 increases monotonically, whereas M_s remains at a constant near-zero value. At a critical degree of disorder, 1-$S \approx 0.6$, a ferromagnetic onset occurs, indicated by a sharp rise in M_s. Beyond this critical point, only fully disordered films are obtained, corresponding to 1-S = 1. In the fully disordered regime, the M_s and a_0 continue to increase. The results in Fig. 3 indicate that a critical degree of disordering is necessary to cause the ferromagnetic onset. In $Fe_{60}Al_{40}$ films possessing the critical degree of disorder, lattice expansion can be used to further increase the M_s. The experimental observations of the lattice relaxation can be a valuable input for theoretical calculations, to help understand the correlation between disorder and ferromagnetism.

For applications in magnetic nano-patterning, the effects of lateral ion scattering must be considered. Section 2.1 describes the method to predict the spatial distribution of magnetization, given the experimental parameters for the ion irradiation.

2.1. *Experimental considerations for disordering using ions*

Films of B2 $Fe_{60}Al_{40}$ of desired thicknesses can be prepared by magnetron sputtering from a stoichiometric target followed by vacuum annealing at 773 K to form the stable B2 structure [5]. Films prepared in this way tend to be polycrystalline. Any ion species can be employed to induce disorder in the alloy, and we consider here the case of Ne^+. Apart from the obvious reason to avoid contaminating the alloy with the implanted ions, Ne^+ is highly compatible for use in focused ion beams. The heavier mass of Ne^+ is more efficient than He^+ in generating disorder, while possessing a sufficiently small radius to penetrate deep inside the alloy film.

The irradiation process occurs as follows: the energetic ion penetrates the film and traverses the crystal field of the lattice, which has a gradual decelerating effect, known as electronic stopping. However, at the typical ion energies employed for disordering thin film alloys, ~100 keV for a 100-nm-thick film, nuclear stopping is much more significant. The ion projectiles undergo nuclear stopping *via* elastic collisions with the host nuclei resulting in ion scattering. At a critical depth, known as the stopping depth, the probability of undergoing a nuclear collision event is maximized. Ions possessing sufficient momentum will displace the atom, thereby forming a vacancy. The ion and the displaced atom are scattered and may subsequently undergo further collision events, forming a collision cascade. Further details on ion–solid interactions can be found in Ziegler *et al.* and references therein [36]. The concomitant vacancies formed by the collision cascades can recombine with thermally diffusing atoms.

In metals, the following assumptions can be made: first, at room temperature it is reasonable to assume that nearly all vacancies recombine, and second, the recombination process is random. Deviations from the above assumptions tend to fall within experimentally observable errors for most alloy systems. Thus, atoms knocked from their B2 ordered sites, for instance, are replaced randomly by diffusing atoms, forming antisite defects, and finally transforming to the fully disordered A2 structure.

Consider the case where the ions are incident along a line perpendicular to the thin film surface. For sufficiently dense collision cascades, the average displacements per atom (*dpa*) is practically isotropic along the film plane, which is a desirable feature for generating homogenous magnetic regions.

However, the *dpa* is inhomogeneous along the film depth; the *dpa* increases from the film surface until the stopping depth, followed by a gradual decay. The inhomogeneity of the antisite defect distribution in the film depth can be estimated using Monte Carlo-type simulations that implement the binary collision approximation [36]. See Fig. 4(a) for examples of calculated depth variation of *dpa* for Ne^+ ions at various ion energies.

Consider an $Fe_{60}Al_{40}$ film of 40 nm thickness prepared on an SiO_2 substrate. This film is irradiated with Ne^+ ions, at various ion energies and ion fluences. Figure 4(a) shows the *dpa* vs. depth

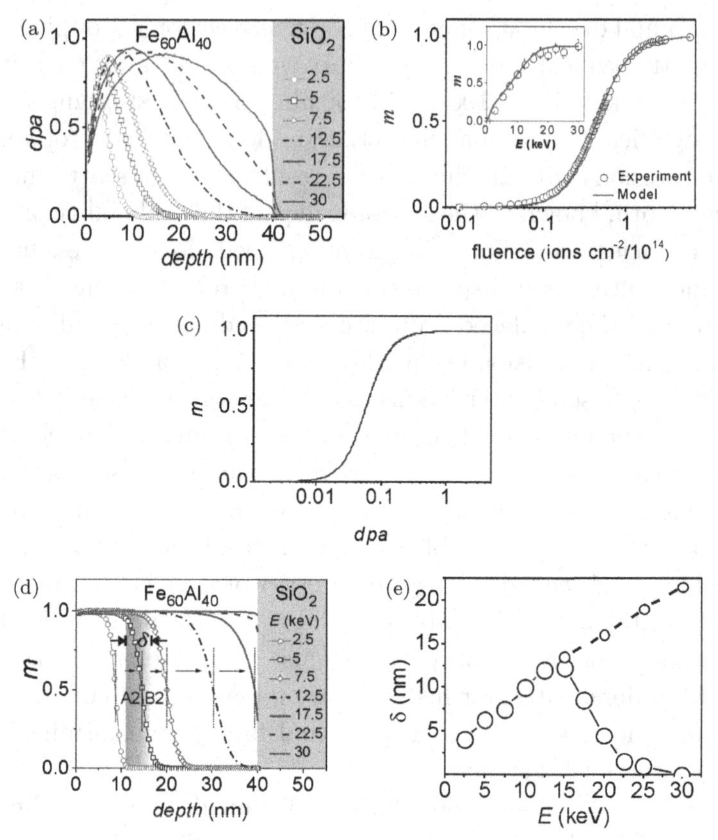

Figure 4. Correlating the atomic displacements due to collisions with penetrating ions, to the induced magnetization, m. (a) Depth variation of the displacements per atom (*dpa*) for given Ne^+ ion energies, shown here for a fixed fluence of 6×10^{14} ions cm^{-2}. (b) Experimentally measured m as a function of ion energy and fluence. (c) Semi-empirical model correlating the *dpa* to the induced magnetization. (d) Semi-empirical depth variation of the m, for the irradiation conditions shown in (a). (e) Variation of the width, δ, of the A2/B2 (ferromagnetic/non-ferromagnetic) interface with the ion energy, for m-variations that are clipped at the film-substrate interface (solid line) assuming a sufficiently thick film with no clipping (dashed line). (b) & (c) Reproduced with permission from Ref. [5]. Copyright 2013 American Chemical Society. (a) & (d) Reproduced with permission from Ref. [38].

calculations obtained using SRIM [36], for Ne^+ ions at energies varying from 2.5 to 30 keV, all at a fixed fluence of 6×10^{14} ions cm^{-2}. As can be expected, with increasing ion energy, the position of the *dpa* maximum is pushed deeper inside the film. It is apparent that the depth distribution is indeed inhomogeneous and raises concerns that the manifested magnetic properties may also be similarly inhomogeneous, which could be undesirable for applications.

In the thin film case, despite the depth-inhomogeneous *dpa*, it is nevertheless possible to obtain depth-homogeneous M_s distributions.

Figure 4(b) shows the variation of the magnetic signal as a function of the ion fluence, at a fixed energy of 20 keV. The magnetic signal is shown here in terms of the magneto-optical Kerr effect (MOKE), which is surface sensitive. The larger the surface magnetization, the greater the Kerr rotation of a linearly polarized laser beam, where the laser penetrates ~10 nm below the film surface.

The normalized magnetization, m, follows a sigmoidal variation, achieving near saturation above 3×10^{14} ions cm^{-2} (Fig. 4b). The saturating effect can be confirmed by fixing the fluence at say, 6×10^{14} ions cm^{-2} and varying the ion energy. As seen in the inset of Fig. 4(b), saturation is achieved for films irradiated at 20 keV and above. The plateau in M_s, despite increasing fluence, implies that the film has been homogeneously magnetized to a maximum value. To understand this effect, it is necessary to establish a relationship between the calculated *dpa* and the corresponding observed increase of m.

To correlate m to the *dpa*, the integral *dpa* from the SRIM calculation should be plotted against the observed magnetization and fitted using a sigmoid [5]. The sigmoidal variation is a typical case of phase transition from phase B2 to A2, with the two phases co-existing during the transition and moreover is also experimentally observed. A logistic function can be applied, for instance

$$m = \left(\frac{dpa}{dpa_0}\right)^p \left[1 + \left(\frac{dpa}{dpa_0}\right)^p\right]^{-1}, \qquad (1)$$

where dpa_0 is the *dpa* at which $m = 0.5$ and p is related to the slope of the quasi-linear region of the sigmoid.

When applying the model in Eq. (1) to the observed m, it is necessary to consider the probe depth, t_i, of the measurement method. Vibrating sample magnetometry tends to probe the entire sample volume, and t_i is the film thickness, measuring over the entire *dpa* distribution. Magneto-optical probes tend to be confined to the surface restricting t_i to ~10 nm, thereby limiting the probed *dpa* distribution.

To consider the effect of depth of probe, Eq. (1) is applied to the $dpa(t)$ in Fig. 4(a) and integrated over the appropriate t_i:

$$\langle m \rangle_t = \frac{1}{t_i} \int_0^{t_i} m(t)dt, \tag{2}$$

to obtain the depth-averaged magnetization, $\langle m \rangle_t$. The constants dpa_0 and p are iterated until the experimentally observed m matches the simulated $\langle m \rangle_t$. For $Fe_{60}Al_{40}$, $dpa_0 = 0.06$ and $p = 2.3$ yields the fits shown in Fig. 4b.

The generalized m vs. dpa relationship for $Fe_{60}Al_{40}$ films is depicted in Fig. 4(c), obtained by plotting Eq. (1) using the dpa_0 and p determined above. The sensitivity of the transition can be evaluated; the maximum M_s can be achieved in regions that undergo 0.3 displacements per atom and above.

Note that in the ~10-keV energy range, each ion can cause ~100 displacements, significantly reducing the ion fluence necessary to magnetize the film in a depth-homogeneous manner. SRIM calculations show that in $Fe_{60}Al_{40}$ irradiated with Ne^+ ions at 10 and 30 keV, each ion causes ~200 and 500 total displacements, respectively. Furthermore, a depth-homogeneous M_s can be achieved while maintaining a fixed ion energy, without the need for a multi-step irradiation process [37].

Applying Eq. (1) to the dpa distribution of Fig. 4(a), the magnetization distributions for the given ion-energies are obtained, shown in Fig. 4(d). It is seen that at the fixed fluence of 6×10^{14} ions cm^{-2}, an irradiation with 30 keV ions leads to a completely depth-homogeneous magnetization. At 20 keV, the magnetization is largely homogeneous through the depth, except for a narrow region at the interface with the SiO_2. Further lowering of ion energy can result in partially magnetized films. Thus, adjusting the ion energy can be used to design partially or fully magnetized films.

In case of limited ion penetration, an interface between the A2 and B2 regions is formed. The A2–B2 interface is not sharp and possesses a certain width, δ, over which the magnetization is expected to decay from maximum to zero. In Fig. 4(d), the δ has been indicated for the case of 12.5 keV Ne^+ ions. Here, δ has been defined as the width over which m decays from 0.95 to 0.05. In cases where the *dpa* distribution is clipped by the film–substrate interface, δ is taken as the width from $m = 0.95$ to the interface.

It is the truncation of the *dpa* distribution in thin films that effects the removal of the depth inhomogeneity in m, expelling the A2–B2 interface and realizing a spatially homogeneous magnetization.

In Fig. 4(e), clipping of the *dpa* curves commences at 15 keV and above. It is seen that at 15 keV, the clipped δ for the 40-nm-thick films and the unclipped δ in thicker films diverge. The unclipped δ shows a quasi-linear increase with the ion energy, since the Gaussian *dpa* distribution tends to broaden at higher ion energies. For 40-nm-thick films, the *dpa* distributions are truncated by the film–substrate interface, thereby narrowing the A2–B2 interface. The clipped *dpa* distribution can be exploited in thin film systems to achieve nearly discrete nanomagnets. The varying δ itself can be used as a means to tune the response of the film to microwave excitations [38].

The $m(dpa)$ function is a powerful way to estimate the magnetization distribution for any given irradiation parameters. The model described by Eq. (1) can be applied to estimate the effect of lateral scattering at the edges of shadow masks as well as to the case of irradiation with a focused ion beam; the magnetization spread due to lateral scattering can be estimated *a priori.*

Magnetic structures can be fabricated using broad-beam irradiation by covering the continuous film with a resist layer, followed by routine lithography steps of exposure, development, ion irradiation and lift-off. For such a shadow mask to be effective, the resist layer must be of sufficient thickness to stop the ions over regions where no magnetization is desired. This can be estimated using SRIM, for instance for an ion energy of 30 keV, a typical PMMA resist of approximately $\approx$200 nm thickness is required. The simulation does not consider the fact that the irradiation tends to burn the resist; therefore, a margin should be provided above the predicted resist thickness. Furthermore, oxygen plasma treatment may be necessary for the complete removal of the resist debris after the ion irradiation process.

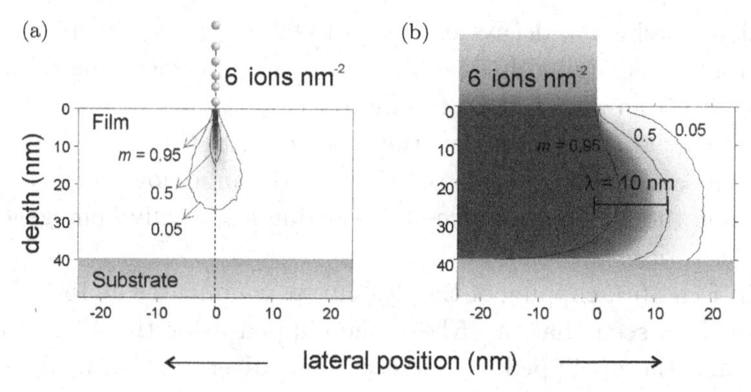

Figure 5. Magnetization distribution generated by Ne^+ irradiation (a) along a 1-dimensional path, i.e. along a line that is perpendicular to the film cross-section shown in the schematic diagram and (b) over a semi-infinite plane. Contour lines indicate the normalized magnetization, m. The Ne^+ energy is 25 keV and fluence 6×10^{14} ions cm^{-2}. Adapted with permission from Ref. [28].

At the edges of the shadow mask, lateral scattering from the material exposed to the ions penetrates the region under the mask, causing disordering and blurring the intended pattern. Lateral scattering of ions ultimately limits the resolution of the magnetic patterning.

The magnetic effect of lateral scattering can be illustrated for two limiting cases. Figure 5(a) depicts the 1D case of a line drawn using a 25-keV Ne^+ ion point source. Applying Eq. (1) to the simulated *dpa* shows that at a fluence of 6 ions nm^{-2}, a non-uniform pear-shaped magnetized region is formed. In the form of a broad beam, this fluence is sufficient to uniformly magnetize a 40-nm-thick film, whereas in the 1D case, the magnetic region extends up to ~30 nm depth. Achieving a uniform magnetization relies on the cumulative effect of a high density of pear-shaped magnetized regions, realized during broad-beam irradiation. The case of a Ne^+-ion point source is applicable in the use of focused ion beams for magnetic patterning. To write nanomagnets using a focused ion beam, it is typical to raster several lines and integrate over the m-distribution shown in Fig. 5(a). This is illustrated experimentally in Section 2.2.

Pertaining to broad-beam irradiation, a limiting case is that of a semi-infinite plane, simulating the edge of a shadow mask. Figure 5(b) depicts a region in the vicinity of a shadow mask edge.

Positive values on the x-axis indicate the region covered by a shadow mask that is assumed to completely stop the ions. On the negative side of the x-axis lies the region exposed to the ions. Figure 5 shows the magnetization distribution resulting from the irradiation of 25 keV Ne^+ ions at 6×10^{14} ions cm^{-2}. The contour lines indicate the induced m, normalized from 0 to 1. As seen in Fig. 5(b), lateral scattering of the ions at the mask edge results in magnetization being induced underneath the mask, with a full-width half-maximum of about 10 nm.

Knowledge of the magnetic effects of the lateral scattering of ions is crucial for designing magnetic structures with sharp interfaces [38, 39]. Lateral scattering is also an important consideration when using focused ion beams, as discussed below.

2.2. *Localized magnetic modifications using focused ion beams*

An ideal tool to achieve focused beams of Ne^+ or He^+ ions, is the gas field ion source (GFIS). This state-of-the-art instrument deploys a tip composed of a triad of Pt atoms, with one of the atoms acting as the point of origin from which the noble gas ion is accelerated. A technical description of the process can be found in Hlawacek *et al.* [40]. The GFIS is capable of producing a Ne^+ ion beam of ~2 nm diameter, at an ion energy of up to 30 keV, with even narrower beam diameters being realized for He^+ ions.

This highly focused ion beam can be deployed for writing magnetic patterns, illustrated here for a pattern of 500-nm-wide magnetic stripes separated by 100 nm of non-ferromagnetic spacer [27, 28]. The stripe pattern was written directly onto a B2 $Fe_{60}Al_{40}$ template using a narrow 25-keV Ne^+ beam. As with the broad-beam case, the fluence was kept fixed at 6 ions nm^{-2}. Transmission electron microscope-based holography has been performed to obtain the structural as well as magnetic information on the magnetic stripes. Figure 6(a) shows the amplitude image of a cross-section of the magnetic stripes, after patterning. The obtained structural contrast shows a polycrystalline film of ~50-nm grain size, with the surface topographically remaining flat, despite the ion irradiation. The magnetic phase image (Fig. 6(b)) shows the distribution of the magnetic

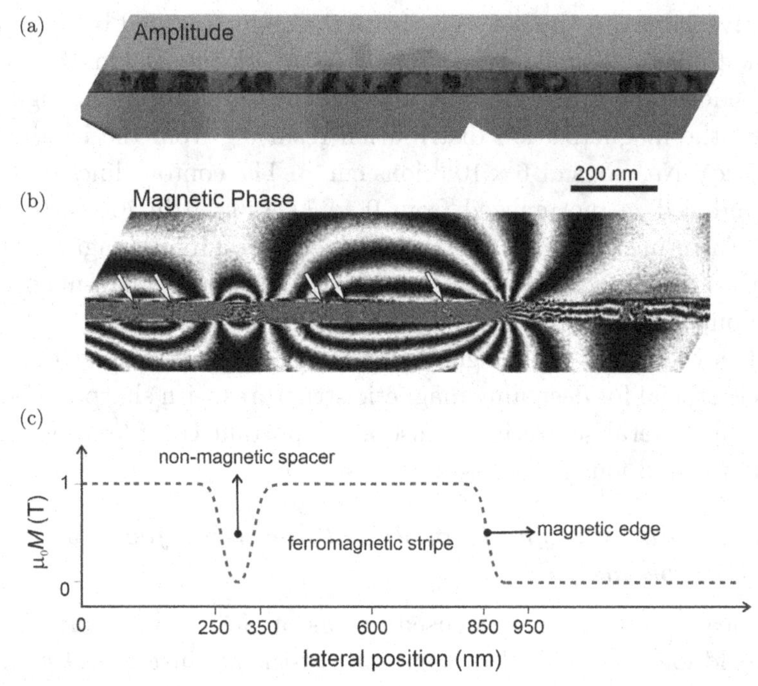

Figure 6. Using a focused 25-keV Ne^+-beam as a magnetic writing stylus. (a) Amplitude image obtained using transmission electron microscopy, showing a polycrystalline $Fe_{60}Al_{40}$ film. The magnetic stripe pattern does not appear in the amplitude contrast. (b) Magnetic phase image, obtained by plotting the cosine of the phase shift, reveals the distribution of magnetic flux lines, emerging from the magnetic pattern embedded within the $Fe_{60}Al_{40}$ film. Arrows indicate disturbances in the flux distribution. (c) Schematic diagram of the magnetic pattern. Adapted with permission from Ref. [27].

flux lines, visualized by plotting the cosine of the magnetic phase shift. The magnetic flux lines seen in the figure are consistent with the intended pattern of two elongated bar magnets separated by a narrow non-ferromagnetic gap, shown schematically in Fig. 6(c). The magnetic flux density within the stripes reaches a value of $\mu_0 M \sim 1\,\mathrm{T}$, giving an $M \approx 800\,\mathrm{kAm}^{-1}$, which is close to the known saturation magnetization for the given irradiation conditions [5].

Inspection of the magnetic phase image in Fig. 6(b) also shows regions where the magnetic flux lines appear to deviate from an ideal distribution, i.e., magnetic poles are located within the lengths of the magnetic stripes, indicated by block arrows. These additional magnetic poles are likely due to rastering errors, causing sharp dips in the magnetization. These dips are confined to regions of 25 nm widths, suggesting that the material is amenable to patterning at this length scale. Moreover, as seen in Fig. 6(b), vertical A2–B2 interfaces can be introduced in a planar film at the desired locations using ion beams.

For obtaining plane-view images of the magnetization, B2 $Fe_{60}Al_{40}$ films of 40 nm thicknesses were prepared on 20-nm-thick Si–N membranes. Scanning transmission electron microscopy was applied, so as to obtain point-by-point diffraction as well as direct-beam data, thereby simultaneously recording the in-plane local variation of structural as well as magnetization distributions in ion-irradiated regions [41]. Using this technique, the magnetization, rather than the magnetic flux density can be imaged. Linear stripes of 10 μm lengths and widths between 30 nm and 4 μm were patterned using a focused Ne^+ ion beam at 26 keV, keeping the fluence fixed at 6 ions nm^{-2}. Stray-field minimization in the high-aspect ratio stripes is expected to drive the magnetization to form single domains, with the moments aligned parallel to the long axis. However, the expected magnetization distribution is contradicted in the experiment, whereby in stripes of widths down to sub-200 nm, periodic magnetic domains are observed. In Fig. 7(a), the example of a 295-nm-wide stripe is shown, where periodic magnetic domains, with domain sizes comparable to the width of the stripe, are observed. The occurrence of periodicity implies that an additional uniaxial magnetic anisotropy (K_U) drives a fraction of moments to align parallel to the short axis, depicted schematically in Fig. 7.

Estimation of the lattice parameter using STEM shows that the embedded A2 $Fe_{60}Al_{40}$ stripes are structurally anisotropic. The inset of Fig. 7(a) shows that the disorder-induced lattice expansion is anisotropic in plane. The lattice parameter measured parallel to the short axis, a_{0w} is larger than that along the long axis, a_{0l} i.e., $a_{0w} > a_{0l}$. Structural anisotropy of the unit cell can originate in embedded structures from the strain field surrounding the structure, where the shape of the object may induce a symmetry breaking. The occurrence of structural and magnetic anisotropies coincide,

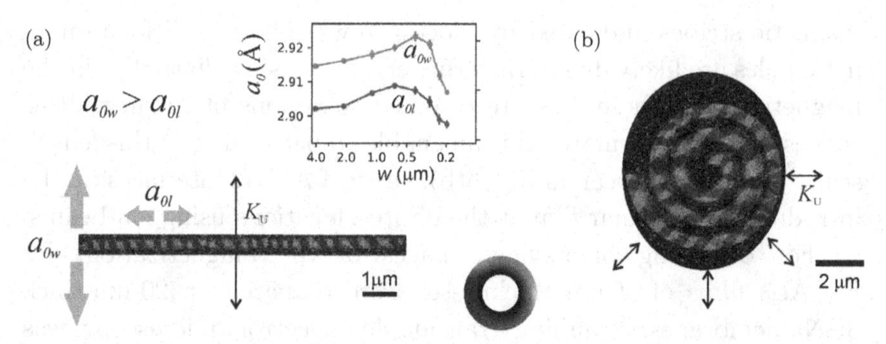

Figure 7. Lattice relaxation and periodic magnetic domains in linear as well as curved embedded magnets. (a) An embedded magnetic stripe of 295 nm width and 10 μm length shows periodic magnetic domains, consistent with the occurrence of a uniaxial magnetic anisotropy, K_U, inducing a tendency for the magnetic moments to align perpendicular to the linear stripe. A lattice distortion is observed, such that the lattice expansion along the short axis is larger than that along its length, shown schematically with block arrows. The inset shows the estimated lattice parameters, along the short axis (a_{Ow}) as well as the long axis (a_{Ol}), as a function of stripe width, w. (b) Domain periodicity can be written onto an embedded archimedean spiral, consistent with a K_U causing a tendency for the moments to align radially at each point along the curve. The magnets have been generated in a 40-nm-thick B2 $Fe_{60}Al_{40}$ film using a focused beam of 26 keV Ne^+ ions at a fluence of 6 ions nm^{-2}. The magnetization directions are indicated by the color wheel. Adapted with permission from Ref. [41].

providing a pathway to modulate strain as well as magnetic properties in embedded nanomagnets.

In addition to linear structures, curved objects have been investigated as well. A magnetic spiral with 500-nm arm width and 500-nm arm spacing was embedded in B2 $Fe_{60}Al_{40}$, where the magnetization also breaks into periodic domains. Micromagnetic simulations show that the occurrence of periodic domains in the linear as well as curved structures is consistent with the presence of an additional magnetic anisotropy component, K_U. The K_U counters the effect of the object's shape by tending to orient the magnetization along the narrow dimension of the object. The net effect of the two anisotropies causes the periodic distribution of the local magnetization. These results demonstrate that focussed ion beams can be used to engineer the strain and magnetic anisotropy of linear as well as curved nanomagnetic objects.

Irradiation-induced magnetic patterning therefore can be used to generate modulations of the magnetization and magnetic domain structure as well as lattice strain. These microscopic observations provide the basis for scaling up the process to large areas, as described in Section 2.3.

2.3. *Scaling-up irradiation-induced magnetic patterning*

Scaling-up is an essential step toward making irradiation-induced magnetic patterning relevant to applications. Krupinski *et al.* [42] demonstrated the use of self-assembled layers of polystyrene nano-spheres (Nano-sphere lithography), with the sphere diameter varying from 800 down to 200 nm, as shadow masks. Self-assembly occurs at the surface of an aqueous suspension of monodisperse nanospheres, and the monolayer is gently transferred to the B2 $Fe_{60}Al_{40}$ film via slow evaporation. The hexagonal close-packed arrangement of the nanospheres acts as a shadow mask (see Fig. 8) through which ion irradiation is performed. Depending on requirement, the ion energy can be selected such that the ions penetrate a certain thickness through the nano-spheres and form interconnected — instead of isolated — magnetic regions. In Fig. 8, 40-nm-thick B2 $Fe_{60}Al_{40}$ films were covered with self-assembled spheres of diameters 784 and 202 nm, respectively. To achieve interconnected magnetic regions, Ne^+ ions at energies of 150 and 30 keV, respectively, for the two sphere diameters were performed. To achieve the maximum M_s in regions directly exposed to the ions, a fluence of 6×10^{14} ions cm^{-2} was used in both cases.

The insets in Fig. 8 show the corresponding magnetic modulations embedded within the flat film topography, after irradiation and removal of the mask. Imaging has been performed by detecting the stray fields emerging from the magnetic modulations, by scanning with a read-head typically used in hard disk drives [42]. The selection of the ion energies allowed the ions to penetrate the polystyrene sphere up to a certain inner diameter, casting a translucent shadow of the spheres onto the films and rendering an inter-connected magnetic

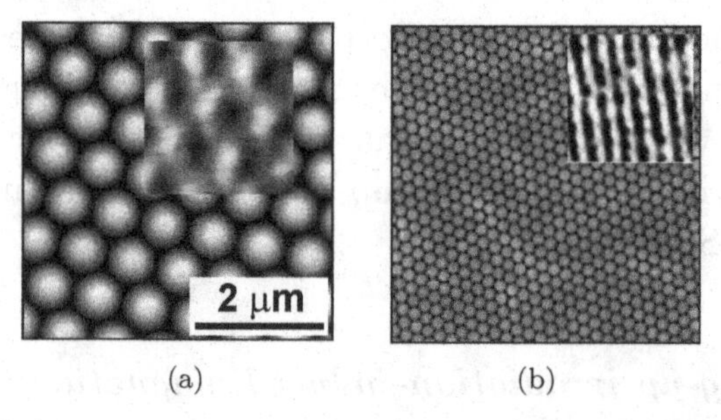

Figure 8. Scaled-up magnetic patterning using self-assembled shadow masks. The masks are composed of a single layer of polystyrene nanospheres of (a) 784 and (b) 202 nm diameters. The nano-sphere masks were transferred onto 40-nm-thick B2 $Fe_{60}Al_{40}$ films and irradiated with 150 and 30 keV Ne^+ ions, respectively, both at a fluence of 6×10^{14} ions cm^{-2}. Insets show the corresponding magnetic contrast, formed by the ions passing through selected locations of the nano-sphere mask and disordering the film. A saturating magnetic field was applied along the horizontal during imaging of the magnetic stray fields emerging from the pattern. Adapted from Ref. [42] — Published by the Royal Society of Chemistry.

pattern over approximately square centimeter regions. The magnetic fraction is estimated to be 59% and 43% for the 784 and 202 nm diameter spheres, respectively, as compared to 9% for perfect shadowing. The role of lateral scattering can be estimated using the previously described semi-empirical model showing for instance that in the case of the 784 nm sphere diameter, A2–B2 interfaces of ∼100 nm widths are formed surrounding 200 nm diameter tubes of non-ferromagnetic material, corresponding to the footprint of the self-assembled nanospheres [42]. A judicious selection of sphere diameters and irradiation conditions can be used to achieve highly ordered 2D magnetic patterns over large areas.

2.4. *Open questions and current research on irradiation-induced antisite disorder*

Research into the mechanism of disorder-induced ferromagnetism in alloys is ongoing. A precise understanding of the mechanism of the M_s induced by antisite disorder has so far been elusive, since varying the ordering also causes changes to the lattice parameter. Interpretations of the M_s vs. lattice expansion and M_s vs. disorder relationships under mechanical stress-induced disordering processes have been without consensus; the induced M_s has been considered purely a disordering effect [34, 43] and contradicted by claims of an M_s contribution from the lattice expansion [15–17].

Theoretically, understanding ferromagnetism induced via antisite disorder is non-trivial, not only due to the difficulties in modeling the randomness but also due to a lack of reliable experimental observations of the lattice relaxation sensitively varying with the disorder [44–46].

A sensitive interplay between the M_s, lattice expansion and defect concentration can result from an evolving electronic structure originating from the varying antisite disorder. Variation in the DOS due to disorder has been calculated [25, 44–46] and is evidenced by the decrease in the resistivity that accompanies the transition of B2 $Fe_{60}Al_{40}$ to the A2 structure [47]. Literature on the variation of transport properties during the B2 $\rightarrow$ A2 transition is lacking, and transport studies may provide an effective way to probe the material, particularly during the initial stages of disordering. The recent experimental observation of a critical disorder degree at which the ferromagnetic onset occurs [35], may motivate theoretical investigations into the evolution of the DOS at intermediate degrees of disordering.

From an applications perspective, the B2 $\rightarrow$ A2 transition can be used for manipulating the size of magnetic domains formed due to magnetic defects occurring in partially magnetized films. Tahir *et al.* demonstrated this by varying the ion energy, achieving distributions similar to those of Fig. 4(d) [48, 49]. At low ion energies (<5 keV Ne^+) the A2–B2 interface forms a significant fraction of the total magnetized region, and lateral inhomogeneities present within the magnetically decaying region result in an increased density of domain nucleation sites. Increasing the ion energy causes an increase of the magnetic domain size, consistent with the truncation of the A2–B2 interface discussed in previous sections. Moreover, the position of the A2–B2 interface in the thin film significantly affects the magnetic response under microwave excitation. The A2–B2 interface can act as a pinning site for precessing spins and can be used as a switch to tune the spin wave modes of disordered films under dynamic excitation

[50]. Schneider *et al.* [38] showed that a fine-tuning of the resonant response in case of uniform precession (excitations with wave-vector $k = 0$) can be realized by gradually shifting the depth position of the A2–B2 interface. These results on the static as well as dynamic response of ion-irradiated thin films open the way for further investigations into ion-irradiated nanomagnets, both as single objects and as nanomagnets in close proximity to induce dipolar coupling, mediated by the paramagnetic surrounding.

2.5. *The role of open-volume defects in A2 → B2 re-ordering*

Until now we have focused only on the antisite defects in the B2 lattice; however, $Fe_{60}Al_{40}$ also possesses a range of vacancy defects, from missing single atoms (mono-vacancies) to vacancy clusters and voids (Fig. 9) [51, 52]. Triads of defects may occur, whereby a vacancy pair is anchored by an antisite, and are known as triple defects [53]. Positrons can act as ideal probes to identify as they dwell within the open volumes for characteristic timescales, depending on the defect size and chemical surrounding, prior to annhilation with electrons and subsequent emission of detectable gamma particles.

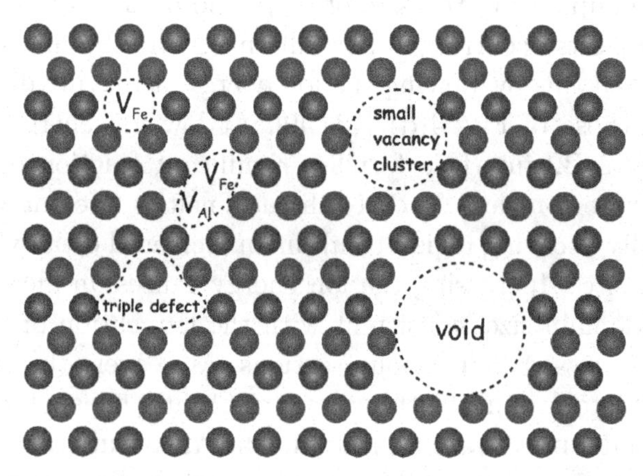

Figure 9. Types of open-volume defects that can occur in $Fe_{60}Al_{40}$. Adapted with permission from Ref. [55].

Using Doppler broadenening spectroscopy, an averaged measure of the open-volume defect concentration was obtained on thin $Fe_{60}Al_{40}$ films possessing varying degrees of antisite disorder [54]. Whereas the M_s varied drastically with the antisite defects, the open-volume defects appeared qualitatively unchanged within the experimental errors, suggesting that the latter defect type does not influence the intrinsic magnetic property.

Through time-resolved observations of the positron annihilation process to precisely identify the defect types, it has been recently shown that open-volume defects can nevertheless significantly influence the re-ordering kinetics during annealing and thereby determine the thermal process necessary for achieving the required degree of disorder and manipulating the M_s [55]. The dependence of the positrons lifetime on the open-volume size and chemical surrounding was modelled using DFT providing a means to identify the defect distribution within each sample. Using positron annhilation lifetime spectroscopy (PALS) on $Fe_{60}Al_{40}$ films prepared from a variety of processes, including the as-sputtered films as well as films ordered to the B2 by vacuum annealing at 773 K, and ion-irradiated films, it has been shown that individual open-volume defect types vary drastically during each process step [55]. For instance, the as-sputtered films possess triple defects as well as vacancy clusters and voids; thermal annealing can suppress these defects. However, ion irradiation is the most effective way to eliminate the above defects, while simultaneously increasing the mono-vacancy concentration as well as unlocking mono-vacancies from the triple defects. The effect of an *in situ* post-annealing at a reduced temperature of 423 K on the positrons lifetime was measured, to reveal the roles of various open-volume defect types on thermal re-ordering. Films possessing higher mono-vacancy concentrations exhibit faster re-ordering kinetics. This shows that triple defects and large vacancy clusters tend to hinder the atomic diffusion and re-ordering to the B2 structure. These defects are prevalant in the as-grown films, sputtered at 300 K. Vacuum annealing at 773 K suppresses the triple defects. Ion irradiation of the as-grown films is highly effective in destroying the triple defects, as well as voids, thereby lowering the temperature necessary for re-ordering, to well below 773 K. Ion irradiation can therefore act as preparation treatment to lower re-ordering temperatures where needed for applications.

So far, the roles of various defect types in determining magnetic properties and the interaction of these defects with incoming ions have been discussed. In addition to ion irradiation, laser pulsing provides an alternative path to defect modification in alloys, as demonstrated in Section 3.

3. Manipulation of Disorder using Laser Pulses

The applicability of B2 ordered materials can be widened with the use of laser pulsing for the local manipulation of disorder. Recently, Ehrler *et al.* have shown that laser pulses can reversibly modify lattice order at the surface of B2 $Fe_{60}Al_{40}$ thin films [20]. Single-shot laser pulses, of fluences above a threshold induce lattice disorder at the alloy surface, whereas below-threshold pulsing re-orders the system. The laser pulse-induced disorder (order) therefore writes (deletes) ferromagnetism at the film surface. The re-writeable (rw) effect is observed under conditions similar to those described in Fig. 10(a), where the laser pulse passes through the transparent substrate and is incident at the film–substrate interface, and magnetization is observed at the film surface.

Laser pulse-induced magnetic contrast changes were observed *in situ* using synchrotron-based XMCD combined with photo-electron emission microscopy (XPEEM). This method is highly surface sensitive, with a probe depth limited to ~5 nm from the surface.

The laser pulse-induced magnetic rw process is depicted in Figs. 10(b)–10(e). A 40-nm-thick B2 $Fe_{60}Al_{40}$ film has been deposited on a single crystalline MgO substrate. The alloy disordering and re-ordering processes can be observed *via* changes to the magnetization. Regions with their magnetization oriented along the direction of X-ray incidence appear in red, whereas oppositely oriented magnetization appears in blue.

Figure 10(b) shows the magnetized region formed by irradiating with a single-shot laser pulse of 500 mJ cm^{-2} fluence, onto the

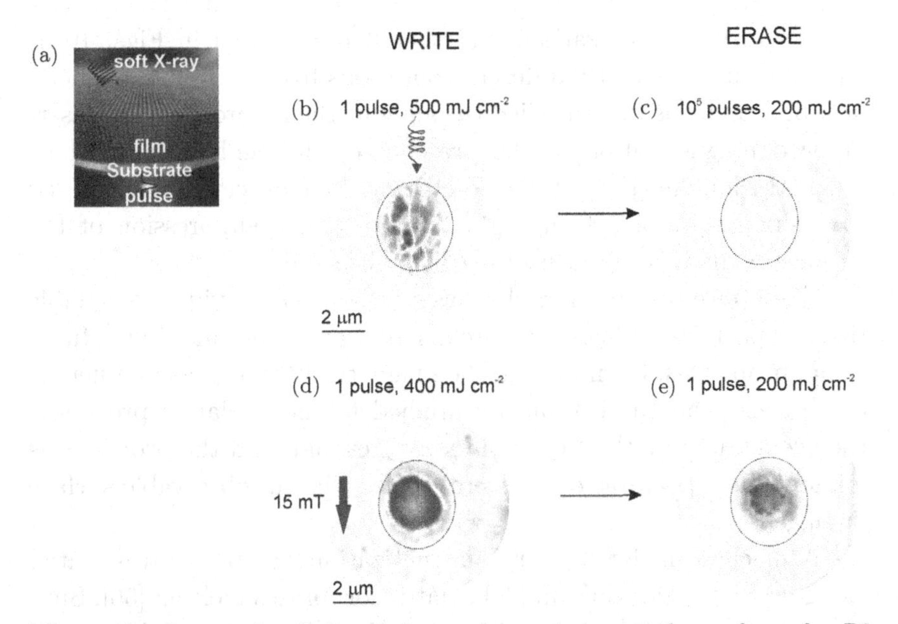

Figure 10. Laser-induced disordering and re-ordering at the surface of a B2 $Fe_{60}Al_{40}$ thin film. (a) Experimental scheme, with the laser pulses incident at the interface of the 40 nm thick film with a transparent MgO substrate. A soft X-ray probe is used to image the top surface. (b) A single-shot pulse of fluence above the magnetic writing threshold leads to formation of A2 $Fe_{60}Al_{40}$ at the surface and the appearance of surface magnetization. A multi-domain magnetic region is formed corresponding to the laser spot. (c) A laser pulse train of fluence below the writing threhold deletes the magnetization. (d) A single-shot writing pulse applied in the presence of a small dc magnetic field aligns the magnetization into a single domain. (e) Suppression of the laser-written magnetization by a single shot of fluence below the writing threshold. The color scale, red-white-blue, indicates the magnetization direction, parallel-orthogonal-antiparallel, with respect to the direction of X-ray incidence (shown here from top to down). Adapted with permission from Ref. [20]. Copyright 2018 American Chemical Society.

film–substrate interface of the B2 $Fe_{60}Al_{40}$ film. The laser wavelength was 800 nm with an ~100-fs pulse width. A magnetized region is observed at the film–vacuum surface consisting of multiple domains. The same spot was subsequently irradiated with a train of 10^6 pulses, at a fluence lowered to 200 mJ cm^{-2}, thereby strongly suppressing the surface magnetization, as seen in Fig. 10(c). The process is observed to be repeatable (tested ~5 times) at the same spot, demonstrating a laser-induced re-writable magnetization effect.

Laser rw magnetization is further demonstrated in Figs. 10(d) and 10(e) under slightly different conditions from above. A magnetic field of 15 mT has been applied throughout the rw process. The laser-induced magnetization in the presence of the field forms a single magnetic domain (Fig. 10(d)). The single-domain region was exposed to a single-shot of 200 mJ cm^{-2} fluence, and a suppression of the surface magnetization of up to 40% is observed.

The above results show that laser pulses can provide a reversible thermal path for realizing antisite disorder and hence modifying functional properties. Even though the magnetic writing–erasing fluences are higher than those typically applied for non-ablative processes, they remain lower than the ablation threshold, and the process was found to be repeatable over several cycles without observable surface damage.

A mechanism for laser rw magnetization has been investigated using a two-temperature model, that incorporates melting [56]. Simulations suggest that above a threshold laser fluence, the alloy melts and enters a supercooled state that tends to persist for several nanoseconds (ns), restricting subsequent solid-state diffusion and re-ordering. The formation of antisite disorder due to persistent supercooling can be pictured as follows. The threshold fluence coincides with the melting of the film, from the point of incidence at the substrate–film interface, upward until the film–vacuum surface. A lack of crystallite nucleation under these conditions tends to drive the molten pool into a supercooled state, thereby avoiding formation of the solid state, even as the temperature drops well below the melting point (T_{m}). Simulations suggest that the supercooled state can persist in excess of 4 ns, as the temperature continues to drop to as low as $0.5T_{\mathrm{m}}$ [20]. Nucleation finally occurs at a temperature that is well below the T_{m}, rapidly re-soldifying the alloy within ~1 ns. Although a limited rise in temperature occurs during re-soldification due to the released crystallization enthalpy, the temperature nevertheless remains well below T_{m}, slowing down diffusion kinetics necessary for re-ordering to B2 $Fe_{60}Al_{40}$. Solid-state diffusion is severely restricted and antisites are trapped within the re-solidified alloy, thereby freezing-in the ferromagnetic A2 $Fe_{60}Al_{40}$ structure.

Conversely, for laser fluences below the complete melting threshold, the film is not melted over its entire thickness, forming two molten regions, separated by the unmelted film. One molten region forms at the point of pulse incidence and a second molten region can appear due to heat accumulation directly above the point of incidence, at the film–vacuum surface. Below the threshold fluence, supercooling of the molten region at the latter surface does not persist, and re-soldification can occur instantaneously as the temeperature drops below T_m. The solid state is achieved at an elevated temperature close to T_m, providing a wide time-temperature window for atomic diffusion to occur, thereby forming re-ordered non-ferromagnetic B2 $Fe_{60}Al_{40}$.

Note that the threshold fluence for magnetic writing applies to the surface region only. This means that even though laser pulses of fluences below the writing threshold cause magnetic erasure at the film surface, magnetic writing may nevertheless occur within a region in the vicinity of the point of laser incidence at the film–substrate interface. An optimized film thickness is necessary, such that the laser re-writeable surface region is distanced from the interface, while still allowing sufficient heat penetration to the film surface. In comparison, magnetic writing (i.e., without laser-erasing) is less restrictive in terms of incidence and read-out geometry.

Further investigations are necessary to reveal laser rw phase transitions in other alloy thin film systems. Binary systems in which drastic property changes can be realized *via* irradiation-induced disorder may also be potential candidates to investigate laser-induced transitions.

4. Disorder in B2 $Fe_{50}Rh_{50}$

Antisite disorder induces property changes in a wide variety of alloy systems. Although B2 $Fe_{60}Al_{40}$ is an ideal prototype material, its potential applications in devices are yet to be proven. Other B2 ordered alloys that exhibit a variety of functional properties, such as an antiferromagnetic (AFM) ground state in B2 $Fe_{50}Rh_{50}$, can broaden the scope for applications. In B2 $Fe_{50}Rh_{50}$, the AFM state undergoes a temperature-driven phase transition to the FM state at 380 K. This temperature-driven meta-magnetic phase transition has been studied extensively, for its potential applications in heat-assisted magnetic recording [57]. The large entropy change associated with the meta-magnetic transition makes B2 $Fe_{50}Rh_{50}$ a candidate magneto-caloric material [58].

The magnetic behavior of B2 $Fe_{50}Rh_{50}$ thin films is highly sensitive to ion irradiation. The red line in Fig. 11(a) shows the magnetization as a function of temperature, for a 35-nm-thick B2 $Fe_{50}Rh_{50}$ film. At 300 K, spins in the antiferromagnetic state perfectly compensate each other, showing near-zero net magnetization, M. As the temperature increases above 370 K, the known first-order AFM to FM phase transition is observed. In Fig. 11(a), the red line does not reach the maximum M as the temperature was limited to 400 K.

The B2 $Fe_{50}Rh_{50}$ film shows a visible magnetic response to irradiation with 20 keV Ne^+ ions, even at a relatively low fluence of 5×10^{12}

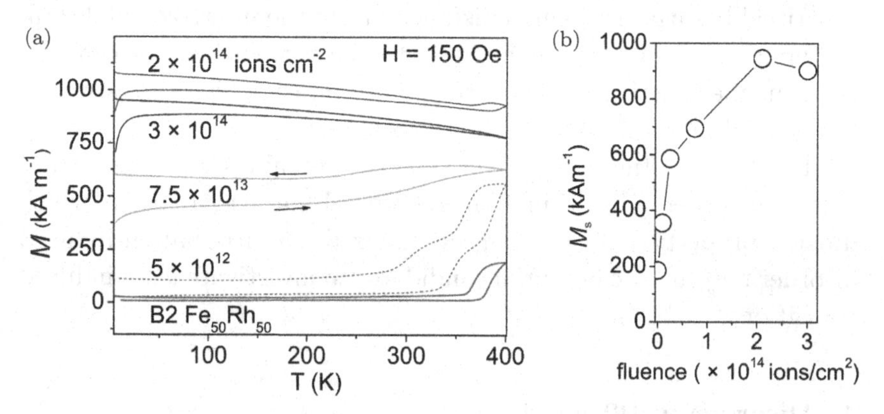

Figure 11. Effect of disordering in B2 $Fe_{50}Rh_{50}$ thin films. Films of 35 nm thicknesses were irradiated with 20 keV Ne^+ ions at given fluences. (a) The magnetization as a function of temperature under a fixed external magnetic field of 150 Oe. (b) Variation of the M_s as a function of the fluence at a temperature of 5 K. The arrows in (a) indicate the direction of temperature change. Adapted with permission from Ref. [9].

ions cm^{-2}. The dotted black line shows that a residual M is observed at 300 K, suggesting the presence of spontaneously magnetized FM regions in the AFM film. Sweeping the temperature shows that the meta-magnetic phase transition occurs at a lower temperature, close to 350 K. The step-like transition behavior indicates the occurrence of a distribution of the ferromagnetic regions due, for instance, to differences in the near-surface and bulk-like defects. Furthermore, a wide temperature hysteresis is observed, resulting in an opening between the increasing and decreasing temperature segments. The temperature hysteresis is also observed on the well-ordered B2 $Fe_{50}Rh_{50}$, however with a closed temperature loop.

Further increase of Ne^+ fluence increases the two effects *viz.*, the increasing residual M at 300 K as well as the decreasing temperature of the meta-magnetic phase transition. Figure 11(b) shows the M_s as a function of Ne^+ fluence measured at 5 K. The M_s increases initially with fluence, until a maximum of 950 kA m^{-1} for 2×10^{14} ions cm^{-2} Ne^+ fluence and begins to decrease thereafter.

The initial increase in magnetization with fluence, followed by a decrease, seen in Fig. 11, suggest the occurrence of three magnetic states: the first two are the known AFM ground state and the observed FM phase for Ne^+ fluences up to 2×10^{14} ions cm^{-2}. The suppression of M_s at higher ion fluences points to the formation of a paramagnetic (PM) phase. Austenitic A1 $Fe_{50}Rh_{50}$ is known to be paramagnetic, and a transformation to the A1 can explain the observed magnetization suppression [59]. It has been shown that the B2 structure can be recovered by vacuum annealing the A1 $Fe_{50}Rh_{50}$ [8].

A precise structural characterization of irradiation-disordered ferromagnetic $Fe_{50}Rh_{50}$ is lacking. It is necessary to establish the nature of disordering that induces the onset of room temperature ferromagnetism in $Fe_{50}Rh_{50}$.

An interesting observation is that unlike in $Fe_{60}Al_{40}$ films, where the occurrence of the ferromagnetic state corresponds to the destruction of B2 ordering; in case of $Fe_{50}Rh_{50}$, the 100 superstructure reflection is only suppressed, but does not vanish, even in films that have been sufficiently irradiated to achieve the maximum M_s [9, 31]. This can imply that although the B2 ordering may have decayed during the irradiation process, the ferromagnetic film can nevertheless be described as a disordered-B2 rather than the A2 structure. Along with a suppression of the superstructure reflection, a lattice expansion is observed in the disordered-B2 [31], which may play a role in the ferromagnetic onset.

In a recent work, Eggert *et al.* [60] showed that the ferromagnetic onset coincides with the occurrence of small deviations of the Fe and Rh atoms from their equilibrium lattice positions, known as static disordering. Using EXAFS, non-irradiated B2 $Fe_{50}Rh_{50}$ films of the AFM state were compared to the films post-irradiation, in their FM state. No significant changes in the Fe–Fe nn environment were observed, suggesting that antisite defects were not present in sufficient concentration to cause the ferromagnetic onset. The irradiated films however possessed static disordering. Interestingly, the static disorder was found to be comparable to the dynamic disorder observed in thermally excited B2 $Fe_{50}Rh_{50}$ films, above the metamagnetic transition temperature, forming the temperature-driven FM state. The two types of disorders are schematically shown in Fig. 12. The implication is that in B2 $Fe_{50}Rh_{50}$, slight deviations of the Fe and Rh atoms from equilibrium lattice positions are sufficient to induce a ferromagnetic onset. These lattice deviations can be dynamic, as realized through thermal excitations above the

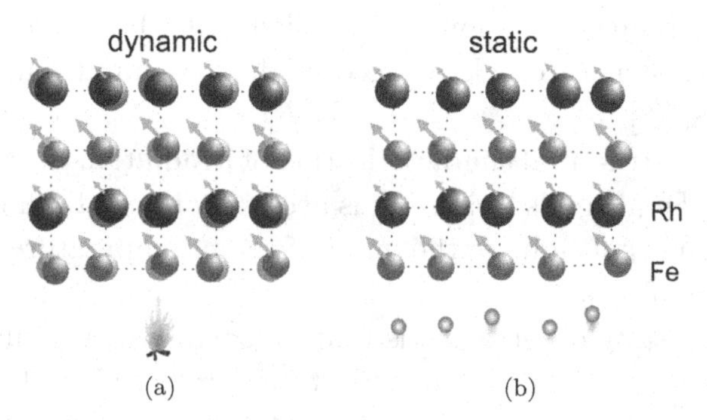

Figure 12. Dynamic and static disorder in FeRh. Dynamic disorder occurs in the thermally excited system (a), whereas ion-irradiation induces static disorder (b). Adapted with permission from the Royal Society [60].

meta-magnetic transition. The ferromagnetic onset can also be realized through irradiation-induced static disorder, which is a snapshot equivalent of the thermally driven dynamic disorder.

It is reasonable that the three structural states in $Fe_{50}Rh_{50}$ thin films viz., B2, disordered-B2 and A1, may co-exist during the irradiation process. Beginning from a B2 ordered structure, ion irradiation will gradually disorder the film, achieving maximum disordering at the peak of the quasi-Gaussian *dpa* curve. The disordering at the *dpa*-peak may cause the onset of ferromagnetism in the vicinity of the *dpa*-peak, thereby inducing FM–AFM interfaces. Further irradiation can thereafter drive these magnetic interfaces away from the *dpa*-peak eventually expelling them from the film, as observed in $Fe_{60}Al_{40}$ (Fig. 4). In the case of $Fe_{50}Rh_{50}$, however, it is conceivable that a further interface formation occurs, where the region around the *dpa*-peak transforms to the A1 structure, thereby inducing PM–FM interfaces that also are driven outward with further irradiation. Finally, the film will transform fully to the PM state. Focussed ion beams may therefore be useful in determining the locations of the FM–AFM and PM–FM magnetic interfaces, which can then propagate radially outward traversing laterally through the film plane.

The sensitivity of the material and the wide magnetic property variations in B2 $Fe_{50}Rh_{50}$ point to a tremendous scope for applied research in the nanoscale manipulation of its magnetic behavior. So far, focused beams of Ga ions have been applied for this purpose [61], whereas studies using focused beams of noble gas ions are limited. B2 $Fe_{50}Rh_{50}$ may also be amenable to laser pulse manipulation of disorder, an aspect which is yet to be rigorously investigated.

5. Conclusions and Future Perspectives

The role of defects in causing ferromagnetic onsets in B2 ordered alloys remains to be fully understood. A microscopic understanding of the phenomenon of disorder-induced ferromagnetism in a variety of alloy systems necessarily requires the systematic and subtle introduction of defects at the nanoscale combined with state-of-the-art spectroscopic and magnetic probes. Ion irradiation of thin films using noble gas ions provides the cleanest path for introducing the defects that cause ferromagnetic onsets. In B2 $Fe_{60}Al_{40}$, further investigations are necessary to shed light on the interplay between antisite defect generation and the lattice expansion and its effect on the magnetization. Studies on irradiated films have helped clarify the role of open-volume defects in $Fe_{60}Al_{40}$, where the open-volume defect distribution significantly influences the re-ordering process. Less is known on the mechanism of ferromagnetic onset in B2 $Fe_{50}Rh_{50}$, where the correlations between the degree of static disorder, the lattice parameter and the magnetic behaviour are yet to be determined. The role of microscopic defects on the magnetism of other alloys, for instance $Fe_{60}V_{40}$ and other systems, such as Heusler alloys [62], as well as layered systems wherein ion irradiation can be used to finely tune the interfaces [63] remain to be fully investigated.

Despite the open questions on the mechanism of disorder-induced ferromagnetism, there is a vast scope for applications. Focused ion beams as well as laser pulsing can be used to generate nanomagnets embedded within paramagnetic or antiferromagnetic films. The dynamic properties of these nanomagnets — as single objects or dipolar coupled arrays — can open avenues for microwave devices, as suggested by ferromagnetic resonance observations [38, 64]. Laser pulses can be used to achieve re-writeable magnetic regions at thin film surfaces in B2 $Fe_{60}Al_{40}$, with other B2 ordered alloys as potential candidates for further investigations.

The fact that highly resolved magnets can be generated at desired locations on non-magnetic thin film templates may enable a wide variety of applications and continue to drive fundamental investigations on defects and magnetism.

Acknowledgments

The author thanks the staff of the Ion Beam Center, HZDR and Jürgen Fassbender for the immense support received over the course of many years. Funding from the German Research Foundation through grant BA5656/1-2 is acknowledged.

References

[1] A. J. Bradley, A. H. Jay, The formation of superlattices in alloys of iron and aluminium, *Proc. Roy. Soc. A* **136**, 210 (1932).

[2] G. P. Huffman, R. M. Fisher, Mössbauer studies of ordered and cold-worked Fe–Al alloys containing 30 to 50 at. % Aluminum, *J. Appl. Phys.* **38**, 735 (1967).

[3] P. A. Beck, Some recent results on magnetism in alloys, *Metall. Trans.* **2**, 2015 (1971).

[4] E. Menéndez, M. O. Liedke, J. Fassbender, T. Gemming, A. Weber, L. J. Heyderman, K. V. Rao, S. C. Deevi, S. Suriñach, M. Dolors Baró, J. Sort, J. Nogués, Direct magnetic patterning due to the generation of ferromagnetism by selective ion irradiation of paramagnetic FeAl alloys, *Small* **5**, 229 (2009).

[5] R. Bali, S. Wintz, F. Meutzner, R. Hübner, R. Boucher, A. A. Ünal, S. Valencia, A. Neudert, K. Potzger, J. Bauch, F. Kronast, S. Facsko, J. Lindner, J. Fassbender, Printing nearly-discrete magnetic patterns using chemical disorder induced ferromagnetism, *Nano Lett.* **14**(2), 435–441 (2014).

[6] G. Shirane, C. W. Chen, P. A. Flinn, R. Nathans, Mössbauer study of hyperfine fields and isomer shifts in the Fe–Rh alloys, *Phys. Rev.* **131**, 183 (1963).

[7] J. M. Lommel, J. S. Kouvel, Effects of mechanical and thermal treatment on the structure and magnetic transitions in Fe–Rh, *J. Appl. Phys.* **38**, 1263 (1967).

[8] A. Tohki, K. Aikoh, A. Iwase, K. Yoneda, S. Kosugi, K. Kume, T. Batchuluun, R. Ishigami, T. Matsui, Effect of high temperature annealing on ion-irradiation induced magnetization in FeRh thin films, *J. Appl. Phys.* **111**, 07A742 (2012).

[9] A. Heidarian, R. Bali, J. Grenzer, R. A. Wilhelm, R. Heller, O. Yildirim, J. Lindner, K. Potzger, Tuning the antiferromagnetic to ferromagnetic phase transition in FeRh thin films by means of low-energy/low fluence ion irradiation, *Nucl. Instrum. Methods Phys. Res. B* **358**, 251 (2015).

[10] S. P. Bennett, A. Herklotz, C. D. Cress, A. Ievlev, C. M. Rouleau, I. I. Mazin, V. Lauter, Magnetic order multilayering in FeRh thin films by He-Ion irradiation, *Mater. Res. Lett.* **6**(1), 106 (2018).

[11] J. C. Krause, J. Schaf, M. I. da Costa, Jr., C. Paduani, Effect of composition and short-range order on the magnetic moments of Fe in $Fe_{1-x}V_x$ alloys, *Phys. Rev. B* **61**, 6196 (2000).

[12] X. He, W.-C. Wang, B.-X. Liu, Ferromagnetic states in Fe−Ru systems studied by *ab initio* calculation and ion-beam mixing, *Phys. Rev. B* **77**, 012401 (2008).

[13] G. F. Zhou, H. Bakker, Influence of mechanical milling on magnetic properties of intermetallic compounds (Overview), *Mater. Trans., JIM* **36**, 329 (1995).

[14] J. Fassbender, M. O. Liedke, T. Strache, W. Möller, E. Menéndez, J. Sort, K. V. Rao, S. C. Deevi, J. Nogués, Ion mass dependence of irradiation-induced local creation of ferromagnetism in $Fe_{60}Al_{40}$ alloys, *Phys. Rev. B* **77**, 174430 (2008).

[15] E. Menéndez, J. Sort, M. O. Liedke, J. Fassbender, S. Suriñach, M. D. Baró, J. Nogués, Two-fold origin of the deformation-induced ferromagnetism in bulk $Fe_{60}Al_{40}$ (at.%) alloys, *New J. Phys.* **10**, 103030 (2008).

[16] J. Nogués, E. Apiñaniz, J. Sort, M. Amboage, M. d'Astuto, O. Mathon, R. Puzniak, I. Fita, J. S. Garitaonandia, S. Suriñach, J. S. Muñoz, M. D. Baró, F. Plazaola, F. Baudelet, Volume expansion contribution to the magnetism of atomically disordered intermetallic alloys, *Phys. Rev. B* **74**, 024407 (2006).

[17] S. Gialanella, X. Amils, M. D. Barò, P. Delcroix, G. Le Caër, L. Lutterotti, S. Suriñach, Microstructural and kinetic aspects of the transformations induced in a FeAl alloy by ball-milling and thermal treatments, *Acta Mater.* **46**, 3305 (1998).

[18] M. Fujii, K. Saito, K. Wakayama, M. Kawasaki, T. Yoshioka, T. Isshiki, K. Nishio, M. Shiojir, Ferromagnetism and structural distortions induced in atomized Fe-Al (35–42 at. % Al) powder particles by cold milling, *Philos. Mag. A* **79**, 2013 (1999).

[19] J. Sort, A. Concustell, E. Menéndez, S. Suriñach, K.-V. Rao, S.-C. Deevi, M.-D. Baró, J. Nogués, Periodic arrays of micrometer and sub-micrometer magnetic structures prepared by nanoindentation of a nonmagnetic intermetallic compound, *Adv. Mater.* **18**, 1717 (2006).

[20] J. Ehrler, M. He, M. V. Shugaev, N. I. Polushkin, S. Wintz, V. Liersch, S. Cornelius, R. Hübner, K. Potzger, J. Lindner, J. Fassbender, A. A. Ünal, S. Valencia, F. Kronast, L. V. Zhigilei, R. Bali, Laser-rewriteable ferromagnetism at thin-film surfaces, *ACS Appl. Mater. Interfaces* **10**(17), 15232 (2018).

[21] J. Ehrler, R. Bali, F. Kronast, L. Zhigilei, Laser schaltet magnet an und aus, *Physik in unserer Zeit* **49**(4), 163 (2018).

[22] J. Fassbender, J. McCord, Magnetic patterning by means of ion irradiation and implantation, *J. Magn. Magn. Mater.* **320**(3–4), 579 (2008).

[23] J. Fassbender, D. Ravelosona, Y. Samson, Tailoring magnetism by light-ion irradiation, *J. Phys. D: Appl. Phys.* **37**, R179 (2004).

[24] B. D. Terris, T. Thomson, Nanofabricated and self-assembled magnetic structures as data storage media, *J. Phys. D: Appl. Phys.* **38** (12), R199 (2005).

[25] E. La Torre, A. Smekhova, C. Schmitz-Antoniak, K. Ollefs, B. Eggert, B. Cöster, D. Walecki, F. Wilhelm, A. Rogalev, J. Lindner, R. Bali, R. Banerjee, B. Sanyal, H. Wende, Local probe of irradiation-induced structural changes and orbital magnetism in $Fe_{60}Al_{40}$ thin films via an order-disorder phase transition, *Phys. Rev. B* **98**(2), 024101 (2018).

[26] R. Soma, Y. Saitoh, M. Sakamak, K. Amemiya, A. Iwase, T. Matsui, Irradiation effect on magnetic properties of FeRh thin films with energetic C60 cluster ion beam, *AIP Advances* **8**, 056433 (2018).

[27] F. Röder, G. Hlawacek, S. Wintz, R. Hübner, L. Bischoff, H. Lichte, K. Potzger, J. Lindner, J. Fassbender, R. Bali, Direct depth- and lateral-imaging of nanoscale magnets generated by ion impact, *Sci. Rep.* **5**, 16786 (2015).

[28] S. A. Cybart, R. Bali, G. Hlawacek, F. Röder, J. Fassbender (2016), Focused helium and neon ion beam modification of high-T_C superconductors and magnetic materials, In Hlawacek G., Gölzhäuser A. (eds.), *Helium Ion Microscopy*. NanoScience and Technology. Springer, Cham.

[29] G. Hlawacek, R. Bali, F. Röder, Y. Aleksandrov, A. Semisalova, S. Wintz, K. Wagner, H. Schultheiss, S. Facsko, J. Fassbender, Tailoring magnetic nanostructures with neon in the ion microscope, *Microsc. Microanal.* **22**(S3), 1716 (2016).

[30] M. Fallot and R. Horcart, Sur l'apparition du ferromagnétisme par élévation de température dans des alliages de fer et de rhodium, *Rev. Sci.* **77**, 498 (1939).

[31] S. Cervera, M. Trassinelli, M. Marangolo, C. Carrétéro, V. Garcia, S. Hidki, E. Jacquet, E. Lamour, A. Lévy, S. Macé, C. Prigent, J. P. Rozet, S. Steydli, D. Vernhet, Modulating the phase transition temperature of giant magnetocaloric thin films by ion irradiation, *Phys. Rev. Mater.* **1**, 065402 (2017).

[32] M. B. Stearns, Variation of the internal fields and isomer shifts at the Fe sites in the FeAl series, *J. Appl. Phys.* **35**, 1095 (1964).

[33] A. Hernando, X. Amils, J. Nogués, S. Suriñach, M. D. Baró, M. R. Ibarra, Influence of magnetization on the reordering of nanostructured ball-milled Fe-40 at. % Al powders, *Phys. Rev. B* **58**, R11864(R) (1998).

[34] L. E. Zamora, G. A. Pérez Alcázar, G. Y. Vélez, J. D. Betancur, J. F. Marco, J. J. Romero, A. Martínez, F. J. Palomares, J. M. González, Disorder effect on the magnetic behavior of mechanically alloyed $Fe_{1-x}Al_x$ $(0.2 \leq x \leq 0.4)$, *Phys. Rev. B* **79**, 094418 (2009).

[35] J. Ehrler, B. Sanyal, J. Grenzer, S. Zhou, R. Böttger, H. Wende, J. Lindner, J. Fassbender, C. Leyens, K. Potzger, R. Bali, Magneto-structural correlations in a systematically disordered B2 lattice, *New J. Phys.* (2020). https://doi.org/10.1088/1367-2630/ab944a.

[36] J. F. Ziegler, M. D. Ziegler, J. P. Biersack, SRIM — The stopping and range of ions in matter, *Nucl. Instrum. Methods Phys. Res. B* **268**(11–12), 1818 (2010).

[37] A. Varea, E. Menéndez, J. Montserrat, E. Lora-Tamayo, A. Weber, L. J. Heyderman, S. C. Deevi, K. V. Rao, S. Suriñach, M. D. Baró, K. S. Buchanan, J. Nogués, J. Sort, Tuneable magnetic patterning of paramagnetic $Fe_{60}Al_{40}$ (at.%) by consecutive ion irradiation through pre-lithographed shadow masks, *J. Appl. Phys.* **109**, 093918 (2011).

[38] T. Schneider, K. Lenz, A. Semisalova, J. Gollwitzer, J. Heitler-Klevans, K. Potzger, J. Fassbender, J. Lindner, R. Bali, Tuning ferromagnetic resonance via disorder/order interfaces, *J. Appl. Phys.* **125**(19), 195302 (2019).

[39] G. L. Causer, D. L. Cortie, H. Zhu, M. Ionescu, G. J. Mankey, X. L. Wang, F. Klose, Direct measurement of the intrinsic sharpness of magnetic interfaces formed by chemical disorder using a He^+ beam, *ACS Appl. Mater. Interfaces* **10**(18), 16216 (2018).

[40] G. Hlawacek, V. Veligura, R. van Gastel, B. Poelsema, Helium ion microscopy, *J. Vac. Sci. Technol. B* **32**, 020801 (2014).

[41] M. Nord, A. Semisalova, A. Kákay, G. Hlawacek, I. Maclaren, V. Liersch, O. M. Volkov, D. Makarov, G. W. Paterson, K. Potzger, J. Lindner, J. Fassbender, D. McGrouther, R. Bali, Strain anisotropy and magnetic domains in embedded nanomagnets, *Small*, 1904738 (2019).

[42] M. Krupinski, R. Bali, D. Mitin, P. Sobieszczyk, J. Gregor-Pawlowski, A. Zarzycki, R. Böttger, M. Albrecht, K. Potzger, M. Marszałek, Ion induced ferromagnetism combined with self-assembly for large area magnetic modulation of thin films, *Nanoscale* **11**(18), 8930 (2019).

[43] S. Takahashi and Y. Umakoshi, The influence of plastic deformation on the magnetic properties in Fe–Al alloys, *J. Phys. Condens. Matter* **2**, 4007 (1990).

[44] N. I. Kulikov, A. V. Postnikov, G. Borstel, J. Braun, Onset of magnetism in B2 transition-metal aluminides, *Phys. Rev. B* **59**, 6824 (1999).

[45] E. Apiñaniz, J. Garitaonandia, F. Plazaola, Influence of disorder on the magnetic properties of FeAl alloys: theory, *J. Non-Cryst. Solids* **287**(1–3), 302 (2001).

[46] S. K. Bose, V. Drchal, J. Kudrnovský, O. Jepsen, and O. K. Andersen, Theoretical study of ordering in Fe–Al alloys based on a density-functional generalized-perturbation method, *Phys. Rev. B* **55**, 8184 (1997).

[47] B. V. Reddy, P. Jena, S. C. Deevi, Electronic structure and transport properties of Fe–Al alloys, *Intermetallics* **8**(9–11), 1197 (2000).

[48] N. Tahir, R. Gieniusz, A. Maziewski, R. Bali, K. Potzger, J. Lindner, J. Fassbender, Evolution of magnetic domain structure formed by ion-irradiation of B2-$Fe_{0.6}Al_{0.4}$, *Opt. Express* **23**(13), 16575 (2015).

[49] N. Tahir, R. Gieniusz, A. Maziewski, R. Bali, M. P. Kostylev, S. Wintz, H. Schultheiss, S. Facsko, K. Potzger, J. Lindner, J. Fassbender, Magnetization reversal of disorder-induced ferromagnetic regions in $Fe_{60}Al_{40}$ thin films, *IEEE Trans. Magn.* **50**(11), 6101304 (2014).

[50] N. Tahir, R. Bali, R. Gieniusz, S. Mamica, J. Gollwitzer, T. Schneider, K. Lenz, K. Potzger, J. Lindner, M. Krawczyk, J. Fassbender, A. Maziewski, Tailoring dynamic magnetic characteristics of $Fe_{60}Al_{40}$ films through ion irradiation, *Phys. Rev. B* **92**(14), 144429 (2015).

[51] J. Wolff, M. Franz, A. Broska, R. Kerl, M. Weinhagen, B. Köhler, M. Brauer, F. Faupel, T. Hehenkamp, Point defects and their properties in FeAl and FeSi alloys, *Intermetallics* **7**(3–4), 289 (1999).

[52] J. Čížek, F. Lukáč, I. Procházka, R. Kužel, Y. Jirásková, D. Janičkovič, W. Anwand, G. Brauer, Characterization of quenched-in vacancies in Fe–Al alloys, *Physica B* **407**(14), 2659 (2012).

[53] R. J. Wasilewski, Structure defects in CsCl intermetallic compounds — I. Theory, *J. Phys. Chem. Solids* **29**, 39 (1968).

[54] M. O. Liedke, W. Anwand, R. Bali, S. Cornelius, M. Butterling, T. T. Trinh, A. Wagner, S. Salamon, D. Walecki, A. Smekhova, H. Wende, K. Potzger, Open volume defects and magnetic phase transition in $Fe_{60}Al_{40}$ transition metal aluminide, *J. Appl. Phys.* **117**(16), 163908 (2015).

[55] J. Ehrler, M. O. Liedke, J. Čížek, R. Boucher, M. Butterling, S. Zhou, R. Böttger, E. Hirschmann, T. T. Trinh, A. Wagner, J. Lindner, J. Fassbender, C. Leyens, K. Potzger, R. Bali, The role of open-volume defects in the annihilation of antisites in a B2-ordered alloy, *Acta Mater.* **176**, 167 (2019).

[56] X. Sedao, M. V. Shugaev, C. Wu, T. Douillard, C. Esnouf, C. Maurice, S. Reynaud, F. Pigeon, F. Garrelie, L. V. Zhigilei, J.-P. Colombier, Growth twinning and generation of high-frequency surface nanostructures in ultrafast laser-induced transient melting and resolidification, *ACS Nano*, **10**, 6995 (2016).

[57] J.-U. Thiele, S. Maat and E. E. Fullerton, FeRh/FePt exchange spring films for thermally assisted magnetic recording media, *Appl. Phys. Lett.* **82**, 2859 (2003).

[58] O. Gutfleisch, T. Gottschall, M. Fries, D. Benke, I. Radulov, K. P. Skokov, H. Wende, M. Gruner, M. Acet, P. Entel, M. Farle, Mastering hysteresis in magnetocaloric materials, *Philos. Trans. A Math. Phys. Eng. Sci.* **374**, 2074 (2016).

[59] N. Fujita, S. Kosugi, Y. Saitoh, Y. Kaneta, K. Kume, T. Batchuluun, N. Ishikawa, T. Matsui, A. Iwase, Magnetic states controlled by energetic ion irradiation in FeRh thin films, *J. Appl. Phys.* **107**, 09E302 (2010).

[60] B. Eggert, A. Schmeink, J. Lill, M. O. Liedke, U. Kentsch, M. Butterling, A. Wagner, S. Pascarelli, K. Potzger, J. Lindner, T. Thomson, J. Fassbender, K. Ollefs, W. Keune, R. Bali, H. Wende, Static and dynamic disorder effects on the magnetism in FeRh, RSC Adv. 10, 14386 (2020).

[61] K. Aikoh, S. Kosugi, T. Matsui, A. Iwase, Quantitative control of magnetic ordering in FeRh thin films using 30 keV Ga ion irradiation from a focused ion beam system, *J. Appl. Phys.* **109**, 07E311 (2011).

[62] F. Hammerath, R. Bali, R. Hübner, M. R. D. Brandt, S. Rodan, K. Potzger, R. Böttger, Y. Sakuraba, S. Wurmehl, Structure-property relationship of Co_2MnSi thin films in response to He^+-irradiation, *Sci. Rep.* **9**(1), 2766 (2019).

[63] D. W. Clark, S. J. Zinkle, M. K. Patel, C. M. Parish, High temperature ion irradiation effects in MAX phase ceramics, *Acta Mater.* **105**, 130 (2016).

[64] A. Heidarian, S. Stienen, A. Semisalova, Y. Yuan, E. Josten, R. Hübner, S. Salamon, H. Wende, R. A. Gallardo, J. Grenzer, K. Potzger, R. Bali, S. Facsko, J. Lindner, Ferromagnetic resonance of MBE-grown FeRh thin films through the metamagnetic phase transition, *Phys. Status Solidi B* **254**(10), 1700145 (2017).

CHAPTER 8

Defects-Induced Magnetism in SiC

YU LIU*,†

*Microsoft Quantum Materials Lab Copenhagen,
2800 Lyngby, Denmark
†Center for Quantum Devices, Niels Bohr Institute,
University of Copenhagen,
2100 Copenhagen, Denmark
yu.liu@nbi.ku.dk

The room-temperature ferromagnetism is unexpectedly observed in semiconductors without transition metal doping. Due to its amazing potential in spintronic application, it is important to understand the mechanism behind this phenomenon and to know whether the itinerant electrons interact with local spins. Here, we reveal the answers to these issues by intentionally creating divacancies to induce magnetism in SiC single crystals. The local moments are confirmed to be intrinsic, which come from p states of the nearest-neighbor carbon atoms around divacancies. The experimental evidence also supports the fact that magnetic moments and itinerant carriers can interact with each other in SiC. These findings will facilitate the development of spintronic devices.

1. Introduction

It is well known that localized magnetic moments and the coupling between them are two indispensable factors to induce long-range spin ordering in solids, exhibiting ferromagnetism (FM), antiferromagnetism (AFM), ferrimagnetism, etc. The local spins usually come from the elements containing the partially filled 3d or 4f subshells, where the electron configuration favors the high-spin states according to the Hund's rule. Then, spontaneous spin coupling will favor ferromagnetic order if complying with the Stoner criterion; i.e., its exchange energy multiplying the density of states larger than unity [1, 2]. The coupling strength between the spins depends on the exchange integral that is sensitive to the separation of the spins. Coupling over a large separation is possible as evidenced by diluted magnetic semiconductors, where the concentrations of 3d metals are only of several percents. The itinerant carriers are thought to play a role in mediating spin orientations between magnetic atoms, known as the Ruderman–Kittel–Kasuya–Yosida (RKKY) theory [3–5]. Recently, more interest is focused on the magnetism in wide band-gap (WBG) semiconductors as the Curie temperatures can be attained at around room temperature (RT) by doping, which is important for applications in spintronics [6–12]. Limited solubility of 3d metals in WBG semiconductors, however, often leads to the precipitation of second phases, thwarting the attempts to get the unambiguous experimental results [7, 13]. Thus, study on the low-solubility regime is preferred even though the separation of contributions to the observed magnetism from magnetic elements and other sources still remains difficult.

On the contrary, there have been increasing evidences that traditional magnetic elements are not the sole source in inducing intrinsic magnetism; for a recent review, see Refs. [14–16]. RT FM was observed in highly oriented pyrolytic graphite (HOPG) [17–19], in HfO_2 [20], in Al-doped SiC [21], in Li-doped ZnO [22], and in various nano-sized compounds [23, 24] that are otherwise nonmagnetic in their bulk states. Theoretical studies revealed that the local moment can form from defects and the extended tails of their wave functions mediate long-range magnetic coupling [25–30].

Convincing experimental evidence requires clean and simple systems, so SiC single crystals are the perfect candidate. In our early work, we found that the magnetism in Mn-doped SiC can be attributed to the transition metal doping [31]. The ferromagnetism brought by Al doping further inspired us to initiate the study on defect-induced magnetism in SiC [21]. In this chapter, we review the experimental evidence and the theoretical explanation on defect-induced magnetism in SiC, which are achieved by us in recent years [32–39]. Here, we would like to address the following questions:

(1) How does the magnetism behave in SiC after defect inducing? (Phenomenon)
(2) How to understand the emergence of the defect-induced magnetism in SiC? (Mechanism)
(3) Are we able to control the local moments in this material or to output signal with it? (Application Initiation)

Therefore, we first take the SiC single crystal after neutron irradiation as a typical example to show defect inducing and characterization in SiC followed by its magnetic measurement [32]. Then, we explore the origin of the magnetism in noble gas-implanted SiC [37, 39]. We also indicate the interaction between magnetic moments and itinerant carriers in SiC with defect-induced ferromagnetism after Al or N implantation [39]. At last, the theoretical calculations are presented, which is helpful to comprehend the obtained experimental results [32].

2. Experimental Evidence

2.1. *Material and defect inducing*

Commercially available 2-inch SiC wafers (TankeBlue, Beijing) were cut into pieces for performing neutron irradiation and ion implantation. The measurement of X-ray rocking curve confirmed the high crystalline quality of the wafer (full-width half-maximum, namely FWHM, 21.6 arcsec). The total concentration of main magnetic impurities (Fe, Co, Ni, Cr, and Mn, etc.) measured by secondary ion mass spectroscopy is below 3×10^{14} cm^{-3}. The maximum saturation magnetization that all magnetic impurities can contribute is 4.3×10^{-6} emu/g (assuming each atom with magnetic moment of 5 μ_B). Other main impurities are B ($5.66 \times 10^{17}\,\mathrm{cm}^{-3}$), Al ($1.33 \times 10^{16}\,\mathrm{cm}^{-3}$), N ($2.99 \times 10^{17}\,\mathrm{cm}^{-3}$), and V ($7.82 \times 10^{15}\,\mathrm{cm}^{-3}$).

At a temperature less than 50°C, the irradiations were carried out with neutrons at a dose rate of $2.65 \times 10^{13}\,\mathrm{n/cm^2\,s}$ (the dose rate due to thermal neutrons and fast neutrons were $2.0 \times 10^{13}\,\mathrm{n/cm^2\,s}$ and $6.5 \times 10^{12}\,\mathrm{n/cm^2\,s}$, respectively). Samples were divided into four groups and irradiated for varying durations, with the corresponding doses in the ranges of 1.91×10^{17} to $2.29 \times 10^{18}\,\mathrm{n/cm^2}$. Each piece was irradiated only once and pristine pieces taken from the same wafer were kept for the purpose of comparison.

Implantation was performed at room temperature in the ion beam center at Helmholtz–Zentrum Dresden–Rossendorf (HZDR). Xenon ions were implanted with the fluence of $5 \times 10^{12}\,\mathrm{cm}^{-2}$ at an energy of 500 keV (labeled as 5E12) to investigate the origin of the observed FM. The incident energies for Al and N were 180 and 110 keV, respectively. Samples were implanted with fluences from 5×10^{13} to $5 \times 10^{16}\,\mathrm{cm}^{-2}$ for Al and 1×10^{14} to $1 \times 10^{17}\,\mathrm{cm}^{-2}$ for N. Each piece was irradiated only once, and pristine pieces taken from the same wafer were kept as reference. Pulsed laser annealing was performed in ambient air with a 308-nm excimer laser with a pulse duration of about 30 ns. The samples for annealing have the fluences of $1 \times 10^{16}\,\mathrm{cm}^{-2}$ for Al and $5 \times 10^{16}\,\mathrm{cm}^{-2}$ for N. The energy density for the implanted SiC is 0.2, 0.4, 0.6, and 0.8 $\mathrm{J/cm^{-2}}$.

2.2. *Structural information*

The crystallinity of the samples is changed during defect inducing. Raman spectra in Fig. 1(a) show that the patterns of 6H–SiC crystals irradiated by varying neutron doses are similar to those of the pristine one. There are no other SiC polytypes or secondary phases detected under the sensitivity of Raman characterization. The main difference among the patterns is the decrease in intensity for each mode with increasing irradiation dose. Figure 1(b) displays the relative intensity variation of the folded longitudinal optical (FLO) mode given by $(1\text{–}I/I_0)$ versus irradiation dose, where I_0 and I are the intensities

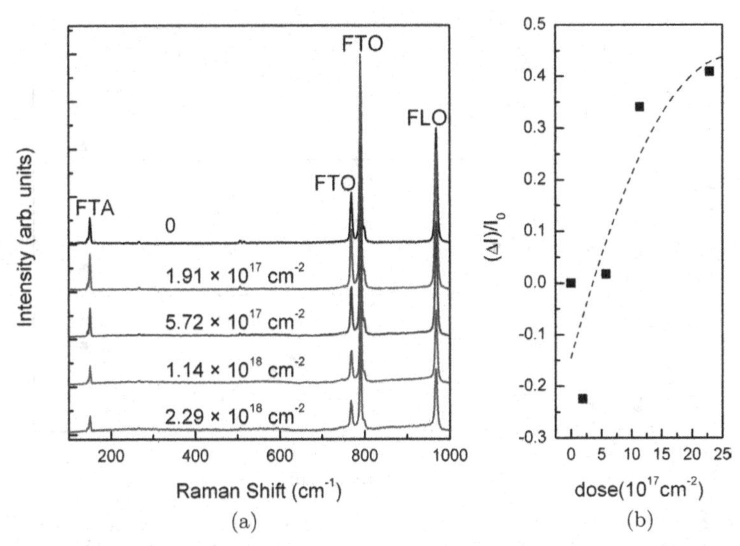

Figure 1. (a) Raman spectra from 100 cm^{-1} to 1000 cm^{-1} for irradiated and pristine SiC samples. (b) The relative intensity of the FLO mode versus irradiation dose. The dash line is a guide to the eyes. Reprinted with permission from Ref. 32. Copyright 2011 American Physical Society.

of the peak measured in the pristine and irradiated samples, respectively. Such a change in relative intensity suggests the existence of defects or lattice damages [40] in the irradiated samples though there is no noticeable change in FWHM of X-ray rocking curves.

2.3. *Defect characterization*

To investigate the defect types and their concentrations created by the neutron beam, we performed a series of measurements by positron annihilation lifetime spectroscopy (PALS). A lifetime spectrum is a linear combination of exponential functions corresponding to different annihilation sites. As a rule, we fit all the measured spectra into an exponential function of three components. A long lifetime of about 2000 ps corresponding to annihilations at voids or the sample surface is found to be very small in its fraction as usual [41] and we do not consider it in the following analysis. The fitted two remaining positron lifetimes τ_1, τ_2, and the fraction of the longer lifetime (τ_2) component I_2 as functions of irradiation dose are shown in Fig. 2. It is found that the lifetimes τ_1 and τ_2, taking values of

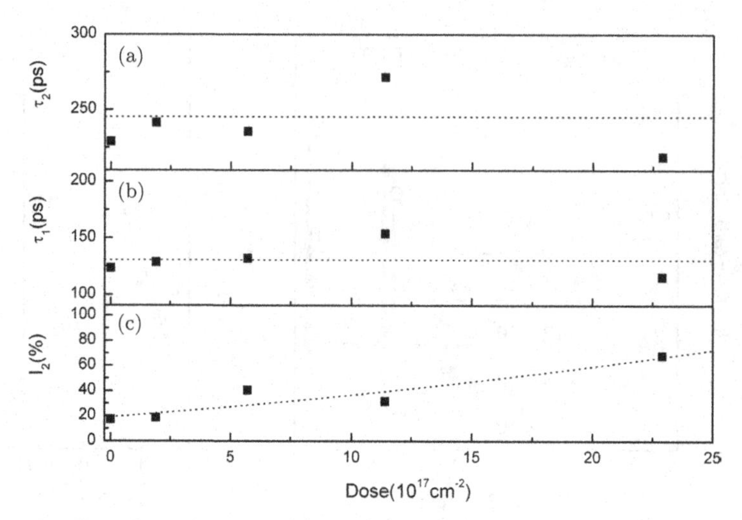

Figure 2. Fitted results of τ_2 (a), τ_1 (b) and I_2 (c) as functions of irradiation dose. τ_1 and τ_2 are found to be independent of dose. I_2 is proved to have positive correlation with dose. The dotted line shows the average values in (a) and (b) while a guide to the relationship between I_2 and irradiation dose is shown in (c). Reprinted with permission from Ref. 32. Copyright 2011 American Physical Society.

130 ± 14 ps and 239 ± 20 ps, are independent of irradiation dose, while I_2 is closely correlated with the irradiation dose. The lifetime value of 130 ± 14 ps, similar to the prior experimental [42–44] and theoretical values [43], is attributed to the bulk lifetime of 6H–SiC. The positron trapping center at 239 ± 20 ps, close to 253 ± 4 ps [42] and 225 ± 11 ps [44], is defect related and corresponds to $V_{Si}V_C$, a typical divacancy, in p-type 6H–SiC crystals. The dependence of I_2 on irradiation dose reveals that the concentration of $V_{Si}V_C$ increases with the increasing irradiation dose, in accordance with the Raman scattering data. The PALS measurements show that $V_{Si}V_C$ is created by neutron irradiations and confirm that its concentration is enhanced by increasing dose. In our case, the concentration of $V_{Si}V_C$ is roughly estimated to be 4.3×10^{17} cm^{-3} at the largest dose of 2.29×10^{18} n/cm^2 assuming that the pristine crystal's concentration is 4.1×10^{16} cm^{-3} [44].

Defects in SiC after ion implantation are also dominated by $V_{Si}V_C$. Another technique using positrons as the probe, called positron annihilation Doppler broadening spectroscopy (PADBS), was applied to clarify the nature of defects. The details can be found in our works [33, 38, 39].

2.4. *Magnetic properties*

The magnetization measurements were performed with a superconducting quantum interference device vibrating sample magnetometer which features a sensitivity of 10^{-7} emu. Figure 3(a) shows the variations of magnetization with the applied field in the range of $-5\,\text{kOe} < \text{H} < 5\,\text{kOe}$ in neutron-irradiated SiC at 5 K. It is noticed that only diamagnetic (DM) features can be observed in the pristine sample. After a low-dose irradiation of 1.91×10^{17} n/cm^2, the sample's DM features are becoming weaker and a minor hysteresis loop can be seen in the low-magnetic field range. With increasing irradiation dose, the magnetic order gradually enhances and DM M-H features further weaken, accompanied by an anticlockwise rotation of the hysteresis loops. Finally, at a dose of 2.29×10^{18} n/cm^2, a distinct hysteresis loop with ferromagnetic features as indicated in Fig. 3(b) can be clearly seen and the saturation magnetization reaches 1×10^{-4} emu/g, which is ~23 times larger than 4.3×10^{-6} emu/g, the maximum value possibly induced by the magnetic impurities in the used SiC crystals mentioned in Section 2.1. Even at 300 K, hysteresis loop can still be observed as shown in Fig. 3(c) with saturation magnetization of about 2×10^{-5} emu/g. A detailed investigation covering wider range of neutron doses shows the relation between ferromagnetism and paramagnetism in SiC [38]. The ion implantation is able to increase saturation magnetization, but it is still difficult to obtain the magnetism in bulk as strongly as predicted [33, 39]. It is understood that the FM is not homogeneously distributed but probably exists in domains, so the magnetization values obtained by dividing the magnetic moment by the total mass do not correspond to the expected intrinsic values. However, there is neither direct evidence to verify this assumption nor feasible measure to make improvements. Anyway, the observed hysteresis loops demonstrate that intrinsic magnetism exists after defect inducing.

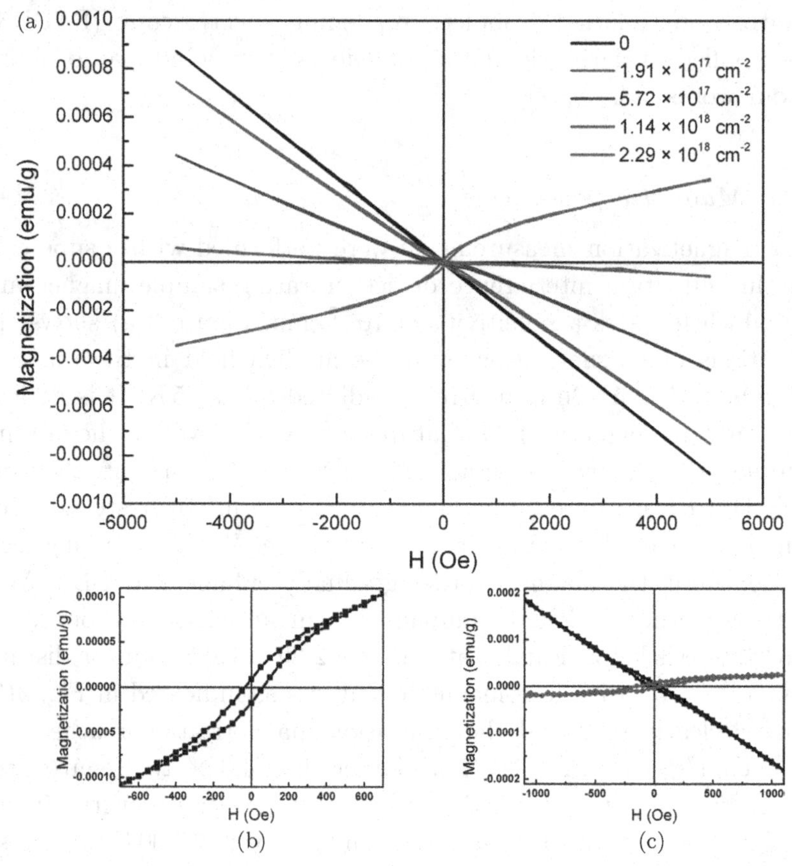

Figure 3. (a) The magnetization (in units of 1 emu $= 10^{-3}$ Am^2) measured at $T = 5\,\text{K}$ as a function of magnetic field ($1\,\text{Oe} = 10^3/4\pi$ Am^{-1}) in the range $-5\,\text{KOe} < \text{H} < 5\,\text{KOe}$ for irradiated and pristine SiC samples. (b) The magnified hysteresis loop for the sample with a dose of 2.29×10^{18} n/cm^2 at $T = 5\,K$. (c) The magnetic signals with (black) and without (red) the DM contribution in the low magnetic field range for the sample with a dose of 2.29×10^{18} n/cm^2 at $T = 300\,K$. Reprinted with permission from Ref. 32. Copyright 2011 American Physical Society.

2.5. *The origin of magnetism*

X-ray magnetic circular dichroism (XMCD) spectroscopy as an element-specific technique has been used to measure the magnetic contribution from different elements with partially occupied 3d or 4f subshells [46, 47] Ohldag *et al.* [48] successfully applied this

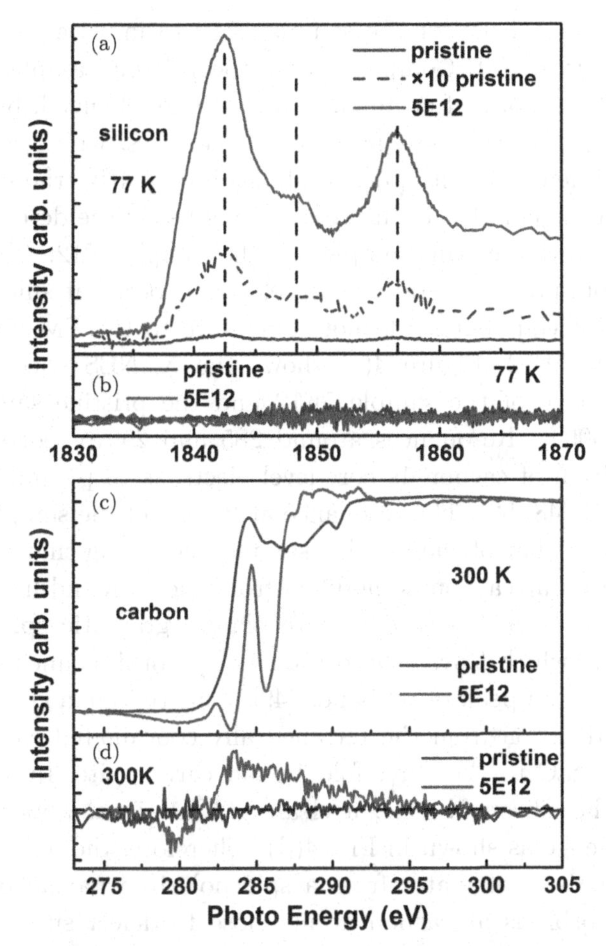

Figure 4. X-ray absorption spectra measured in EY (electron yield) mode for the sample 5E12 and the pristine sample: (a) XANES of the silicon K-edge at 77 K, (b) XMCD at the silicon K-edge at 77 K, (c) XANES of the carbon K-edge at 300 K. (d) XMCD at the carbon K-edge at 300 K [37].

technique to investigate the magnetism at the carbon K-edge in proton-irradiated HOPG. As it is possible to investigate the bonding state in SiC single crystals using X-ray absorption near-edge structure (XANES) spectroscopy [49, 50], it is also possible to explore the magnetic contribution in defect-induced ferromagnetism in SiC with soft X-ray spectroscopy. Figure 4(a) shows the XANES spectra of the silicon K-edge for the sample 5E12 to investigate the origin of the observed FM. Compared with the pristine sample, the peak positions of samples after implantation are not changed, but the relative strength of the peak at 1848 eV decreases, which suggests an increase of defect density [51]. As shown in Fig. 4(b), the strength of the XMCD signal at the silicon K-edge is below the detection noise level in both the pristine sample and the sample 5E12. We conclude that no spin-polarized states of 3p electrons occur at silicon atoms, and thus, silicon centers do not contribute to the FM observed in the sample 5E12. Figure 4(c) shows the XANES spectra at the carbon K-edge of the sample 5E12 and the pristine sample measured at 300 K. Resonances around 285 and 290 eV correspond to the transition of carbon 1s core-level electrons to p* and s* bands, respectively [48, 52]. The resonance at 285 eV of the sample 5E12 is sharper than that of the pristine sample, indicating that the orbital hybridization at carbon is modified from the diamond-like sp3-type carbon in pure SiC toward a more planar, graphitic sp2-type carbon center, which leaves the orthogonal p_z orbital unchanged and gives rise to the peak of p* bands [49]. This reflects a change of the local coordination from the tetrahedrally coordinated carbon atom in pristine SiC to the three-fold bound carbon site. In sharp contrast to the silicon K-edge, a clear XMCD signal appears at the carbon K-edge as shown in Fig. 4(d). Therefore, the defect-induced ferromagnetism originates from a spin-polarized partial occupancy of the p_z orbitals at carbon atoms close to defect sites in SiC. It is worth noting that an XMCD peak at around 280 eV in Fig. 4(d) appears well below the onset of the p* resonance. This peak was also observed in graphite [52]. This intriguing feature is not yet fully understood.

Figure 5 shows typical XANES and XMCD signals obtained from an Al-implanted SiC sample after annealing. Two peaks at ~286 and ~290 eV appear in the XMCD spectrum at the carbon K-edge. The corresponding peak of Xe-implanted (undoped) samples is located at ~284 eV mentioned above, and its position is at ~291 eV in Al-doped SiC polycrystalline film [53]. Therefore, both cases exist in Al-implanted samples after annealing. The emergence

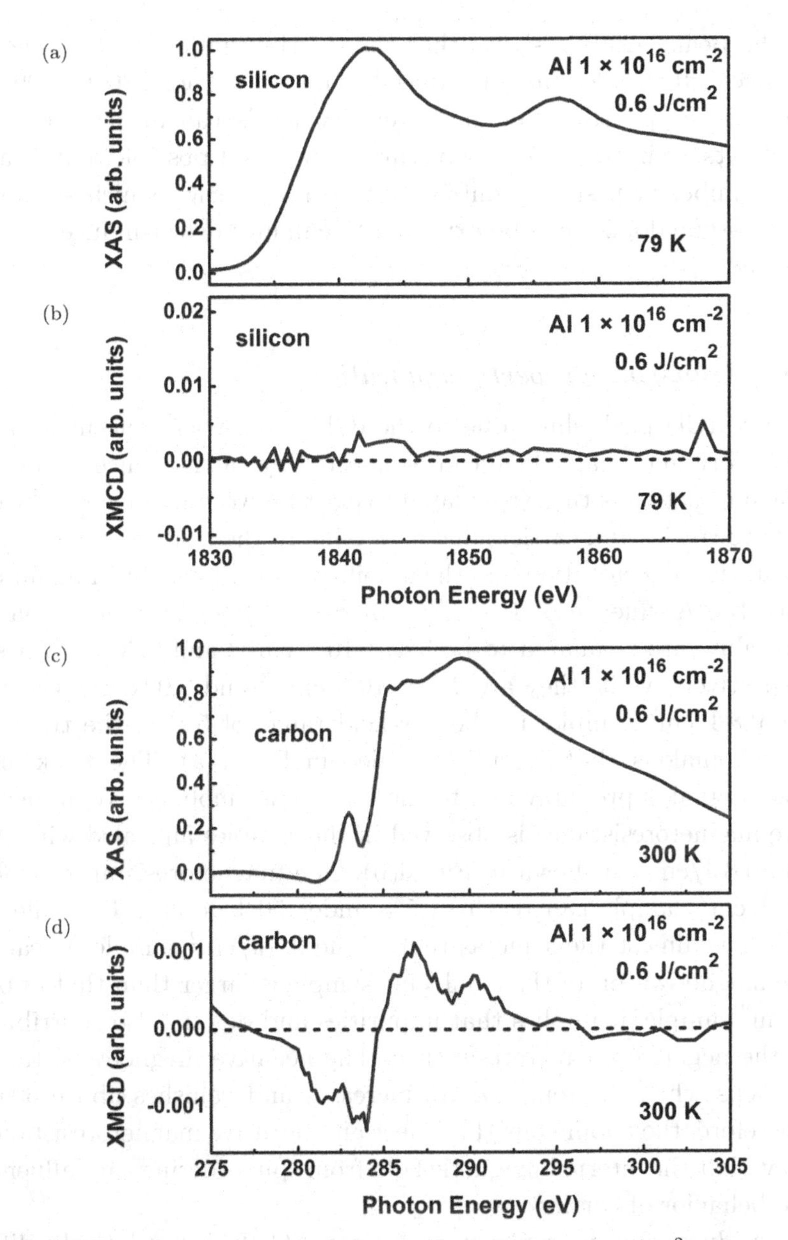

Figure 5. XANES and XMCD of Al-implanted SiC after 0.6 J/cm^2 annealing at the silicon K-edge at 79 K and at the carbon K-edge at 300 K. (a) XANES at the Si K-edge, (b) XMCD at the Si K-edge, (c) XANES at the C K-edge, and (d) XMCD at the C K-edge. Reprinted with permission from Ref. 39. Copyright 2017 American Physical Society.

of the peak ~290 eV shows that the distribution of local moments changes after holes are introduced. The shift from 286 to 290 eV means that p∗ states in higher energy levels can also form local moments, so introducing p-type carriers makes it possible to increase the number of p states contributing to FM in SiC, which suggests that p-type doping can be expected to enhance defect-induced magnetism.

2.6. *Transport property exploration*

The XMCD peak shifts due to the different carrier concentrations, so the origin of magnetism can be affected by carrier concentrations. The interaction is thus speculated to exist between moments induced by divacancies and carriers due to doping. In the Hall effect measurement, the concentration and the mobility of carriers in SiC implanted with the Al fluence of 1×10^{16} cm^{-2} after 0.6 J/cm^2 pulsed laser annealing are estimated to be 3.4×10^{19} cm^{-3} and 0.40 cm^2/(Vs), respectively, while they are 2.3×10^{20} cm^{-3} and 0.02 cm^2/(Vs) for the 0.8 J/cm^2 sample. In the low-field range of 5 kOe, the trace of the anomalous Hall effect is revealed in Fig. 6(a). The weak and noisy signal is probably due to the low carrier mobility. Weak negative magnetoresistance is observed in the samples annealed with 0.6 and 0.8 J/cm^2 as shown in Fig. 6(b). The magnetoresistance of the 0.6 J/cm^2 sample can reach –1.5% under 50 kOe at 2 K, while at 5 K, it is almost the same as that of the 0.8 J/cm^2 sample. Because the magnetization of the 0.6 J/cm^2 sample is larger than that of 0.8 J/cm^2 sample, it implies that impurities and defects also contribute to the negative magnetoresistance. The negative magnetoresistance weakens when the temperature increases and vanishes above 50 K. Therefore, the anomalous Hall effect and negative magnetoresistance show that the internal magnetic field from spin ordering can influence the behavior of carriers.

Additionally, in contrast to p-type SiC, FM is relatively difficult to be induced in n-type SiC [39]. This phenomenon is common in diluted magnetic semiconductors, which elucidates that the

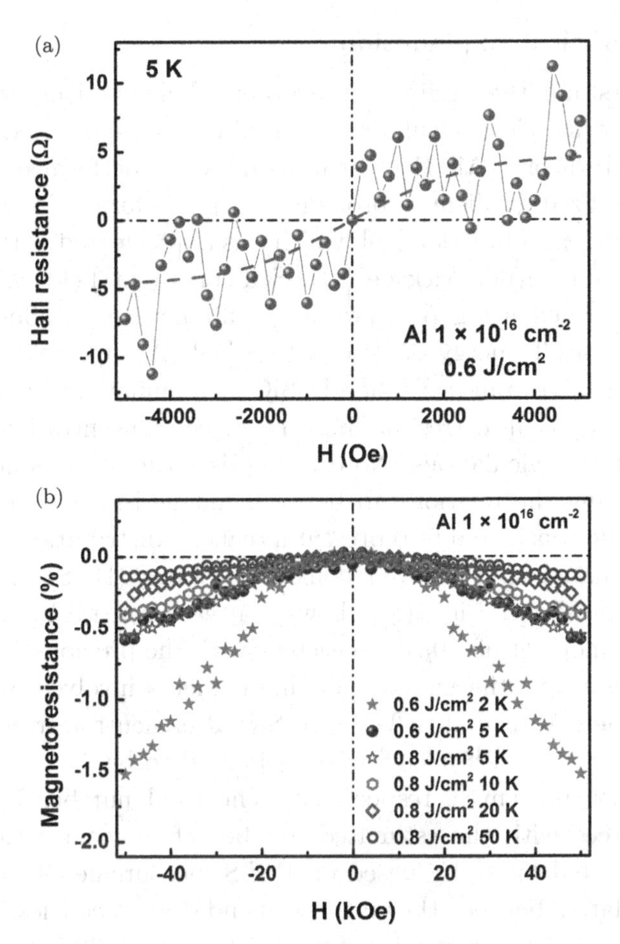

Figure 6. The transport properties of Al-implanted SiC after pulsed laser annealing. (a) The Hall resistivity of the 0.6 J/cm^2 sample with the field of 5 kOe at 5 K. The pink dashed line is a guide to the eye. (b) The magnetoresistance of 0.6 and 0.8 J/cm^2 in the field range of 50 kOe. Reprinted with permission from Ref. 39. Copyright 2017 American Physical Society.

electrons in p-type semiconductors are more localized than those in n-type semiconductors [6, 54, 55]. Based on the results from XMCD, Hall effect, magnetoresistance and the different magnetic behaviors between p-type and n-type SiC, it is confirmed that carriers play a crucial role in defect-induced magnetism.

3. Theoretical Explanation

To understand the origin of the observed magnetism, we carried out first-principles calculations based on spin-polarized density-functional theory. All the calculations were performed by using the generalized gradient approximation in the form of the Perdew–Burke–Ernzerhof function [56], which was implemented in the Vienna *ab initio* simulation package [57]. Self-consistent field calculations were performed using projector-augmented wave pseudopotentials with the cutoff energy set to 310 eV [58]. A supercell consisting of $4 \times 4 \times 1$ unit cells of 6H–SiC containing one axial $V_{Si}V_C$ [$Si_{95}(V_{Si})C_{95}(V_C)$], corresponding to a defect concentration of 0.5%, was built for calculations. Note that this structure is studied only for obtaining the relationship between magnetism and defects and shall not be considered to represent a realistic distribution of defects. The calculated spin-resolved density of states (DOS) of the 192-atom supercell (see Fig. 7(a)) shows that each neutral $V_{Si}V_C$ yields a magnetic moment of $2.0\mu_B$, consistent with the previous report [59]. Using this value, the numbers of the moments involved in FM and paramagnetism related to $V_{Si}V_C$ in SiC after neutron irradiation at a dose of 2.29×10^{18} n/cm^2 were estimated to be 3.5×10^{16} cm^{-3} and 6.7×10^{17} cm^{-3}, respectively. The total number 7.0×10^{17} cm^{-3} agrees with the estimated number of moments (8.6×10^{17} cm^{-3}) created by $V_{Si}V_C$ based on PALS measurements, confirming the correlation between the magnetism and the divacancies. The spin-polarization energy, which is defined as the energy difference between the spin-polarized and spin-unpolarized states, was calculated to be 1.90 eV, suggesting the spin-polarized state stable well above RT [27]. The spin polarization induced by the neutral $V_{Si}V_C$ leads to a 0.50-eV splitting between the majority- and minority-spin states. It should be pointed out that the splitting energy correction is not included here due to the consistency between the current calculation and the observation [60]. Figure 7(b) shows the charge density isosurface of defect states associated with a neutral $V_{Si}V_C$ in the 192-atom supercell. It demonstrates both the localized nature and the extended tails of the

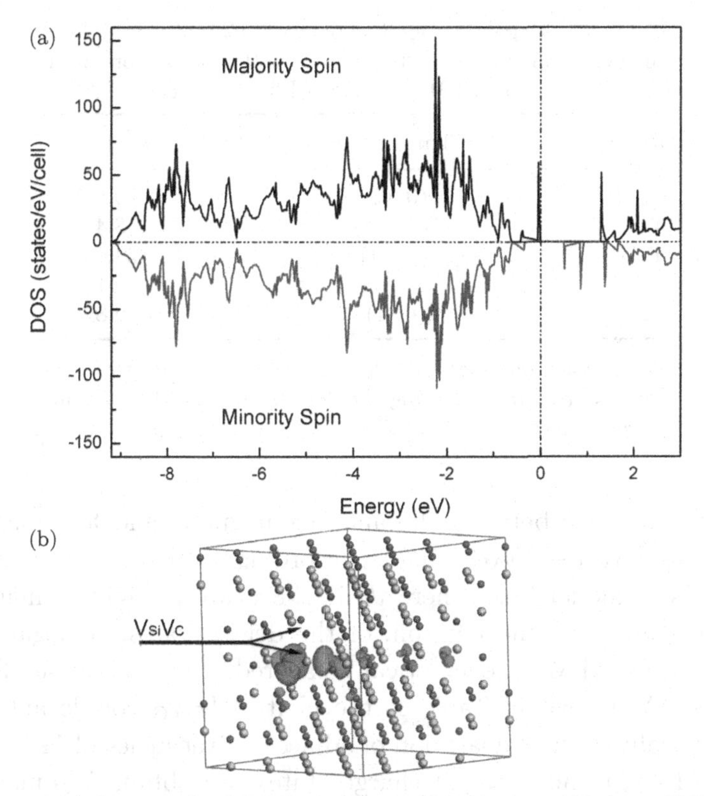

Figure 7. (a) Spin-resolved DOS of a neutral $V_{Si}V_C$ in a 192-atom SiC supercell. The spin polarization leads to a splitting between two kinds of spin states. (b) Isosurface charge density plot (isovalue is 0.02 e/Å^3) of the total spin states about a neutral $V_{Si}V_C$ in a 192-atom SiC supercell, showing both the localized nature and the extended tails of the defect wave functions. Si atoms are shown in yellow and C atoms in gray. Arrows indicate the location of $V_{Si}V_C$. Reprinted with permission from Ref. 32. Copyright 2011 American Physical Society.

defect wave functions beyond the supercell, similar to the calculated results in vacancy-containing III-nitrides [27].

The extended tails of the defect wave functions will induce the long-range coupling between the moments caused by $V_{Si}V_C$. Following the calculating scheme by Dev *et al.* [27], we doubled the size of the supercell by putting two 192-atom cells side by side and attempted to study the magnetic coupling (FM or AFM) between the $V_{Si}V_C$-induced local moments. Note that this antiferromagnetic structure is designed only for obtaining the magnetic interaction. The

Table 1. Magnetic coupling between $V_{Si}V_C$-induced local moments in 6H–SiC. Reprinted with permission from Ref. 32. Copyright 2011 American Physical Society.

dV(Å)	Charge state	J_0 (meV)
12.3	(0, 0)	−9.8
12.3	(+1, +1)	8.4
12.3	(-1, -1)	34.1
12.3	(0, +1)	−19.1
12.3	(0, -1)	9.8

Note: The coupling is antiferromagnetic between two neutral $V_{Si}V_C$. Upon charging the defects entirely, the coupling becomes ferromagnetic.

energy difference between the antiferromagnetic and ferromagnetic phases is $E_{\mathrm{AFM}} - E_{\mathrm{FM}} = 8J_0\mathrm{S}^2$ according to the nearest-neighbor Heisenberg model [61], where J_0 is the nearest-neighbor magnetic coupling and S is the net spin of the defect states. A negative J_0 means that AFM is energetically favored, and otherwise, FM is favored. As shown in Table 1, the neutral $V_{Si}V_C$ couple antiferromagnetically with a separation of adjacent divacancies of 12.3 Å. For the (+1, +1) and (-1, -1) charge states, we obtain ferromagnetic couplings with the identical separation. The varying values of J_0 for different charge states represent the different localized levels of the wave function, while the much larger value of J_0 for the (-1, -1) charge state suggests that the electronic structure is more delocalized. Charging the divacancies is found to be favorable for promoting the magnetic coupling between the $V_{Si}V_C$-induced local moments. In our case, charging is probable as B acceptors and N donors with a level of $\sim 10^{17}$ cm^{-3} exist in the samples, with the concentration of the former a little bit higher. In addition, if the charge concentration is less than the divacancy density, this means that $V_{Si}V_C$ will be partially charged. Charged $V_{Si}V_C$ and neutral $V_{Si}V_C$ may exist side by side in some domains. We calculated the exchange interaction and found J_0 is -19.1 meV for (0, +1) and 9.8 meV for (0, -1) if we assume that the spins of $V_{Si}V_C$ are all the same in this case, suggesting that partially charged divacancies also favor inducing spin ordering.

The strong antiferromagnetic interaction for (0, +1) charge state implies $V_{\mathrm{Si}}V_{\mathrm{C}}$ charged by this way is not significant, if any, in our samples.

4. Outlook

Nowadays, the defect-induced magnetism is accepted as a physical phenomenon without doubt. Based on our previous works, good progress has been achieved in SiC. Similar to noble gas implantation, proton irradiation was also used to induce magnetic moments on the SiC surface [62]. The d0 magnetism was realized in V-doped SiC, for both single crystals [63] and polycrystalline films [64, 65]. The observed magnetism is well understood with calculations in Cu-implanted SiC [66, 67] as well as Zn-implanted one [68]. The wide variety of preparation methods guarantees the possibility for the future research on defect-induced magnetism.

The mobility in SiC is low, and the scattering of defects makes it even lower, which brings difficulties in its spintronic applications. It would be a good idea to search more suitable materials. Recently, we successfully used Weyl semi-metals in this field [69]. The breakthrough was also made in transition metal dichalcogenides [70]. As the thin films are very useful in device fabrication, the $LaAlO_3/SrTiO_3$ interface has great potential in the future works [71, 72]. Further studies on the spintronic device applications, such as spin transistors [73], spin solar cells [74], spin light emitting diodes [75], and spin amplifiers [76], are expected on these novel materials.

5. Conclusions

In this chapter, we review our recent works about the experimental evidence and the theoretical explanation on defect-induced magnetism in SiC. After neutron irradiation or ion implantation, the structure of SiC is changed due to defect inducing. The defects are dominated by $V_{\mathrm{Si}}V_{\mathrm{C}}$ in all SiC samples, and their concentrations increase along with an increase in the fluence. RT FM was observed in SiC single crystals with $V_{\mathrm{Si}}V_{\mathrm{C}}$. The p electrons of the nearest-neighbor carbon atoms of $V_{\text{Si}}V_{\text{C}}$ are mainly responsible for the observed ferromagnetism. The results from the difference of XMCD between semi-insulating and conductive SiC samples, anomalous Hall effect, negative magnetoresistance and the different magnetic behaviors between p-type and n-type SiC confirmed the interaction between magnetic moments and itinerant carriers in SiC with defect-induced ferromagnetism. These findings will facilitate the design and optimization of spintronic devices utilizing d0 ferromagnetism.

References

[1] E. C. Stoner, Collective electron ferromagnetism, *P. Roy. Soc. Lond. A Mat.* **165**(922), 372 (1938).
[2] E. C. Stoner, Collective electron ferromagnetism. II. Energy and specific heat, *P. Roy. Soc. Lond. A Mat.* **169**(938), 339 (1939).
[3] M. A. Ruderman, C. Kittel, Indirect exchange coupling of nuclear magnetic moments by conduction electrons, *Phys. Rev.* **96**(1), 99 (1954).
[4] T. Kasuya, A theory of metallic ferro- and antiferromagnetism on Zener's model, *Prog. Theor. Phys.* **16**(1), 45 (1956).
[5] K. Yosida, Magnetic properties of Cu-Mn alloys, *Phys. Rev.* **106**(5), 893 (1957).
[6] T. Dietl *et al.*, Zener model description of ferromagnetism in zinc-blende magnetic semiconductors, *Science* **287**(5455), 1019 (2000).
[7] J. Y. Kim *et al.*, Ferromagnetism induced by clustered Co in Co-doped anatase TiO2 thin films, *Phys. Rev. Lett.* **90**(1), 017401 (2003).
[8] H. Saito *et al.*, Room-temperature ferromagnetism in a II–VI diluted magnetic semiconductor Zn1-xCrxTe, *Phys. Rev. Lett.* **90**(20), 207202 (2003).
[9] S. B. Ogale *et al.*, High temperature ferromagnetism with a giant magnetic moment in transparent Co-doped SnO2-delta, *Phys. Rev. Lett.* **91**(7), 077205 (2003).
[10] M. Venkatesan *et al.*, Anisotropic ferromagnetism in substituted zinc oxide, *Phys. Rev. Lett.* **93**(17), 177206 (2004).
[11] S. Dhar *et al.*, Colossal magnetic moment of Gd in GaN, *Phys. Rev. Lett.* **94**(3), 037205 (2005).
[12] W. Pacuski *et al.*, Observation of strong-coupling effects in a diluted magnetic semiconductor Ga1-xFexN, *Phys. Rev. Lett.* **100**(3), 037204 (2008).
[13] D. C. Kundaliya *et al.*, On the origin of high-temperature ferromagnetism in the low-temperature-processed Mn-Zn-O system, *Nat. Mater.* **3**(10), 709 (2004).
[14] O. Volnianska, P. Boguslawski, Magnetism of solids resulting from spin polarization of p orbitals, *J. Phys.: Condens. Matter* **22**(7), 073202 (2010).

[15] P. Esquinazi *et al.*, Defect-induced magnetism in solids, *IEEE Trans. Magn.* **49**(8), 4668 (2013).
[16] S. Q. Zhou, Defect-induced ferromagnetism in semiconductors: A controllable approach by particle irradiation, *Nucl. Instrum. Meth. B* **326**, 55 (2014).
[17] P. Esquinazi *et al.*, Induced magnetic ordering by proton irradiation in graphite, *Phys. Rev. Lett.* **91**(22), 227201 (2003).
[18] H. H. Xia *et al.*, Tunable magnetism in carbon-ion-implanted highly oriented pyrolytic graphite, *Adv. Mater.* **20**(24), 4679 (2008).
[19] X. M. Yang *et al.*, Correlation between the vacancy defects and ferromagnetism in graphite, *Carbon* **47**(5), 1399 (2009).
[20] M. Venkatesan, C. B. Fitzgerald, J. M. D. Coey, Thin films: Unexpected magnetism in a dielectric oxide, *Nature* **430**(7000), 630 (2004).
[21] B. Song *et al.*, Observation of glassy ferromagnetism in Al-doped 4H-SiC, *J. Am. Chem. Soc.* **131**(4), 1376 (2009).
[22] J. B. Yi *et al.*, Ferromagnetism in dilute magnetic semiconductors through defect engineering: Li-doped ZnO, *Phys. Rev. Lett.* **104**(13), 137201 (2010).
[23] M. A. Garcia *et al.*, Magnetic properties of ZnO nanoparticles, *Nano Lett.* **7**(6), 1489 (2007).
[24] A. Sundaresan, C. N. R. Rao, Ferromagnetism as a universal feature of inorganic nanoparticles, *Nano Today* **4**(1), 96 (2009).
[25] P. O. Lehtinen *et al.*, Irradiation-induced magnetism in graphite: A density functional study, *Phys. Rev. Lett.* **93**(18), 187202 (2004).
[26] T. Chanier *et al.*, Magnetic state around cation vacancies in II–VI semiconductors, *Phys. Rev. Lett.* **100**(2), 026405 (2008).
[27] P. Dev, Y. Xue, P. H. Zhang, Defect-induced intrinsic magnetism in wide-gap III nitrides, *Phys. Rev. Lett.* **100**(11), 117204 (2008).
[28] H. W. Peng *et al.*, Origin and enhancement of hole-induced ferromagnetism in first-row d(0) semiconductors, *Phys. Rev. Lett.* **102**(1), 017201 (2009).
[29] H. Lee, Y. Miyamoto, J. Yu, Possible origins of defect-induced magnetic ordering in carbon-irradiated graphite, *Phys. Rev. B* **79**(12), 121404 (2009).
[30] M. W. Zhao, F. C. Pan, L. M. Mei, Ferromagnetic ordering of silicon vacancies in N-doped silicon carbide, *Appl. Phys. Lett.* **96**(1), 012508 (2010).
[31] B. Song *et al.*, Magnetic properties of Mn-doped 6H–SiC, *Appl. Phys. Lett.* **94**(10), 102508 (2009).
[32] Y. Liu *et al.*, Defect-induced magnetism in neutron irradiated 6H-SiC single crystals, *Phys. Rev. Lett.* **106**(8), 087205 (2011).
[33] L. Li *et al.*, Rise and fall of defect induced ferromagnetism in SiC single crystals, *Appl. Phys. Lett.* **98**(22), 222508 (2011).
[34] L. Li *et al.*, Defect induced ferromagnetism in 4H-SiC single crystals, *Nucl. Instrum. Meth. B* **275**, 33 (2012).
[35] Y. T. Wang *et al.*, Disentangling defect-induced ferromagnetism in SiC, *Phys. Rev. B* **89**(1), 014417 (2014).
[36] Y. T. Wang *et al.*, Structural and magnetic properties of irradiated SiC, *J. Appl. Phys.* **115**(17), 17c104 (2014).

[37] Y. T. Wang *et al.*, Carbon p electron ferromagnetism in silicon carbide, *Sci. Rep.* **5**, 8999 (2015).
[38] Y. T. Wang *et al.*, Defect-induced magnetism in SiC: Interplay between ferromagnetism and paramagnetism, *Phys. Rev. B* **92**(17), 174409 (2015).
[39] Y. Liu *et al.*, Interaction between magnetic moments and itinerant carriers in d0 ferromagnetic SiC, *Phys. Rev. B* **95**(19), 195309 (2017).
[40] A. Perez-Rodriguez *et al.*, Analysis of ion beam induced damage and amorphization of 6H-SiC by Raman scattering, *J. Electron. Mater.* **25**(3), 541 (1996).
[41] C. C. Ling, C. D. Beling, S. Fung, Isochronal annealing studies of n-type 6H-SiC with positron lifetime spectroscopy, *Phys. Rev. B* **62**(12), 8016 (2000).
[42] W. Puff *et al.*, An investigation of point-defects in silicon-carbide, *Appl. Phys. A* **61**(1), 55 (1995).
[43] S. Dannefaer, D. Craigen, D. Kerr, Carbon and silicon vacancies in electron-irradiated 6H-Sic, *Phys. Rev. B* **51**(3), 1928 (1995).
[44] C. C. Ling *et al.*, Positron lifetime spectroscopic studies of as-grown 6H-silicon carbide, *Appl. Phys. A* **70**(1), 33 (2000).
[45] G. Brauer *et al.*, Evaluation of some basic positron-related characteristics of SiC, *Phys. Rev. B* **54**(4), 2512 (1996).
[46] G. Schütz *et al.*, Absorption of circularly polarized X-rays in iron, *Phys. Rev. Lett.* **58**(7), 737 (1987).
[47] J. Stöhr *et al.*, Principles of X-ray magnetic dichroism spectromicroscopy, *Surf. Rev. Lett.* **5**(6), 1297 (1998).
[48] H. Ohldag *et al.*, π-Electron Ferromagnetism in Metal-Free Carbon Probed by Soft X-Ray Dichroism, *Phys. Rev. Lett.* **98**(18), 187204 (2007).
[49] X. Y. Gao *et al.*, Disorder beneath epitaxial graphene on SiC(0001): An x-ray absorption study, *Phys. Rev. B* **78**(20), 201404 (2008).
[50] M. Ohkubo *et al.*, X-ray absorption near edge spectroscopy with a superconducting detector for nitrogen dopants in SiC, *Sci. Rep.* **2**, 831 (2012).
[51] Y. S. Dedkov, M. Fonin, Electronic and magnetic properties of the graphene–ferromagnet interface, *New J. Phys.* **12**(12), 125004 (2010).
[52] H. Ohldag *et al.*, The role of hydrogen in room-temperature ferromagnetism at graphite surfaces, *New J. Phys.* **12**(12), 123012 (2010).
[53] M. He *et al.*, Study on spin polarization of non-magnetic atom in diluted magnetic semiconductor: The case of Al-doped 4H–SiC, *Solid State Commun.* **197**, 44 (2014).
[54] T. Dietl, H. Ohno, F. Matsukura, Hole-mediated ferromagnetism in tetrahedrally coordinated semiconductors, *Phys. Rev. B* **63**(19), 195205 (2001).
[55] T. Dietl, H. Ohno, Dilute ferromagnetic semiconductors: Physics and spintronic structures, *Rev. Mod. Phys.* **86**(1), 187 (2014).
[56] J. P. Perdew, K. Burke, M. Ernzerhof, Generalized gradient approximation made simple, *Phys. Rev. Lett.* **77**(18), 3865 (1996).
[57] G. Kresse, J. Furthmuller, Efficient iterative schemes for ab initio total-energy calculations using a plane-wave basis set, *Phys. Rev. B* **54**(16), 11169 (1996).

[58] P. E. Blochl, Projector augmented-wave method, *Phys. Rev. B* **50**(24), 17953 (1994).
[59] N. T. Son *et al.*, Divacancy in 4H-SiC, *Phys. Rev. Lett.* **96**(5), 055501 (2006).
[60] J. A. Chan, S. Lany, A. Zunger, Electronic correlation in anion p orbitals impedes ferromagnetism due to cation vacancies in Zn chalcogenides, *Phys. Rev. Lett.* **103**(1), 016404 (2009).
[61] P. Dev, P. H. Zhang, Unconventional magnetism in semiconductors: Role of localized acceptor states, *Phys. Rev. B* **81**(8), 085207 (2010).
[62] R. W. Zhou *et al.*, Ferromagnetism in proton irradiated 4H-SiC single crystal, *AIP Adv.* **5**(4), 047146 (2015).
[63] S.-Y. Zhuo *et al.*, Defects mediated ferromagnetism in a V-doped 6H — SiC single crystal, *Chin. Phys. B* **21**(6), 067503 (2012).
[64] H. Wang *et al.*, Influence of mole ratio of Si: C on the magnetic property of undoped and vanadium carbide doped 3C-SiC, *Chem. Phys. Lett.* **556**, 142 (2013).
[65] H. Wang *et al.*, Effect of vanadium on the room temperature ferromagnetism of V-doped 6H—SiC powder, *Chin. Phys. B* **22**(2), 027505 (2013).
[66] H. W. Zheng *et al.*, Room-temperature ferromagnetism in Cu-implanted 6H-SiC single crystal, *Appl. Phys. Lett.* **102**(14), 142409 (2013).
[67] Y. L. Yan *et al.*, Effect of vacancies on magnetic behaviors of Cu-doped 6H-SiC, *Appl. Phys. A* **117**(2), 841 (2014).
[68] Q. Li *et al.*, Carbon p-electron induced magnetic ordering in Zn-implanted 6H-SiC: experimental observation and theoretical calculation, *Mater. Res. Express* **3**(5), 056103 (2016).
[69] Y. Liu *et al.*, Towards diluted magnetism in TaAs, *Phys. Rev. Mater.* **1**(4), 044203 (2017).
[70] S. Mathew *et al.*, Magnetism in MoS2 induced by proton irradiation, *Appl. Phys. Lett.* **101**(10), 102103 (2012).
[71] A. Brinkman *et al.*, Magnetic effects at the interface between non-magnetic oxides, *Nat. Mater.* **6**(7), 493 (2007).
[72] Ariando *et al.*, Electronic phase separation at the LaAlO3/SrTiO3 interface, *Nat. Commun.* **2**, 188 (2011).
[73] C. Betthausen *et al.*, Spin-transistor action via tunable Landau–Zener transitions, *Science* **337**(6092), 324 (2012).
[74] B. Endres *et al.*, Demonstration of the spin solar cell and spin photodiode effect, *Nat. Commun.* **4**, 2068 (2013).
[75] G. Kioseoglou, A. Petrou, Spin Light Emitting Diodes, *J. Low Temp. Phys.* **169**(5–6), 324 (2012).
[76] Y. Puttisong *et al.*, Room-temperature electron spin amplifier based on Ga(In)NAs alloys, *Adv. Mater.* **25**(5), 738 (2013).

CHAPTER 9

Ferromagnetism in ZnO-based Materials and Its Applications

MUHAMMAD YOUNAS

Physics Division, PINSTECH, P.O. Nilore,
Islamabad 45650, Pakistan
Department of Physics, The University of Hong Kong,
Pokfulam, Hong Kong, China
chuhan.pieas@gmail.com

The observation of ferromagnetism (FM) in transition metal-doped ZnO has fueled much interest in this material, thus making it a potential diluted magnetic semiconductor (DMS). The search for a room-temperature uniform DMS is an active field of research and an important area for exploring novel phenomena, functionalities, and concepts at the intersection of semiconductor physics and magnetism. Controlled FM with intrinsic origin in DMS is debatably a stimulating research topic in materials science and condensed-matter physics. Due to its amazing potential in spintronic applications, a clear understanding of the principal mechanisms responsible for the FM in DMS will definitely offer the much-needed direction for the material design. The striking properties and functionalities of DMS not only influence semiconductor physics but to a large extent would also extend the new horizons for spreading the spintronic research area over many other materials families. In the current scenario, although much work has been done in the past and further research is ongoing on to find out whether the FM in DMS is intrinsic or extrinsic; the picture is still ambiguous in this development.

1. Introduction

In many of the ferromagnetic metals (such as Iron, Nickel, Cobalt), a set number of free electrons makes them electrical conductors. Magnetism in these materials is originated by the difference in the occupancy of spin-up and spin-down states of the electrons. The collective spin alignment of these electrons leads to the spontaneous magnetism that persists up to the Currie temperature (T_C). Ferromagnets have non-volatile memory and their magnetization can be swapped swiftly at ambient temperature [1]. On the contrary, electrical conduction in semiconductors occurs via different type of charge carriers and most of the semiconductors are non-magnetic [2]. Semiconductors have a potential advantage over metals because of their easy manipulation to fabricate an applicable heterostructures. The physical properties of semiconductors can be controlled by many factors such as doping, irradiation, and electric field (amplification by electrostatic gates) [1]. A coupling between ferromagnetic and semiconducting properties can be worthwhile from a practical point of view. A material that simultaneously shows ferromagnetic and semiconductor properties is normally termed as a "magnetic semiconductor". Long spin coherence time in semiconductors [3] and swift improvement of the information storage technology by employing ferromagnetic metals have amplified the research activities in magnetic semiconductors [4]. Magnetic semiconductors can be very exciting from the view point of merging ferromagnets and conventional semiconductor electronics (by integrating nonvolatile magnetic memory and logic computing) in a single device [1, 2].

Magnetic semiconductors have many potential applications in "spintronics". In the field of spintronics, solid-state physics and materials research are strongly intermingled and play a decisive role in a matchless, fast-paced, technological jazz. In spintronics technology, charge and spin degrees of freedom of the charge carriers are being exploited for device functionality [5]. Rapid efforts are being devoted in developing the spintronics technology after the ground-breaking discovery of giant magnetoresistance effect in the multilayered magnetic materials [6, 7]. It is believed that high T_C and controlled spin-dependent phenomena in conventional semiconductors could form the basis for spin-, charge-based, or mixed spin- and charge-based devices [1] such as spin light-emitting diodes (spin-LEDs), spin field effect transistors (spin-FETs), and the spin qubits for quantum computers [8, 9].

Most of the currently available electronic devices are based on transport, manipulation, and storage of electrical charges. These devices will definitely cease to exist in the near future due to the heat generation and miniaturization problems, and such devices would not meet the needs of rapidly increasing power consumption. These issues clearly demand a fast and best substitute for the technology currently in use [1]. Spintronic devices, if successfully manufactured in the future, would be faster due to the fact that flipping of the electron spin is 10 times faster than transporting it and the device would consume 10–50 times less power [4, 10]. In spintronic technology, it is of vital importance that a device must have an element, which produces spin-polarized carriers with reliable transport for it to process the information. Such type of devices must have both ferromagnetic as well as semiconductor materials for the electron spin storage and traditional device operations, respectively. It seems simple to fabricate a device that possesses both ferromagnetic and semiconductor properties. One such example is a heterostructure device with the ferromagnetic metal–semiconductor junction having 100% spin polarization from ferromagnetic metal to semiconductor. However, there are some limitations to these devices. The conductivities between the two materials should match closely for effective injection of spin-polarized carriers from the ferromagnetic metal into the semiconductor material. However, due to the large conductivity mismatch between the two joining materials, formation of interfacial layers and the non-Ohmic nature of the contacts, the degree of spin polarization is reported to be reduced in these structures [11, 12]. For effective incorporation of spin into semiconductor material, numerous technical issues such as proficient injection, detection and manipulation of spin-polarized charge carriers should be considered seriously [4].

Dilute magnetic semiconductors (DMS) are the best replacement to resolve these kinds of issues. DMS materials solve the problem of conductivity mismatch at the interfacial layer of metal and semiconductor materials [13]. In these materials a fraction of transition metal element is doped into the host material to achieve spin-polarized charge carriers and high Curie temperature (T_C) [12] For the successful implementation for device applications, these materials should be of single phase and FM should be carrier mediated with T_C at or above the room temperature. However, many DMS become ineffective due to their lower T_C, hampering their practical applications. Room-temperature operation of spintronics devices is highly anticipated for the pragmatic and economic feasibility. Although extensive research is ongoing in order to find a true DMS with a high T_C for realistic function in devices, it is still a striking challenge to develop such a DMS material that could be employed in spintronics devices operating at high temperature [14]. Controlled ferromagnetism and realistic origin in DMS are debatably the greatest stimulating research topics in materials science and condensed-matter physics in these days.

Wide-bandgap DMS materials with transition metal doping are considered to show high T_C at and above room temperature [11]. A great surge in these materials came with the theoretical prediction of room temperature ferromagnetic ordering in GaN (III–V) and ZnO (II–VI) for 5 at.% Mn doping [15]. These materials find interesting applications in transparent electronics, to add spin functionality in FETs, UV light emission, low-threshold spin lasers, gas sensing, varistors and surface acoustic wave properties [16, 17] Generally, for the true DMS material, it is obligatory that it should be of single phase (free from secondary phases, clusters, magnetic contamination) with sufficient population of spin-polarized charge carriers to mediate the ferromagnetic interactions. If secondary phases, precipitates, small size clusters, and possible magnetic contamination are responsible for the room-temperature ferromagnetism (RTFM), then the efficiency of such DMS materials for the useful device applications would be doubtful. Observation of RTFM in the DMS does not simply mean that it is suitable for device applications; it should pass multiple tests to explore the true nature of the ferromagnetism. Most of the secondary phases or precipitates of the TM elements in the wide-bandgap materials are magnetic and some of the isolated magnetic clusters are found to be superparamagnets. These kinds of potential sources should be fairly excluded prior to being used in the carrier-mediated RTFM interactions [12].

In the current scenario, although much research has been done in the past and still continues to find out whether the RTFM is intrinsic or extrinsic; the picture is still ambiguous in this development. However, the observation of FM in Mn–GaN [18], Mn–ZnO [19], Co–ZnO [20–22], and V–ZnO [23] has fueled much interest in these materials. A clear understanding of the principal mechanisms responsible for the RTFM in DMS will definitely offer the much-needed direction for material design [4]. The study in this context is fairly a wide field and a large number of outstanding researches have been conducted so far. Unluckily, only a small part of these works could be covered in this review chapter. We sincerely apologize the authors of much admirable work, which we have had to skip inadvertently.

2. Exchange Interaction Models in DMS

The physical properties of DMS continue to challenge our understanding of these materials. As a result, several different models and mechanisms have been proposed to explain their magnetism. Among the most frequently quoted is the work of Dietl [15] in which Zener's theory for FM, in particular, the so-called p–d exchange, was utilized. This theory considers a coupling between carrier spins and local atomic moments of magnetic impurities and in parallel to mean-field theory provides an estimate of the ordering temperatures of several DMS materials [24]. The theory has predicted several systems with potential for large ordering temperatures, for example, ZnO and GaN [25]. Mean-field theory assumes ferromagnetic correlations mediated by holes from shallow acceptors in a matrix of localized spins. According to mean-field theory, it is hard to achieve FM in n-type semiconductors due to the generally smaller s–d interaction. Mean-field theory generally overestimates the stability of the ordered phases, which may lead to overly optimistic predictions of the critical temperature [26].

In the absence of ferromagnetic clusters or second phases, different models have been proposed to explain robust FM. The mechanisms relevant to magnetism are direct super exchange (antiferromagnetic), indirect super exchange (could be ferromagnetic), carrier-mediated exchange (ferromagnetic) including the much celebrated double-exchange mechanism, and bound magnetic polarons [27]. According to the mean-field theory proposed by Pierre Weiss [28] and later modified by Heisenberg [29], a direct exchange interaction between atoms inducing a ferromagnetic or antiferromagnetic character depends on the distance between the two atoms in the crystal of the material. If the atoms are close enough to each other, the dominant direct interactions are antiferromagnetic and when they are at certain distances, the material develops ferromagnetic character. But in 1951, Zener [30] stated that the direct exchange interactions between neighboring atoms always lead to antiferromagnetic coupling. However, an indirect exchange coupling of d-electrons spins and the conduction electron spins leads to ferromagnetic ordering. A competition between direct and indirect exchange interactions determines the overall ferromagnetic or antiferromagnetic character of the material. A few important exchange mechanisms are described in the ensuring sections.

2.1. *Super exchange mechanism*

In the mean-field theory approach, it is assumed that FM occurs through interactions between the local moments of the Mn atoms mediated by free holes in the material. The spin–spin coupling is also supposed to be a long-range interaction, and the effective spin–density is calculated on the basis of Mn ion distribution. In the Zener model approach, the direct super exchange coupling between d shells of the adjacent Mn–Mn ions (with similar charge state) leads to an antiferromagnetic configuration because of the half-filled Mn-d shell. Similarly, the indirect super exchange interactions between d shells of the Mn atoms through the conduction electrons (possibly O–p band valence electrons in case of ZnO) tend to align the spins in a ferromagnetic configuration. Ferromagnetism is only possible if indirect superexchange interactions between d shells of the Mn atoms dominate over the direct superexchange coupling between adjacent d shells. Therefore, T_C in the DMS is determined by a competition between the ferromagnetic and antiferromagnetic interactions [27].

2.2. *Double exchange mechanism*

In the Zener's double-exchange mechanism [31], magnetic ions in different charge states couple with each other by virtual hopping of the "extra" electron from one ion to the other. Spin ordering assists the hopping of a charge carrier over the *d* states by lowering the carrier energy and giving rise to the ferromagnetic ordering. The best example is the manganite material in which the ferromagnetic state is stabilized by the double-exchange mechanism originated from hopping of the electron between the Mn^{3+} and Mn^{4+} levels [12]. In the DMS material, the so-called double-exchange mechanism occurs if neighboring TM ions have parallel spins and the 3d electron hops in the partially occupied 3d orbitals of the adjacent TM ions. The TM-d band in this case is widened by the hybridization between the up-spin states, and the d-electron lowers its kinetic energy by hopping in the ferromagnetic state [27].

2.3. *RKKY interactions*

In 1954, Ruderman and Kittel [32], Kasuya [33] and Yosida [34] independently presented an extended model, which later became the well-known RKKY interaction mechanism for magnetic materials. The RKKY type interaction is based on the exchange coupling between the localized d-spins of the magnetic ion and the conduction electrons/holes. According to this model, the conduction band s-electrons are magnetized due to the polarization in the locality of the magnetic ion through s–d exchange interaction. This polarization decays with distance from the magnetic ion in an oscillatory fashion. The oscillation produced in this way causes an indirect superexchange interaction between the two magnetic ions on the nearest-or-next nearest magnetic neighbors. Depending on the separation of the interacting atoms, this coupling may result in an anti-parallel (antiferromagnetic) or parallel (ferromagnetic) configuration. It should also be kept in mind that s and d wave functions are orthogonal and would not lead to any interaction in a perfect one-electron system. Furthermore, if the carriers originated from Mn-d states are not behaving similar to free electrons, then the RKKY interaction may not be realistic. The RKKY interaction and the original model of Zener [30] are the basis for the mean-field Zener theory.

2.4. *Bound magnetic polaron*

In addition to the above-mentioned mechanisms responsible for the RTFM, there is another mechanism called bound magnetic polaron (BMP) model. A BMP normally exists in compound semiconductors such as ZnO due to point defects such as oxygen vacancies (V_{O}) and zinc interstitials. According to the BMP model, an appropriate amount of dopant and defects concentrations efficiently interact to yield a hydrogenic orbital originating from an electron associated with a particular defect. In BMP configuration, spins of many transition-metal ions are aligned with those of a much lower number of weakly bound carriers such as excitons within a polariton radius. The hydrogenic orbital stretches out adequately to overlap with a large number of BMPs, resulting in correlated clusters of polarons along with few isolated polarons within the host matrix. When the size of such clusters of polarons is identical to the size of the sample, ferromagnetic transition in the sample can be observed. The strength and the field of interaction increase as the temperature decreases. Definite dopant content and a minimum disparity between the dopant and defects concentrations are prerequisites for the appearance of a giant magnetic moment and polaron percolation threshold, respectively.

The BMP model is applicable to both p- and n-type host materials [35] and is more suitable for low-carrier density systems such as 1 at.% Co-doped ZnO, which produces a much lower magnetic moment for weak interaction. In case of high-carrier density systems such as greater than 1 at.% Co-doped ZnO, a prompt reduction in the magnetic moment is observed owing to the increased dopant–dopant interactions with progressive orbital moment reduction [36, 37]. In insulating or semi-insulating material with sufficiently higher concentrations of magnetic impurities such as Mn, ferromagnetic ordering is possible due to direct exchange interaction between BMP.

2.5. *Density Functional Theory Study in TM:ZnO*

In 2000, Dietl [15] theoretically estimated that ZnO doped with 5 at.% Mn and similar TM-doped wide-band gap semiconductors would show RTFM. In their calculations, they used the mean-field approach by assuming hole-mediated ferromagnetism interactions between Mn ions. T_C was evaluated on the basis of competition between the long-range carrier-mediated ferromagnetic interactions and short-range Mn–Mn antiferromagnetic exchange interactions. This prediction motivated much attention towards TM:ZnO research. Following this prediction, Sato and Katayama–Yoshida [38] used the Korringa–Kohn–Rostoker (KKR) Green's function method based on LDA of DFT to study the TMs:ZnO systems (TMs = V, Cr, Mn, Fe, Co, and Ni). Replacing some of the Zn atoms with Ga, electrons are introduced into the system. For the incorporation of extra holes in to the system, O atom is replaced with N [39].

In Mn:ZnO system, the Mn impurity has a d^5 character due to the substitution of Zn^{2+} by Mn^{2+} ion, so that there is no itinerant carrier. Antiferromagnetic superexchange interaction between Mn ions is supposed to stabilize the spin glass state. However, when Mn:ZnO is doped with nitrogen ($1s^2 2s^2 2p^3$), holes are itinerant having d-character owing to the strong hybridization of the N-2p states with the Mn-3d states. Due to this hybridization, the kinetic energy is decreased rapidly and the ferromagnetic state is stabilized by the double-exchange mechanism. On the contrary, when the Mn:ZnO system is doped with donors, the doped electrons never reside on the Mn-3d states but move into the host conduction band. Hence, ferromagnetism mediated by the double-exchange mechanism was most observed in (Zn,Mn)O [27]. Transition from the antiferromagnetic state to the ferromagnetic state only occurs as the holes are introduced, but no transition is observed by n-type doping. In case of doping level up to 25 at.% of TMs:ZnO (TMs = $V^{2+}(d^3)$, $Cr^{2+}(d^4)$, $Fe^{2+}(d^6)$, $Co^{2+}(d^7)$, and $Ni^{2+}(d^8)$), the 3d-band of these elements is not completely occupied. In these TMs:ZnO, ferromagnetic ordering is anticipated to occur without the need for additional charge carriers. Upper limit of the T_C was decided by the energy difference between ferromagnetism and antiferromagnetism ordering [27, 39].

It is suggested that the n-type doping in ZnO can increase the T_C of Fe-, Co-, and Ni-doped samples when the effects of disorder are taken into account by the coherent potential approximation (CPA) [39]. The intrinsic defects such as V_O and Zn-interstitials (Zn_i) form donor states within the n-type ZnO. It was proposed that Fe:ZnO, Co:ZnO, and Ni:ZnO are encouraging candidates for high-T_C ferromagnets and electron doping by regulating the ratio of Mn to Fe, Co, or Ni in the co-doped systems such that (Zn,Mn,Fe)O, (Zn,Mn,Co)O, or (Zn,Mn,Ni)O may show carrier-induced ferromagnetism. Wang and Jena [40] investigated the electronic structure and magnetic properties of a Mn-doped ZnO thin film using first-principles calculations. Magnetic coupling between Mn ions is found to be antiferromagnetic in agreement with the calculations of Sato and Katayama-Yoshida [38]. No free holes were present to facilitate the ferromagnetism when Mn atoms were substituted at the Zn site due to the same valence state of Mn and Zn. At high Mn concentrations, Mn atoms are observed to cluster around O atoms. It is concluded that co-doping or defects are necessary to acquire ferromagnetism in the Zn–MnO system in quite agreement with previous experimental results of Mn:ZnO systems [20, 41, 42].

Spaldin [43] employed the DFT approach with an LSDA exchange-correlation function to recognize possible ferromagnetic ground states within (Zn,Co)O or (Zn,Mn)O. In this approach a 32-atom supercell containing two dopant ions was developed. The two dopant ions were positioned at different arrangements. First configuration was a "close" in which TM atoms were alienated by a single O ion, and the second configuration was a "separated" configuration in which the dopant ions were connected via the –O–Zn–O– bond. In the absence of free carriers and with little meV energy difference between the antiferromagnetic and ferromagnetic spin states, a paramagnetic behavior was observed. As observed in these calculations, total energy of the TM-doped ZnO systems was very close to the ferromagnetic and antiferromagnetic orders; therefore, these systems generally show ferromagnetism, antiferromagnetism, and spin-glass behavior in experiments. This discrepancy may be caused by the fact that these systems might remain in dissimilar metastable states depending on the different experimental conditions, even for similar films.

Wang and coworkers [40, 44] used the first-principles DFT calculations by developing GGA for exchange and correlation to analyze the total energies of a (Zn, Mn)O thin film. The thin film is demonstrated by a slab comprising eight layers. The different configuration of Mn substitution on Zn sites is established to contrast the Mn–Mn separation and the nearest-neighbor environment. In the dilute limit, systems exhibit paramagnetic behavior due to the large distance among the substituted Mn atoms on the Zn site. With the increase of Mn concentration at the Zn site, Mn atoms cluster around the O atom leading to the antiferromagnetic behavior. However, antiferromagnetic to ferromagnetic transition is observed when the thin film was co-doped with N [45]. The N atoms prefer to exist as nearest neighbors to the Mn atom in ZnO and generate carriers in the system. The overlap between Mn 3d and N 2p states in the spin-up bands enhances the significant DOS at the Fermi energy leading to the half-metallic character of N in the co-doped (Zn,Mn)O system. The magnetic moments of the neighboring Mn and N atoms interact antiferromagnetically while Mn atoms couple ferromagnetically with one another as represented by Mn(↑)–N(↓)–Mn(↑). This interaction shows that ferromagnetism is facilitated through the p–d exchange coupling between the carriers and Mn atoms.

The local spin density approximation and coherent potential approximation used in the first-principles calculation may not be suitable to tackle the strong correlation and lattice relaxations. These approaches often fail to describe systems with localized (strongly correlated) d and f electrons. This deficiency can be improved by introducing a strong intra-atomic interaction in the form of onsite Coulomb interaction (U) substitution of the LSDA. This scheme is commonly known as the generalized gradient +U(GGA + U) scheme or the local spin density approximation +U (LSDA + U) scheme.

Based on the GGA and GGA + U approximations, *ab initio* electronic structure calculations have been employed to analyze the ferromagnetism in TM-doped ZnO [46]. The results are totally different from those reported by Sato [38], in which Mn-, Co-, or Cu-doped ZnO were ferromagnetic semiconductors, while other TM-doped ZnO were metallic. Lee and Chang [47] calculated the energy difference between ferromagnetic and antiferromagnetic for two Co atoms substituted in ZnO using the GGA approximation. The total energy difference between the ferromagnetism and antiferromagnetism states is $< 3\,\text{meV/Co}$. It signifies that competition between ferromagnetic and antiferromagnetic coupling occurs in Co-doped ZnO and spontaneous magnetization is not possible in intrinsic Co-doped ZnO. Instead, high concentrations of both Co ions and electron carriers are necessary to achieve FM [47, 48]. This is in agreement with experimental results, which report ferromagnetism at 300 K in some Co-doped ZnO films, but the reproducibility was less than 10% [20].

For both the Co- and Mn-doped ZnO, the ferromagnetic state is stabilized by the introduction of holes, such as the creation of Zn vacancies (V_{Zn}) in (Co,Cu)-doped ZnO. Following this prediction, the effects of V_{Zn} were experimentally and theoretically investigated and some groups demonstrated that V_{Zn} endorse RTFM [49–51]. Furthermore, Yan *et al.* [49] demonstrated that V_{Zn} favor the RTFM in Mn-doped ZnO films using simulations of O K-edge X-ray absorption near-edge structure (XANES) and first-principle calculations. To investigate the effects of co-doping and defects on magnetic ordering in TM-doped ZnO, Gopal and Spaldin [52] systematically studied four cases: (a) single-TM (Cr, Mn, Fe, Co, Ni, and Cu) substitution at Zn sites; (b) substitutional magnetic TM ions combined with additional Cu and Li dopants; (c) substitutional magnetic TM ions combined with V_{O}; and (d) pairs of magnetic ions (Co and Fe, Co, and Mn). The results showed that ferromagnetic ordering of TM ions is not induced by substitutional TM impurities or V_O. Incorporation of interstitial or substitutional Li is favorable for ferromagnetism, as are V_{Zn}, which is consistent with experimental results for (Co, Li) co-doped ZnO [53, 54].

Recently, hybrid and local-density approximation calculations were used to evaluate the Hubbard model parameters for electrons trapped by defects in ZnO [55]. The trapped electrons are described by a one-band Hubbard Hamiltonian, where the Hubbard U is the effective electron–electron repulsion for the pair of electrons in the vacancy. The ferromagnetic moment in the model is generated from electrons trapped at negatively charged vacancies in the n-type oxide. Strong ferromagnetic coupling between defects is found over a range exceeding 10 Å when the defects have a large, positive Hubbard U value. The negatively charged Zn–O divacancy (V_{ZnO}^-) and Zn monovacancy (V_{ZnO}^-) are proposed as the possible sources of magnetic moment in ferromagnetic ZnO films. These vacancies are observed to trap one or two electrons and their charge transition levels lie in the bandgap.

Up-Till now, we cannot draw a definite conclusions regarding intrinsic RTFM from the computational work. Both electron carriers [47, 56] and hole doping [52, 53, 57, 58] seem necessary for ferromagnetic ordering and the calculated magnetic behavior strongly contingent on the computational details and on the choice of exchange correlation function. Although there is a large spread of contradictory magnetic behaviors in literature, most of the reports agree that intrinsic FM does not occur in TM-doped ZnO at reasonable doping concentrations, but in many cases, the addition of carriers or point defects stabilizes the ferromagnetic state [59, 60].

3. Transitional Metal-Doped ZnO

Although a substantial amount of experimental data and related underlying mechanisms have been collected since the first report on RTFM [20, 35, 58–63] the origin of ferromagnetic ordering in DMS, particularly TM-doped ZnO, is still under debate, and whether TM^{2+}

replacement of Zn^{2+} or the formation of secondary phases is responsible for carrier-mediated RTFM is still under study [64, 65]. It is certainly fair to state that there is a great deal of controversy and discrepancy concerning the actual ferromagnetic T_C and the source of FM in transition metal-doped wide-bandgap semiconductors, especially in ZnO. Although there are still controversies, in this section we will discuss what has been reported experimentally about the ferromagnetism and its origin in TM-doped ZnO.

The Lande g-factor and total angular momentum (J) can be used to evaluate the effective magnetic moment of TM ions in TM:ZnO films. The Lande g-factor and effective magnetic moment μ_{eff} can be represented by the following relations [66]:

$$g = \frac{3}{2} - \frac{L(L+1) - S(S+1)}{2J(J+1)} \qquad \mu_{\text{eff}} = g\mu_B\sqrt{J(J+1)}.$$

Here J is the total angular momentum, including the spin (S) and orbit (L) angular momentum, μ_{eff} is the effective magnetic moment, and μ_B is the Bohr magnetron. In case of Co^{2+} in Co:ZnO, the Lande g-factor is 4/3, the effective magnetic moment is 6.6 μ_B/Co, and the net spin moment (calculated from the spin-only formula $\mu_{\max} = g\mu_B\text{S}$) is $3\,\mu_B$. Likewise, for Mn^{2+} the value of both an effective and net spin magnetic moment is $5\,\mu_B$. These diverse values of ideal magnetic moments for TM ions are most probably one of the reason for the different magnetic moments of TM:ZnO. Knowing that ZnO can be made ferromagnetic by doping with transition metals, numerous groups have fabricated (Zn,TM)O (TM = Sc, Ti [67, 68], V [23, 67], Cr [20, 67, 68], Mn [19, 20, 65, 67, 69–72], Fe [23], Co [20, 22, 67, 68, 73] Ni [20, 67, 68, 74], and Cu [67, 68], films by different techniques.

Several reports specify that TM ions at Zn sites are a necessary but not sufficient condition for ferromagnetism. Some ZnO films doped with Ti [75], Mn [76], Co [77] and Cu [78] showed different behavior, such as paramagnetism or superparamagnetism as well as spin glass behavior [41]. In ZnO, the solubility limit of 3d transition metals such as Mn (20 at.%) and Co (15 at.%) is much larger than the other TM atoms such as Ti, V, Cr, Fe, Ni, and Cu [51, 68, 69, 79, 80].

The electron effective mass in Mn-doped ZnO is $0.3m_e$, where m_e is free-electron mass. Similarly, ZnO films doped with 5 at.% Co usually show much stronger RTFM than ZnO doped with the other TM elements. The larger amount of injected spins and carriers thus make Mn- and Co-doped ZnO ideal for the fabrication of spintronic devices. Concentration-dependent magnetization shows that the largest magnetic moments is observed for ZnO doped with different TM concentrations such as 3–5 at.% for Co [79, 81, 82], 2 at.% for Mn [19, 83], and 1 at.% for Cr, Ni, and Cu [84–86].

Thin film deposition techniques, such as MBE [87], metalorganic chemical vapor deposition (MOCVD) [88], and pulsed laser deposition (PLD) [89], offer excellent control of the dopant concentration and the ability to grow single-phase layers of ferromagnetic materials. Although the magnetic behavior of TM-doped ZnO films depends on the deposition conditions [12, 61, 62], there is no final conclusion to say which deposition method is best for ferromagnetic ordering of the films. For different growth techniques employed for the DMS material, the magnetic properties can be very different. The results concerning the existence of ferromagnetism have been rather controversial, such that some Co-doped ZnO films deposited by PLD [20] or magnetron sputtering [90, 91], could, respectively, exhibit RTFM or RT paramagnetism. Similarly, several groups have reported ferromagnetism in (Zn,TM)O systems [19, 20, 23, 92] and others reported antiferromagnetic and spin-glass behavior [41].

Ando and coworkers [67] investigated the magneto-optical properties by measuring the magnetic circular dichroism (MCD) spectra of TM:ZnO films grown by PLD. The films doped with Sc, Ti, V, and Cr did not indicate any obvious magneto-optical effect, while the films alloyed with Mn, Co, Ni, and Cu showed prominent negative MCD peaks near the band edge of the host ZnO. The MCD amplitudes for undoped ZnO was almost temperature independent, while it decreased with increasing temperature for thin films doped with Mn, Co, Ni, and Cu. The MCD signal near the semiconductor band gap may be estimated with the effective g factor. For $Zn_{1-x}Co_xO$, the effective g values were estimated to be 4 for $x = 0.012$ and 9 for $x = 0.016$, which are much smaller than the values for other II–VI host magnetic semiconductors. Strong MCD signals designated that ZnO alloyed with Mn, Co, Ni, and Cu was a diluted magnetic semiconductor with strong exchange interaction between sp-band carriers and localized d electrons.

The theory of Dietl [15] for high-temperature ferromagnetism in Mn-doped ZnO was explicitly for p-type materials. However, Pearton [61] obtained primary evidence of ferromagnetism in n-type ZnO, using Mn and Sn as the transition metal and donor impurities, respectively. Similarly, Norton [93] reported the ferromagnetism with a $T_C \sim 250\,K$ in ZnO co-doped with Mn and Sn. From the XRD measurements, no evidence was observed for Mn–O phases and the Sn in these samples acted as a doubly ionized donor introducing deep states in the energy gap. By assessing the difference between the field-cooled (FC) and zero-field-cooled (ZFC) magnetization between 4.2 and 300 K, the paramagnetic and diamagnetic contributions to the magnetization were subtracted to attain the hysteretic ferromagnetic regime. This difference was more pronounced for the 3% Mn-doped samples, and the ferromagnetism was preserved up to 250 K for both the 3% and 5% Mn-doped samples. Within the XRD detection limit, it was reported that only Mn_3O_4 could not be accounted for the ferromagnetic transition temperature of 250 K, since ferromagnetic Mn_3O_4 has T_C of 42 K.

Besides RTFM arising from TM ion substitution at Zn^{2+} site, secondary phases such as TM metal complex and TM-based oxides are frequently found to be responsible for the RTFM observed. The formation of secondary phases mostly relies on the TM solubility limit in ZnO, oxygen partial pressure (PO_2), and the substrate temperature. All these parameters play significant roles in controlling the microstructure of the films. The oxygen partial pressure and the substrate temperature affect the formation of secondary phases, which subsequently affects the magnetic properties [19, 94].

3.1. *Mn:ZnO system*

Primarily, the majority of DMS investigated until now involved Mn^{2+} ion as a genuine member of the transition metal family to be inserted in various II–VI and III–V hosts. The transport and magnetic properties of Mn:ZnO films co-doped with Ga, Cr, and Fe deposited on sapphire substrates via pulsed laser deposition technique were studied by Fan *et al.* [95]. The results indicated that the co-dopants could effectively adjust the carrier densities and resistivities of Mn:ZnO films. The Coulombic attraction between the n-type Ga, Cr, or Fe and p-type Mn:ZnO favorably decreases the energy of the system, thereby preventing dopant aggregation and effectively enhancing the Mn-dopant solubility. The non-compensated n–p co-doping provides a certain amount of carrier density and local spins. This would lead to the RTFM as well as the positive and negative magnetoresistances at low temperatures in ZnO wide-gap semiconductors. Furthermore, the non-compensated n–p co-doping resulted in the metallic Ga-doped Mn:ZnO and the semiconducting behavior of the Cr- or Fe-doped Mn:ZnO. The Ga-doped Mn:ZnO sample has sufficient number of mobile electrons to show temperature-independent magnetization. However, the magnetization of Cr- and Fe-doped samples exhibited strong dependence on the temperature because of the localized d electrons. A redshift at room temperature in the near-band-edge emission peak was observed in Mn:ZnO thin films with 2% Mn doping [96]. It was analyzed that increasing the magnetic moment by increasing the carrier concentration leads to an increase of the redshits. The decrease of the bandgap (redshift) was attributed to the sp–d interaction between the free charge carriers in the semiconductor band and the localized magnetic moments. The magnetic moment was correlated to the concentration of V_O and not to the crystallographic order. The strong increase in magnetic moment upon incorporation of Mn in films with the same concentration of V_O suggested that the ferromagnetism in Mn-doped ZnO was due to the exchange between the Mn cations mediated by the shallow donor electrons. Mn-related secondary phases were not detected and were found not to be responsible for the observed ferromagnetism.

Liu *et al.* [97] has fabricated Mn-doped ZnO nanoparticles by a solution route followed by post-growth annealing in N_2, O_2, and Ar, respectively. The XRD and X-ray photoelectron spectroscopy (XPS) data revealed the single-phase wurtzite structure without any impurity phases. Magnetization loops for the Mn:ZnO samples displayed typical ferromagnetic saturation behavior. Using photoluminescence spectroscopy and first-principle calculations, it was suggested that the V_O, especially singly ionized oxygen vacancies, played a crucial role in mediating ferromagnetism in the Mn-doped ZnO samples. Single-phase ZnO:Mn thin films with wurtzite structure were grown under ambient argon–oxygen admixture to investigate the effects of stoichiometry and interstitial oxygen on magnetic properties [98]. The main focus of the study was to understand the connection between the defect concentration, material composition, and ferromagnetic properties. Extended near-band-edge emission spectra with noticeable decrease of the defect to near-band-edge emission ratio in the photoluminescence (PL) characterization showed the improved optical quality of ZnO:Mn thin films. Magnetic measurements showed that there was optimum argon–oxygen ambient pressure for achieving the maximum RTFM in the samples. With an increase of argon–oxygen partial pressure from 1 to 2 mbar, the saturation magnetization increases from 16.3 to 115.8 μemu. However, when the argon–oxygen partial pressure further increased to 5 mbar, the saturation magnetization decreased to 52.1 μemu. A strong direct correlation between the Zn–O stoichiometry and the saturation magnetization was demonstrated in these thin films. The oxygen-rich stoichiometry was reported to reduce the V_O and favored the p–d exchange coupling of Mn in ZnO host matrix responsible for the enhanced RTFM.

3.2. *Co:ZnO system*

The Co-doped ZnO material has been considered as a role model system for a wide range of DMS [27, 99, 100] after the theoretical prediction of the intrinsic ferromagnetism by the *ab initio* computation [38] and the subsequent experimental evidence of the RTFM [20] in this class of the material. Regardless of this great perspective, the understanding of RTFM in this class of system is far from being realized, since the conventional theory of carrier-mediated magnetism [15] does not hold while the relationship between the carriers and the magnetism is not clear [101]. Experimental and theoretical studies have endorsed RTFM in these systems to a variety of origins including intrinsic defects [102], hydrogen contamination [103], non-homogeneity of the dopant spatial distribution [104], or the presence of secondary phases [105] such as small Co nanoclusters. Furthermore, the observation of RTFM depends upon the sample preparation and growth conditions. Many exciting experimental studies showed that RTFM is absent in the near-perfect epitaxial Co-doped ZnO films [106–110]. Ney and coworkers [108, 109] observed a paramagnetic nature of the Co-doped ZnO films deposited by PLD on Al_2O_3 substrate. They used the X-ray magnetic circular dichroism (XMCD), X-ray linear dichroism (XLD) and electron paramagnetic resonance (EPR) results to justify their arguments. They observed the absence of intrinsic ferromagnetic interactions for the isolated and paired Co dopant atoms in the high structural quality Co-doped ZnO thin films. Using the PLD growth method, Xu and coworkers [110] prepared Co-doped ZnO films under low oxygen pressure for enhancing the V_O concentration but the film was paramagnetic. However, Heald and coworkers [111] observed that the RTFM originated from the intermetallic ferromagnetic Co_{Zn} under the low oxygen partial pressure.

The role of V_O is important in affecting the RTFM in Co-doped ZnO. The isolated V_O is a deep donor [112]. Theoretical calculations [113, 114] indicate that the Co–V_O complex in the Co:ZnO alloy generates an electronic state close to the conduction band which could play a role in the RTFM. Using first-principles calculations, Liu and coworkers [115] proposed that V_O could induce the RTFM, whereas Assadi and coworkers [116] by employing density functional calculations considered that the RTFM in Co:ZnO was mainly contributed by the hydrogen interstitial (Hi) rather than the V_O. However, a substantial experimental demonstration of the role of V_O in activating the RTFM in ZnO:Co is still lacking and the issue is still under debate [52, 117–119]. Several groups have tried to address this issue. The RTFM was observed in $Zn_{0.98}Co_{0.02}O$ thin films fabricated by the PLD method under low oxygen partial pressure [120]. However, after the annealing in air, a dramatic change in the magnetic property from ferromagnetic of the as-grown film to paramagnetic of the annealed film was observed. The XANES spectral features showed an intimate relation to the presence of V_O in their study. Similar results were observed during the investigation of the local structure of the ferromagnetic $Zn_{1-x}Co_xO$ (for $x = 0 : 02$, 0.04, and 0.06) high-quality epi-film by comparing the results of polarization-dependent X-ray absorption spectroscopy and *ab initio* calculations of the selected defect structures [114]. A clear ferromagnetic behavior [114] for all the samples is observed, with particularly high saturation magnetization for the $x = 0.04$ sample (about $0.44\,\mu_B/Co$). However, annealing the same sample ($x = 0.04$) in oxygen would completely remove the ferromagnetism. Considering the Co K-edge XANE spectra and the calculated energy of the several defect structures including V_O, Co–V_O complexes, Co pairs, and Co clusters, a specific Co–V_O defect complex is shown to be present in the samples. The observed correlation between the Co and V_O concentration and the RTFM indicated that V_O plays an active role in the origin of the observed magnetic moment.

The same research group then studied the influence of the surface structure on the ferromagnetic $Zn_{1-x}Co_xO$ wurtzite epilayers using coupling atomic force microscopy and advanced X-ray spectroscopy [121]. It was found that in high-quality epilayers, the formation of Co clusters and iso-space-group Co-rich region would take place at the sample surface with random Co distribution in the bulk. The surface modifications were not the origin of the magnetic properties of the material, and the ferromagnetism in the sample with the lowest concentration of background phases and the smoothest surface is enhanced. The possibility of ferromagnetic behavior originated due to trivial defects, and the poor quality of the samples mentioned in the previous reports [122] is totally nullified in this study. Instead, these results supported the idea of nontrivial intrinsic ferromagnetism due to the formation of cobalt–oxygen vacancy complexes.

Some findings [123] also support the role of V_O in mediating the RTFM in ZnO:Co system. The $Zn_{0.95}Co_{0.05}O$ films deposited on sapphire substrates by PLD are shown to be superparamagnetic and ferromagnetic. The ferromagnetism in these films was correlated to the substitutional Co ions instead of Co cluster or Zn-related defects and confirmed the important role of V_O in the Co–Co ferromagnetic coupling. The superparamagnetic behavior of these films was associated to the nanosize effect or the ferromagnetism due to the non-homogeneous V_O-induced nanoscale aggregation. A detailed analysis of the (Zn,Co)O system using advanced nanocharacterization tools, such as electron probe micro-analysis, XANES, extended X-ray absorption fine structure (EXAFS), high-resolution transmission electron microscopy (HR-TEM), and XMCD, is carried out [124]. The homogeneous and heterogeneous magnetism is observed in a series of (Zn,Co)O layers with different Co contents up to 40% grown by atomic layer deposition. The structures deposited at low temperature (433 K) showed spin-glass freezing for $x = 0.16$ and 0.4. This specific behavior was due to a random distribution of Co_{Zn} ions coupled by strong localized spins having short-range antiferromagnetic interactions. At a higher growth temperature (673 K), a superparamagnetic- and a ferromagnetic-like contributions originated, respectively, from the volume-dispersed nanocrystals and the nanocrystals residing at the (Zn,Co)O/substrate interface. The ferromagnetic behavior is associated with dipolar coupling within the interfacial two-dimensional dense dispersion of nanocrystals, but the discussion about the role of V_O in FM is missing in this study.

In the above-discussed scenario, it has become more and more evident that the presence or absence of ferromagnetism in Co:ZnO is far from being clearly understood. The origin and mechanism of the observed ferromagnetism remains controversial. The ambiguous picture regarding RTFM in these systems is mainly caused by the contradictory experimental and theoretical results. Although the ferromagnetic properties of Co:ZnO were observed, paramagnetic and antiferromagnetic behaviors have been demonstrated in Co:ZnO samples [125–127]. Numerous studies ascertain that the ferromagnetic properties originate during the formation of Co clusters or secondary phases; whereas others suggest that the observed FM is associated with superparamagnetism [128, 129]. Experimentally, it is very perplexing to rule out the formation of the ferromagnetic Co clusters or Co-containing secondary phases. The other important issue is the distribution of the dopants within the host material. It is reported that no ferromagnetism is expected above the temperature of ~10 K [130, 131] with a random distribution of TM dopants of concentration below 10% and in the absence of valence band holes. The nonrandom aggregation of the dopant promotes ferromagnetic-like response up to and above the room temperature [132–134] Therefore, ambiguities are associated with the high-temperature ferromagnetism in DMS irrespective of whether it is a real carrier-mediated effect or originated from highly nonrandom distribution of TM ions. These kinds of issues are creating hurdles in achieving the real ferromagnetic behavior in DMS and imposing constraints to the successful application of these materials in device applications. In order to have a true carrier-mediated RTFM to validate the practical use of these materials for the device applications, detailed systematic studies are required to provide the experimental evidence for the underlying mechanism in the Co:ZnO system. This would require the use of advanced characterization tools in order to study for how the magnetic impurities are actually incorporated and distributed depending on the growth conditions and co-doping. The other possibility is to employ such kind of dopants that have high solubility limits, have high magnetic moments, and at the room-temperature all its secondary phases are non-magnetic.

3.3. *Cu:ZnO system*

The Cu-doped ZnO material is also considered as a potential candidate to qualify for the real spin-based material after the observation of Anomalous Hall Effect (AHE) in n-type Cu-doped ZnO [86]. Recently, Cu-doped ZnO has got much attention due to its mutual ferromagnetic–ferroelectric properties [135] and bipolarity charge storage capabilities [136]. One of the principal motivation for selecting the Cu dopant in ZnO is that it can be helpful to fabricate an unambiguous DMS, since neither metallic Cu nor its oxides (CuO and Cu_2O) are ferromagnetic at room temperature. In addition, the size mismatch between Cu and Zn is very small, resulting in the low formation energy of Cu_{Zn} and the small lattice distortion [137]. The Cu-doped ZnO has been predicted to be ferromagnetic, and it develops as one of the major challenges to build a room-temperature DMS using nonmagnetic dopants. Indeed, both theoretical and experimental studies report the RTFM in Cu-doped ZnO films [78, 138]. In this section we will discuss the results of the previous studies conducted to explore the ferromagnetism and its potential origin in the Cu:ZnO system.

Although there is much hope for Cu:ZnO materials to act as the genuine DMS, similar to other TM-doped ZnO materials, there are also controversies related to the RTFM in Cu-doped ZnO. There are some reports showing the absence of ferromagnetism in Cu-doped ZnO films [139], as well as the lack of MCD signal in the Cu 3d and O 2p states even though the samples exhibited RTFM and were free of contamination [78]. Some researchers suggest that the presence of Cu^{2+} ions together with V_O are necessary for ferromagnetic order in Cu-doped ZnO films while some studies report that the Cu in ZnO is in the Cu^{1+} state [140–142] or mixed 2+/1+ oxidation state. It is also suggested that the ferromagnetism in the Cu-doped ZnO film is induced by the planar nanoscale Cu–O inclusions lying in the basal planes of the ZnO lattice [143]. Considerably, dissimilar RTFM strengths are reported with the same amount of Cu doping by different groups [86, 141, 143].

One of the possible reasons for this controversial data might be the mixed valence state of the Cu doped into the ZnO materials. The configuration of Cu^{3+} is analogous to $3d^84s^0$, while Cu^{2+} is corresponding to $3d^94s^0$. Both these configurations possess non-zero magnetic moments due to the unfilled d electrons. On the contrary, the Cu^{1+} and Cu^0 have the configuration of $3d^{10}4s^0$ and $3d^{10}4s^1$, respectively, and do not possess any local magnetic moment owing to the filled d electrons. The ZnO:Cu system containing $CuZn^+(Cu^{3+})$ or $CuZn^0$ (Cu^{2+}) is found to possess magnetic moments of $2\,\mu_B$/Cu and $1\,\mu_B$/Cu, respectively. In O-rich condition, when the doping concentration of Cu increases, substituting Cu^{2+} or Cu^{1+} assists the p-type defects and the Fermi level would be pinned around

Cu^{2+}/Cu^{1+} as a result of the overall charge neutrality [141, 144]. In this case, most of the Cu would be in the oxidation state of +2 or +1 in the ZnO:Cu material. The ratio of Cu^{2+}/Cu^{1+} is normally determined by the chemical potential of Cu and other growth conditions since the positively charged Zn_i and Cu_i defects would neutralize the negatively charged CuZn defects. Ferromagnetic coupling is favored between the substitutive Cu ions at +2 or +3 oxidation state, which are only available in the *p*-type samples. The Cu^{1+} and Cu^0 defects would form in *n*-type samples. Furthermore, clustering of Cu defects is observed to be dynamically promising for ferromagnetic ordering. Therefore, in the discussed scenario, the *n*-type ZnO:Cu would not show ferromagnetism, while *p*-type ZnO:Cu could have ferromagnetic property [145].

The other reason is the misunderstanding of the role of defects. The defects in the form of oxygen [146, 147] or cation [148] vacancies, interstitials [149] and strain [150] have drawn much attention toward the observed RTFM. Since super-exchange or double-exchange interactions cannot produce long-range magnetic order at the magnetic cation doping level of a few percent, both the V_O and V_{Zn} induce unoccupied states of primarily 2p–3d character [151] and generate charge carriers, which may be holes at the oxygen site [152, 153] or electrons at the Cu or Zn sites. The V_O and V_{Zn} are playing an essential role in the RTFM of Cu-doped ZnO materials. There are observations that the RTFM decreases with increasing copper doping, which may possibly be caused in part by the increased amount of CuO secondary phase [86, 141, 154, 155]. Using soft XMCD, direct evidence of the RTFM in O-deficient Cu-doped ZnO films is observed [140]. The double-exchange coupling between V_O and copper impurities proposes the origin of the RTFM in these films. Similarly, the XPS, XAS, and XMCD studies on 2.2% Cu-doped ZnO nanowires reveal that the ferromagnetic interaction resulted from the Cu^{2+} and Cu^{3+} states in the bulk part of the nanowire, suggesting that the hole doping plays an important role in the ferromagnetism since the Cu^{3+} state is a Cu^{2+} state plus an oxygen p hole [156].

The defects-enhanced FM is observed in semi-insulating Cu-doped ZnO thin films (with Cu conc. ~ 0.05 to 5 at.%) prepared by inductively coupled plasma-enhanced physical vapor deposition. The BMP [157] model explains the ferromagnetism in the defect-rich insulating oxides, suggesting that the carriers are loosely bound to the point defects (V_O) resulting in an extended orbital forming the BMP. The interactions between the Cu^{2+} ions through the BMP makes the Cu^{2+} ions spin in parallel with each other and leads to an effective ferromagnetic coupling among the Cu^{2+} magnetic moments. With the help of local structural and magnetic characterization, the sample having higher concentration of point defects shows the largest saturation magnetization [158]. In a similar study on the 2 at.% Cu-doped ZnO thin films prepared by reactive magnetron sputtering (RMS) and PLD, defect-enhanced ferromagnetism is observed. The valence states of the Cu ions in both films were in the 2+ state. In the film prepared by PLD, the V_O is neighboring around the Zn ions and Cu ions in the hexagonal wurtzite structure. Upon the annealing in oxygen, the V_O population reduces and the RTFM also decreases. On the contrary, for the film prepared by RMS, the V_O around the Cu ions is not detected, and the V_O population around the Zn ions is also smaller than that in the PLD-prepared film. Therefore, the ferromagnetic exchange in the RMS film is most likely mediated through cation vacancies, if point defects play a role in the RTFM [156]. In order to check for the effect of oxygen partial pressure during the growth and the possible role of V_O on the electric and magnetic properties of (Cu,Li) co-doped ZnO thin films, the $Zn_{0.989}Cu_{0.01}Li_{0.001}O$ thin films deposited at 500°C under different PO_2 ranging from 0.04 to 40 Pa are studied. Three conductivity regimes are reported for these films deposited under different PO_2. The n-type film with low resistivity is obtained when the PO_2 is maintained at 0.04 Pa during growth and the p-type films are achieved with PO_2 at 40 Pa. However, with PO_2 between 0.4 and 4 Pa, the insulating behavior of the films is observed. Only thin-film samples fabricated at 0.04Pa having the n-type nature showed the RTFM with moments $\sim 0.25_{\mu B}/Cu$ while the others grown at higher PO_2 were paramagnetic. The V_O

are speculated to play a crucial role for the observed ferromagnetic behavior in this study [159].

Recently, a model is proposed to verify the local ferromagnetic correlation between the Cu moments and the V_O in the Cu:ZnO nanoparticles [160]. According to this model, strong hybridization between the Cu 3d state and the valence band oxygen 2p state push the 3d energy levels inside the bandgap, as well as both the lower Hubbard band (LHB) and the upper Hubbard band (UHB) falling within the band gap. At low temperatures, the superexchange between the LHB electrons through the occupied V_O electrons prefers an antiferromagnetic correlation between the Cu moments near the V_O. With the increase in temperature, the number of promoted carriers in the UHB increased and the low-temperature local antiferromagnetic correlation transits to the higher temperature local ferromagnetic correlation. The large super-paramagnetic clusters of Cu atoms centered on V_O are considered essential to form super-moments and the nucleating centers of such moments are the V_O centers. In high-temperature annealed samples, the short-range ferromagnetic correlation between the Cu moments induced by the V_O center is destroyed due to migration of the V_O centers from the vacancy clusters toward the surface. The observed paramagnetic behavior supported the proposed model. The magnetic hysteresis (finite coercivity and low field saturation) and the irreversibility of temperature-dependent magnetization are observed to be destroyed upon thermal annealing in this study.

4. Anomalous Hall Effect in TM:ZnO System

The initial idea of TM-doped ZnO has some reservations whether the ferromagnetic behavior is intrinsic or extrinsic due to the presence of some extrinsic phases such as nanoscale inter-metallic clusters, magnetic contamination and magnetic secondary phase [161]. One of the most stimulating issues in this regard is misunderstanding of the carrier-mediated RTFM. In case of ZnO, on the one hand do we not have enough hole carriers owing to the *p*-type doping difficulty, and on the other hand, the *p*–*d* exchange interactions between the hole and the dopant d state are not as large as predicted by the mean-field theory. For the n-type ZnO case, the s–d exchange interactions between the electrons and the dopant d state is even weaker [162]. As far as defect-mediated ferromagnetism is concerned, controversies still exist [148, 162]. The problem in this scenario is the serious lack of utilization of the most refined methods to investigate the true picture of the material under observation are EXAFS, NEXAFS, XMCD, secondary ion mass spectrometry (SIMS) and AHE [161].

The problem is that employing all these techniques, especially the use of AHE measurement, is a pain-taking task and frequently skipped by researchers investigating RTFM observed in their samples. Therefore, we still have no factual picture about the origin of RTFM and whether the intrinsic or extrinsic defects are responsible for the RTFM. In order to gain a clear understanding of the charge transport mechanisms and have a full control over the magnetic properties of the ZnO-based DMS, it is of fundamental requirement to have spin-polarized charge carriers free from magnetic contamination and secondary phases. The superconducting quantum interference device (SQUID) measurement or magnetometry can only replicate the overall magnetic state but cannot be employed as a meaningful tool to diagnose whether the ferromagnetism is intrinsic or not. A key issue in understanding and exploiting the FM in TM-doped semiconductors materials is to elucidate the degree of magnetic ordering associated with the carriers, and one such probe is the AHE. The spin-polarized, carrier-mediated ferromagnetism can be best judged by observing the AHE signal [163, 164]. It is an important tool and generally considered as the final fingerprint for the presence of carrier-mediated FM and is regarded as an excellent substitute for M–T or M–H measurements [13].

4.1. *Theory of AHE*

In 1879, Hall [165] made the historic discovery in the antiquity of electricity and magnetism by observing that when a current-carrying conductor is placed in a magnetic field, the Lorentz force "presses" its electrons against one side of the conductor. This observation of solid-state transport experiment is recognized as the Hall effect in solid-state semiconductor physics and electronics and provides a simple sophisticated tool to measure the carrier concentration in non-magnetic conductors. At a later stage, Hall defined that this "pressing electricity" effect was 10 times stronger in ferromagnetic iron than in non-magnetic conductors [166]. This robust effect in ferromagnetic conductors is acknowledged as the anomalous Hall effect (AHE). In the ordinary Hall effect for nonmagnetic materials, the Lorentz force deflects the charges moving in a perpendicularly oriented magnetic field. This deflection generates an electric field transverse to the current. Typically, this effect is linear in field. However, in AHE of magnetic materials, the Hall voltage is a ferromagnetic response of the charge carriers in electrically conductive ferromagnets, which is proportional to the magnetization. The magnetic moment associated with the atoms gives rise to an additive term in the Hall equation [164, 167] as represented below:

$$\rho_{xy} = \frac{E_y}{J_x} = R_0 B + R_A \mu_0 M_S,$$

where ρ_{xy} is the Hall resistivity, E_y is the electric field perpendicular to the current and magnetic field, J_x is the current density, $R_0 = 1/\text{pe}$ is the ordinary Hall coefficient, μ_0 is the permeability of free space, M_S is the field-dependent spontaneous magnetization of the material, and R_A is the anomalous Hall coefficient arises from the spin–orbit interaction. R_A induces the anisotropy between the scattering of the spin-up and spin-down electrons, showing strong temperature dependence and is usually correlated with the electrical resistivity [168].

Presence of the AHE is generally attributed to the asymmetric scattering of the charge carriers by magnetic impurities where the spin–orbit interaction plays an important role. Generally, three mechanisms of scattering act to influence the electron motion in any of the real material, and these are known to be responsible for the AHE: (i) intrinsic deflection (interband coherence induced by an external electric field gives rise to a velocity contribution perpendicular to the field direction), (ii) the side jump (the electron velocity is deflected in opposite directions by the opposite electric fields experienced upon approaching and leaving an impurity, and (iii) the skew scattering (anisotropic amplitude of scattered wave packet in the presence of spin–orbit coupling of the electron or the impurity) [169].

4.2. *AHE in TM:ZnO*

The magnetotransport behavior can be associated to the magnetic scattering of spin-polarized charge carriers. The AHE is normally scaled by the positive and negative magnetoresistance (MR). The positive MR at low temperature is accredited to the spin-splitted conduction band due to the s–d exchange interaction. The negative MR at high field may originates from the magnetic scattering of electrons by isolated magnetic impurities or could be associated with BMP formation, which is not clearly understood and is still under debate [170, 171]. The Hall measurements are normally carried out in an applied magnetic field limit, where the magnetization is saturated (at low temperature and very high magnetic field). At low magnetic fields, the behavior of ρ_{xy} is dominated by the field dependence of M_S. Once the material's magnetization is saturated, the ρ_{xy} field dependence is linear due to the ordinary Hall effect [168]. One of the challenging tasks in the DMS is the successful detection of the AHE signal. Both high resistance of the DMS and diverse behavior of the carriers in the electric conduction, magnetic coupling and the coupling of the conduction electrons to the magnetic degrees of freedom add to the difficulty in measuring AHE signal.

In TM-doped ZnO films, when ZnO is doped in the metallic regime, it behaves as a genuine DMS by displaying clear RT AHE signal. However, in the semiconducting state of the TM:ZnO, no AHE signal was detected [101] due to possible noise present in the sample. In order to measure the weak AHE signal in the semiconducting regime of DMS, a noise reduction circuit combined with a digitizing measurement technique was employed in Co-doped ZnO study. With the utilization of this circuit, RT AHE signal is observed in the semiconducting state (with carrier concentration $\sim 10^{19}$ cm^{-3}) of the

Co-doped ZnO system [163]. In the case of Co:ZnO and Mn:ZnO thin films, the AHE is found favorable only in n-type Co-doped ZnO and is represented by pronounced s-shape Hall curve; while in n-type Mn-doped ZnO, no s-shape Hall curve is observed, suggesting the absence of AHE in this material [171]. The absence of AHE in Mn:ZnO is proposed due to the decrease of spin polarization with increased electron concentration. Both skew scattering ($\propto\rho_{xx}$) and side-jump mechanisms ($\propto\rho_{xx}^2$) are anticipated to contribute to the asymmetric scattering processes of AHE in the form of $\rho_{xy} \propto \rho_{xx}^{1.4}$ exponential dependency [170, 172].

The observation of carrier-mediated RTFM with low transition-metal doping or non-magnetic adatoms are of vital importance in DMS system. Experimental evidences reveal that the defects such as the vacancies with or without the presence of non-magnetic adatoms play an important role in triggering magnetic order. Defect-induced magnetism (DIM) seems to be of general consideration in nominally non-magnetic solids. The implantation of light non-magnetic elements such as hydrogen and lithium is an appropriate way to analyze the DIM in ZnO materials. The low-energy implantation is also a good way to suppress the other significant defects. Hydrogen implantation is a simpler and effective technique to produce magnetic order in ZnO with a strong reduction in the electrical resistance. Magnetotransport properties of H-implanted ZnO single crystals with different hydrogen concentrations (up to ~3 at.% in the first 20-nm surface layer between 10 and 300 K) in the atomic percentage range shows the ferromagnetic-like loops in all H-implanted ZnO samples at room temperature. All the experimental data specify that hydrogen atoms alone in the few percent range activate magnetic order in the ZnO crystalline state. In H-implanted ZnO single crystals, the similar AHE and SQUID hysteresis loops strongly suggest a spin-split band with a non-zero spin–orbit coupling by eliminating impurities as the origin of the observed ferromagnetic [173].

Thus, so for a given low doping level of spin-carrying TM ions, the above-mentioned results specify the strong support for the defect-mediated ferromagnetism in TM-doped ZnO materials. The experimental identification of the possible magnetic contamination and uneven spin distribution has been the extremely stimulating issue in DMS research. As seen in the above discussion, there are still notable variations in the reported magnetic behavior, with some films exhibiting only paramagnetism and even those with ferromagnetism showing a wide range of apparent Curie temperatures. Many arguments and mechanisms have been predicted for the intrinsic magnetic ordering. Apparently, no single model is capable of clarifying the magnetic properties of a wide class of DMS including ZnO. The most plausible reason of the controversy in the magnetic properties of ZnO-based DMS might have its roots in inadequate characterization of the samples and the lack of information regarding distribution of TM elements in the host ZnO [174]. In Section 5, we will discuss the prevailing characterizing techniques and their limitations in detecting the secondary phases.

5. Current Tools in Detecting True FM in TM:ZnO

To a certain extent, the characterization techniques are imperative in analyzing the DMS. It is really hard to analyze experimentally and theoretically the true nature of RTFM because so many factors affect the magnetic interaction, including the microstructure, local structure, electronic structure, coupling between localized magnetization, and trace amounts of TM dopant atoms in the host oxide films [174]. Therefore, the most critical issue for DMS is whether the resulting material truly contains evenly distributed TM elements or contains nanosize clusters, precipitates, or second phases that are liable for the observed magnetic properties.

Several methods have been employed to determine the local structure of the doping elements and to detect the secondary phases so as to identify whether the TM-doped ZnO is a genuine DMS or not. The major characterizing tools used for secondary phase identification are given as follows:

(1) The XRD measurement is a convenient tool to detect the potential impurity phases. Sometimes, the sensitivity may not be good enough to identify the trace amount of nanocrystals when they

are of very small sizes [100, 174]. Sometimes, coherent nanocrystals present in the host matrix are concealed under the high-resolution XRD due to the indistinguishable crystallographic structure and lattice constant to the surrounding matrix [175].

(2) More sophisticated characterization tools, such as SEM, TEM, and SIMS, are normally employed to distinguish the precipitates down to nanoscale and to probe the homogeneity of the dopant distribution in the host material [27]. The combined use of TEM (with appropriate mass and strain contrast) and EDX uncovers the aggregation of magnetic cations without distorting the host wurtzite structure under certain growth conditions [176]. Electron energy loss spectroscopy (EELS) combined with high-resolution transmission electron microscopy (HRTEM) [19, 177] and XPS [79] can be used to determine the precipitates, homogenous dopant distribution, and valence state.

(3) The EXAFS studies provide detailed microscopic structure of the lattice. The transition metal K-edge [49], L-edge [78], and O-K edge [178, 179] are usually monitored in the EXAFS studies. Simultaneous use of the EXAFS and NEXAFS techniques for a single system is a power tool to get information about the chemical bond between the host and the guest atom, the local lattice distortion and the chemical state of the atoms [79, 179].

(4) The XMCD and Optical MCD are considered as excellent tools to gauge the role of individual magnetic element in ferromagnetism, distribution of spin and orbital angular momenta and band structure of the TM-doped ZnO in order to identify whether the TM-doped ZnO is a real DMS [179, 180]. In the XMCD technique, the difference in the absorption of the left- and right-hand circularly polarized X-rays by a magnetized sample offers quantitative information about the distribution of spin and orbital angular momenta. On the contrary, optical MCD determines the relative difference in the circular polarization-dependent optical absorption under an applied magnetic field. The optical absorption of the circularly polarized light under magnetic field is the reflection of Zeeman splitting of the band structures and provides

information about the electronic structure and secondary phases by considering the strong sp–d exchange interactions and the subsequent Zeeman splitting of the optical transitions [181, 182]. The MCD spectra displaying a multiplet structure suggest that the ferromagnetic ordering originates from the TM ion substitution at Zn sites and not from the metallic TM secondary phases. The part of the MCD signal at H = 0 Oe is the ferromagnetic component, while linearly increased MCD signal with the increasing magnetic field (H) signifies the paramagnetic component [180, 183].

(5) The average atomic magnetic moment is an interesting parameter in assessing the magnetic properties. It is therefore important to determine the accurate TM content in samples. The most frequently employed instrument for the investigation of magnetization is SQUID magnetometer with a high sensitivity of 10^{-8} emu. However, trace ferromagnetic contamination of the substrate can lead to false magnetization signals of the samples, which may generate the misinterpretation of the results [174]. To avoid this kind of confusion and to testify the observed FM as a carrier-mediated mechanism, anomalous Hall effect (AHE) measurements are usually employed. However, problems in the detection of small AHE signal due to the high resistivity of DMS systems [163] and the observation of AHE in samples with magnetic clusters embedded in a non-magnetic matrix have raised some serious questions on the implementation of AHE as an authentic tool to confirm the carrier-mediated ferromagnetism [184, 185].

(6) Hall effect measurements are unable to provide sufficient information about the electrical transport properties when resistance is very high. In such a situation, ferroelectric testing and impedance spectroscopy are the best options to analyze the electrical transport properties [79, 177, 186]. Impedance spectroscopy (IS) is an informative and exceptional characterizing tool in fundamental and applied materials research. It can be used to resolve the contribution of different phases to the electrical properties such as contact effects, grains, grain boundaries, and any type of impurity inside a sample. Compared to other

techniques, IS explicitly distinguishes among strongly coupled processes having different proceeding rates as well as the concealed multiple phases having diverse conductivities even if the concentrations of phases are very low [186, 187].

5.1. *Spinodal decomposition in TM:ZnO*

In the above-discussed scenario, the AHE measurements are used to analyze the carrier-mediated ferromagnetism [13, 188] and XMCD establish a dominant element-specific tool for exploiting the inter-link between the ferromagnetism and crystallographic phase separation. However, they both cannot in general express whether the ferromagnetism originates from the diluted spins or is the consequence of the chemical phase separation [129]. Currently, it becomes increasingly clear owing to the use of advanced element-specific nanocharacterization tool that in some DMS, robust ferromagnetism also associates with the presence of nanoscale regions containing a large density of magnetic cations. This large density of magnetic cations forms the condensed magnetic semiconductors (CMSs) immersed in the host matrix. This chemically separated nanophase is known as spinodal decomposition in the literature [133, 175].

A good example of spinodal decomposition is the addition of Fe to (Ga, Fe)N. The incorporation of Fe leads to a broadening of the GaN-related diffraction maxima without revealing any secondary phases according to the standard laboratory high-resolution X-ray diffraction [61]. However, by the application of state-of-the-art element-specific much brighter synchrotron source as the nanocharacterization tool, the detection of chemically separated dominant ferromagnetic precipitate (Fe_3N) is possible, and in some cases, elemental Fe ferromagnetic nanocrystals are also visible [176, 189]. In another study on (Zn,Co)O thin films, a careful examination by HRXRD and HRTEM demonstrated a good wurtzite structure of the films without the evidences for precipitates, such as Co precipitates. However, the observed high-temperature ferromagnetism in these films was justified on the basis of the spinodal decomposition of Co-rich antiferromagnetic nanocrystals embedded in the Co-poor

(Zn,Co)O host. According to this model, the ferromagnetic signature originates from the uncompensated spins at the nanocrystal surface and could be visible below both the Neel temperature T_N and the blocking temperature T_B of the nanocrystals. Although these coherent nanocrystals exist in the host matrix, they normally remain undetected under standard high-resolution XRD or transmission electron microscopy (TEM) owing to the identical crystallographic structure and lattice constant to the surrounding (Zn,Co)O matrix [175]. Similarly, no precipitates of other crystallographic phases were detected and ferromagnetism was linked to the chemically separated wurtzite antiferromagnetic nanocrystals of CoO or Co-rich (Zn,Co)O [190]. Although CMS nanocrystals are reported for DMS, the independent existence of such nanocrystals and their properties are not yet included in the literature. It is still unknown whether they are metallic or insulating as well as whether they exhibit ferromagnetic, or antiferromagnetic spin order. However, their spin ordering temperature is expected to be relatively high, typically above room temperature [188].

The above-discussed characterizing tools with different physical bases have been recognized to be effective in detecting secondary phases. However, it is difficult to determine a trace amount of TM dopant using a random selection of the characterization techniques. To believe in reports of ferromagnetism based on magnetic hysteresis alone is not a fair approach. Prior to considering a true DMS, complete characterization of the sample would require a careful affiliation among the measured magnetic properties and materials analysis methods that are capable of sensing other phases and precipitates. A relatively complete characterization would involve magnetic hysteresis measurements as well as FC and ZFC magnetization measurements to rule out the possibility of small nanoparticles. Magnetotransport properties, XMCD and optical MCD can be used to scale the role of the individual magnetic element, distribution of spins, orbital angular momenta and band structure. The HRTEM, XPS, and EXAFS may be employed to investigate the secondary phase, chemical bonding, and local lattice environments [191].

5.2. *Electric field control of FM*

For a material to qualify as an intrinsic DMS for its successful implementation in spintronic devices, it should pass three basic tests: (i) observation of the AHE [86, 163, 171], (ii) it should possess optical-MCD signal [67, 107, 109], and most importantly (iii) it must show electric field control of magnetization [192]. Observation of AHE, although necessary, is not really a strict test and can be found even in a sample with magnetic clusters embedded in a paramagnetic host matrix [185]. The O-MCD test has been verified in different systems [67, 108, 109]. Therefore, the electric field control of magnetization is the most rigorous test of intrinsic FM [192]. The electrically controlled magnetism of a material is an attractive possibility and has attracted much attention in recent years. In magnetic semiconductors, the application of electric field can significantly modify the distribution profile of the carrier density due to the limited number of conducting electrons and results in the change of the magnetic properties.

Wang *et al.* [193] studied non-volatile electric control of magnetism in 2% Mn-substituted ZnO films sandwiched between two Pt electrodes grown on Al_2O_3 (0001) crystal substrates by using PLD technique. With the application of an electric field, bipolar resistive memory switching was induced in the film sandwiched between the two metallic electrodes. With forward threshold current, the system switches sharply from a low resistive state (LRS) to a high resistive state (HRS), and the hysteresis remains stable over repeated sweeps.

Switching the system in the low resistance state is corresponding to depleting the diode space–charge region from V_O, which are the dominant carriers. Since V_O mediate ferromagnetism in this compound, a non-volatile switching of the resistive state coexists with a non-volatile switching of the magnetic moment in the depletion layer of the diode. Therefore, the bistable switching of the resistive state is accompanied by a bistable switching of the magnetic moment.

In a study by Chang *et al.* [194] the electric field control of FM was demonstrated in a back-gated Mn-doped ZnO nanowire (NW) field effect transistor (FET). The ZnO NWs were synthesized by the thermal evaporation method, and the Mn doping of 1 at.% was subsequently carried out in an MBE system using a gas-phase surface diffusion process. The characteristic $I_{ds} - V_{gs}$ curve at $V_{ds} = 20\,mV$ shows that the majority carriers in Mn–ZnO NWs are n-type. The magnetic hysteresis curves measured under different temperatures ($T = 10$–$350\,K$) clearly reveal the presence of ferromagnetism above the room temperature, proposing that the quantum confinements in NWs can improve the T_c, and meanwhile minimize the crystalline defects. The T_C of the Mn–ZnO NWs was estimated to be about 437 K from the SQUID measurements. Most importantly, the gate modulation of the MR ratio is up to 2.5% at 1.9 K as the gate voltage changes from −40 to +40 V, which indicates the electric field control of ferromagnetism in a single Mn–ZnO NW. The successful demonstration of electrical modulation of ferromagnetism in DMS thin films and NW not only confirms a true DMS with carrier-induced ferromagnetism but also demonstrates a significant step toward its successful implementation in future spin logic devices.

In another study, Younas *et al.* [195] prepared cell structure from the $Ga_{0.01}Zn_{0.99}O$ and $Cu_{0.04}Zn_{0.96}O$ by employing a pulsed KrF excimer laser. First, a buffer layer of $Ga_{0.01}Zn_{0.99}O$ thin film ~250 nm was deposited on the (001) sapphire crystal substrate at 600°C temperature under $P(O_2) = 0.02\,Pa$. Subsequently, the $Cu_{0.04}Zn_{0.96}O$ layer with a thickness of ~250 nm was grown on top of the $Ga_{0.01}Zn_{0.99}O$/(001) sapphire through shadow-masking. Their results showed that the Cu multivalent (Cu^{M+}) ions modulate the magnetic and resistive states of the cells. The magnetic moment reduced by ~30% during the high-resistance state (HRS) to low-resistance state (LRS) switching. X-ray photoelectron spectroscopy results revealed an increase of the oxidation Cu^{+}/Cu^{2+} state ratio during the HRS to LRS transition, which decreases the effective spin-polarized $Cu^{2+} - V_O - Cu^{+}$ channels and thus the magnetic moment. A conduction mechanism involving the formation of conductive filaments from the coupling of the Cu^{M+} ions and V_O has been suggested.

Recently, in a similar study, Li *et al.* [196] prepared oxide thin films by employing the magnetron co-sputtering system in the ultra-high vacuum chamber. The bottom electrode of In:ZnO (In: 2%) films was deposited on c-plane sapphire substrate. Subsequently, the Co:ZnO (Co: 3%) thin films (~100 nm) have been deposited on In:ZnO/sapphire structures at 800°C. Further, they investigated the bipolar resistive switching (RS) behavior, and the intensity of the saturation magnetization in the device in HRS and LRS was observed to be the smallest and largest, respectively. X-ray photoelectron spectroscopy exhibited the presence of oxygen vacancies in the device, the concentration of which in LRS was larger than that in HRS, and a change in the conductive filament was consistent with the resistance states and FM. The results presented in their study showed that the RS effect and FM could be attributable to film defects.

5.3. *Ultrafast imaging tools for the future spintronic*

The currently available third-generation brighter synchrotron sources are the appropriate options to understand the local lattice distortion, chemical state of the atoms, and chemical phase separation. They provide an average photon flux per second originated from about 10^6 electron bunches each radiating about 10^6 photons with a pulse length of about 100 ps. With this intensity (about 10^{12} X-rays/s), a typical diffraction image can be recorded in 1 s. However, these sources are unable to provide information regarding the mechanism of the exchange interaction and angular momentum transfer between the injected spin-polarized itinerant electrons and the localized moments for magnetic switching. On the contrary, new X-ray sources such as the X-ray free electron laser (X-FEL) has a peak brightness factor of about 10^{10}, which is larger than that of the currently available third-generation X-ray sources. These sources will have 10^7 times larger flux per pulse and a 10^3 times shorter pulse length than the currently available sources. A soft X-FEL will emit about 10^{13} X-rays in a single ultrafast burst of about 100 fs and will be capable of recording the complete diffraction images with a single shot. The advanced X-FEL schemes will definitely go beyond the soft X-FEL by delivering even shorter pulses down to about 1 fs or even into the attosecond (10^{-18}s) range [197]. With the revolutionary advancements in these techniques, we will be able to have much higher resolutions down to 1 fs and can have a closer look at the ultimate problem in magnetism like the understanding of the exchange interaction itself. Third-generation storage rings typically produce X-ray pulses of about 50 ps length indicating great research opportunities to help explore and develop faster technologies [197].

6. Current Spintronics Device Concepts

The materials used for high-tech applications are concerned with the real origin of magnetic coupling and the true spin transport across the interfaces and the time dependence of magnetic reversal processes. The future of the magnetic storage and memory technology is related to the new materials patterning and self-assembling down to nanoscale with their magnetic stability at room temperature. The first generation of spintronics devices [197, 198] are based on the ferro- and antiferromagnetic materials [198]. Second-generation spintronic devices having a combination of magnetic materials and semiconductors are quickly gaining importance [8, 199]. In semiconductors, the long spin flip lifetime of nanoseconds or coherence length of 10 of microns are the best features for manipulating and detecting the spin states in small devices.

Therefore, application of spintronics devices demands both a clear understanding and control over the spin-dependent phenomena in semiconductor materials. Devices utilizing spin transport rather than charge transport have the advantages of faster data processing and operation at much less power consumption [8]. The understanding of the spintronics device applications highly depends on the ability to efficiently inject spin-polarized charge carriers into a conventional semiconductor. Electrical spin injection can be achieved either by a spin-polarized source or by spin-filtering unpolarized carriers at the interface. For a typical modern transistor, the spin injection and detection efficiencies have to be of the order of 99.9995% to achieve an on–off conductance ratio of 10^5 [200]. Regardless of the continued efforts, spin injection from a conventional ferromagnetic metal into a semiconductor has proved extremely inefficient. Therefore, experimentally, the desired target is still not achieved. The prime hindrance in this scenario is the failure to achieve high spin injection efficiency at the interface between the source and channel and high spin detection efficiency at the interface between the channel and drain, which are very important for spin FETs. The potential reasons behind the low spin injection and detection efficiencies may be the conductance mismatch and the high chemical incompatibility between the ferromagnetic metal and the semiconductor interface and also the scattering by the defects existing at the interface that can cause spin relaxations [201].

The possible solution to avoid these problems is the use of DMS materials. These materials having T_C above the room temperature would pave the way for triumphing spin polarization with high spin injection and detection efficiency with a long spin coherence needed for realizing the spintronic applications. The ZnO-based DMS is also favorable in realizing high spin-polarized injection efficiencies and carriers due to the high solubility of 3d transition metals (up to ~30% for Mn and Co). The simultaneous unique optical, electrical, and magnetic properties of ZnO could result in many novel spintronic devices with unprecedented functionality. There have been reports of DMS-based ferromagnetic material device structures such as the spin field effect transistors (SFETs) and photo-induced ferromagnets (PIF) [61, 202, 203]. The SFET suggested by Datta and Das [204] and subsequent numerous reports by other researchers have been the motivation behind the extensive studies that have been conducted so far on spin precision-controlled electronic semiconductor devices [200, 205, 206]. In Section 6.1, we will discuss some of the representative devices based on the spintronic concepts.

6.1. *Spin FET*

The most common example of a spintronic device structure is the SFET. The SFET is based on the idea first proposed by Datta and Das [204]. Their scheme shows a drain, a source, a narrow channel, and a gate for controlling the current. The gate allows the current to either flow through (ON): or not flow through (OFF). In the Datta–Das SFET device, carriers are injected into the channel from a spin-polarizing electrode (called polarizer or source), which can be either a ferromagnetic metal or a dilute magnetic semiconductor, through a channel to be collected by another magnetically polarized electrode (called analyzer or drain). The drain injects electrons ballistically through the channel with spins parallel to the transport direction. The current through the transistor depends on the relative orientation of the electron spins and the source and drain magnetic moments. The current is maximum when they are all aligned (ON): otherwise, it is scattered away (OFF) [199]. In an SFET, the application of a relatively low gate voltage causes an interaction between the electric field and the spin precession of the carriers via the so-called Rashba spin–orbit coupling effect [206–208]. This effective magnetic field causes the electron spins to precess. By adjusting the low applied voltage, one can generate the precession to lead to either parallel or antiparallel electron spin at the drain, which results in the effective control over the current. With sufficient control over the spin orientation, the current can be effectively shut off and on at much lower bias than needed in a charge-controlled FET.

The ZnO-based DMS heterostructures can also be used to fabricate the spin FETs with the potential to show a high spin injection efficiency and spin transport coherence. Sato and Katayama–Yoshida [209] suggested a ZnO-based SFET. They did first-principles calculations for the ground state of transition metal-doped ZnO and concluded that Mn:ZnO is of antiferromagnetic ground state, and ferromagnetism can be induced by hole doping. In this structure, the system changes to a half-metallic ferromagnet when holes are injected into Mn:ZnO by applying a negative bias gate voltage. Using ferromagnetic (Zn,Co)O as the source and drain contact material and by manipulating the relative orientation of the magnetization of the injector and detector with a different in-plane external magnetic field and gate voltage, the corresponding current in the detector can be changed due to 100% spin-polarized electron flow in the (Zn,Mn)O channel [210–212].

6.2. *Photo-induced ferromagnets*

The photo-induced ferromagnets idea is based upon the fact that ZnO:MnCr is a half-metallic ferromagnet upon hole doping while ZnO:FeMn is a half-metallic ferromagnet upon electron doping [210, 213]. For photons of suitable energy, electrons and holes created in the GaAs substrate near the interface with the ZnO:MnCr or ZnO:FeMn can transfer into these materials by biasing, initiating them to become half-metallic ferromagnets. The presence of these ordered states can be identified using the magneto-optical effect from another probe of photons beam with energy lower than the ZnO band gap.

6.3. *Spin LEDs*

Spin injection has been effectively verified in all-semiconductor tunnel diode structures by using a spin-polarized DMS as the injector in one case and a paramagnetic semiconductor as a spin filter in the other [188, 212]. In this type of scheme, spin-polarized holes and unpolarized electrons are injected from either side and recombined in a quantum well in zero magnetic field. The polarization of the injected holes can be confirmed by comparing the intensity of the right and left circularly polarized light in the electroluminescence (EL) spectra originated from the recombination of the holes with the injected (unpolarized) electrons. Among such devices, spin LED is the simplest concept of a light emitting diode (LED) with one of the contact layers as ferromagnetic. In this device, MBE-grown p-type Mn-doped GaAs is ferromagnetically ordered at low temperatures (52 K) and is separated from non-magnetic n-type GaAs layer [208]. Under forward bias, spin-polarized holes from (Ga, Mn)As and unpolarized electrons from the n-type GaAs substrate are injected into a non-magnetic (In,Ga)As quantum well [188], producing a polarized EL with the largest polarization at the quantum well (QW) ground state energy of 1.34 eV. The spin-polarized electrons have been detected in the form of circularly polarized light emitted from the QW. The presence of spin polarization has been established by the observation of hysteresis in the polarization of the emitted light as a function of the magnetic field. The emitted light with a specific helicity in order to conserve angular momentum in the recombination process also proves that the holes in the Mn-doped GaAs layer are spin polarized [27, 188].

Although the carried-mediated ferromagnetism in DMS is still not clear, the realization of the above-mentioned devices critically depends on the stability of carrier-mediated ferromagnetism. Provided that the carrier-mediated ferromagnetism in DMS is understood unambiguously, it will propose a range of new control not only over the semiconducting properties but also over magnetism leading to new opportunities for magneto-electronics devices.

7. Summary and Viewpoint

A series of accomplishments in this review have acknowledged a prominent role of DMS, especially TM:ZnO, in associating the science and technology of semiconductors and magnetic materials. A number of findings reviewed here documented the substantial progress in assessing the role played by the competition between ferromagnetic and antiferromagnetic interactions, solubility limits, nanosecondary phases, and the characterizing tools to deal with these issues. One of the striking significances of the solubility limits is a self-organized accumulating of magnetic nanocrystals inside a semiconductor host by the chemical or crystallographic phase separation. These heterogeneous magnetic systems have apparent T_C usually well above the room temperature. However, detailed structural, electronic, and magnetic analysis is still lacking to look for the possible spintronic functionalities of such nanocomposites [214].

From the theoretical point of view, rapid progress in the reliability of *ab initio* methods is expected when dealing with issues such as, the convergence as a function of the energy cutoff, the number of atoms in the supercells, and the density of k points. The future development of *ab initio* methods for the incorporation of the spin–orbit interaction that significantly affects the band structure [215] the developed treatment of the exchange-correlation functional to calculate more realistic values of the band-gap and d-level positions, proper dealing with Mott–Hubbard localization of d electrons [216, 217], and finally the improvement in handling the Coulomb gap in the density of states at the Fermi level (Anderson–Mott localization phenomena) will definitely help in achieving a better agreement between experiment and theory. On the experimental side, the use of advanced element-specific nanocharacterization tools such as XMCD, HRTEM, and high-energy synchrotron radiation will ascertain the more fundamental role of TM elements inside the host matrix. The simultaneous use of these nanocharacterization tools along with the implementation of the three basic tests (i) observation of the AHE (ii) existence of optical-MCD signal and (iii) electric field control of magnetization — will categorically assist to certify the intrinsic DMS and their successful application in spintronic devices.

In the light of this review, the search for a room-temperature uniform DMS will continue to be an active field of research and will constitute an important area for exploring novel phenomena, functionalities, and concepts at the intersection of semiconductor physics and magnetism. The striking properties and functionalities of DMS not only influence semiconductor physics but to a large extent would also extend the new horizons for a spread of spintronic research over many other materials families. In future, in addition to 3d TM impurities in various hosts like ZnO and GaN, other spin dopants will be considered, including 4d TMs and elements with open f shells as well as spin-carrying defects [214]. It might be, therefore, expected that studies of DMS will continue to convey surprising and stimulating discoveries in the upcoming years.

Acknowledgments

Younas would like to thank The University of Hong Kong and Pakistan Atomic Energy Commission for support during the writing of this review. He would also like to thank the Research Grant Council, HKSAR (GRF 17302115).

References

[1] R. Jansen, *Nature Mater.* **11**, 400 (2012).
[2] A. Zunger, S. Lany, H. Raebiger, *Physics* **3**, 53 (2010).
[3] M. L. Reed, N. A. El. Masry, H. H. Stadelmaier, M. K. Ritums, M. J. Reed, C. A. Parker, J. C. Roberts, S. M. Bedair, *Appl. Phys. Lett.* **79**, 3473 (2001).
[4] U. Ozgur, Ya. I. Alivov, C. Liu, A. Teke, M. A. Reshchikov, S. Dogan, V. Avrutin, S. J. Cho, H. Morkoc, *J. Appl. Phys.* **98**, 041301 (2005).
[5] J. Sinova, I. Zutic, *Nature Mater.* **11**, 368 (2012).
[6] G. Binasch, P. Grunberg, F. Saurenbach, W. Zinn, *Phys. Rev. B* **39**, 4828 (1989).
[7] M. N. Baibich, J. M. Broto, A. Fert, F. N. van Dau, F. Petroff, P. Eitenne, G. Creuzet, A. Friedrich, J. Chazelas, *Phys. Rev. Lett.* **61**, 2472 (1988).
[8] S. A. Wolf, D. D. Awschalom, R. A. Buhrman, J. M. Daughton, S. von Molnar, M. L. Roukes, A. Y. Chtchelkanova, D. M. Treger, *Science* **294**, 1488 (2001).
[9] T. Dietl, *Semicond. Sci. Technol.* **17**, 377 (2002).
[10] S. Dogan, V. Avrutin, S. J. Cho, H. Morkoç, *J. Appl. Phys.* **98**, 041301 (2005).
[11] S. Karamat, R. S. Rawat, T. L. Tan, P. Lee, S. V. Springham, Anis-ur-Rehman, R. Chen, H. D. Sun, *J. Supercond. Nov. Magn.* **26**, 187 (2013).
[12] S. J. Pearton, W. H. Heo, M. Ivill1, D. P. Norton, T Steiner, *Semicond. Sci. Technol.* **R59**, 19 (2004).
[13] H. Ohno, *Science* **281**, 951(1998).
[14] J. M. D. Coey, S. A. Chambers, *Meter. Res. Bull.* **33**, 1053 (2008).
[15] T. Dietl, H. Ohno, F. Matsukura, J. Cibert, D. Ferrand, *Science* **287**, 1019 (2000).
[16] R. N. Gurzhi, A. N. Kalinenko, A. I. Kopeliovich, A. V. Yanovsky, E. N. Bogachek, U. Landman, *Appl. Phys. Lett.* **83**, 4577 (2003).
[17] J. Rudolph, D. Hagele, H. M. Gibbs, G. Khitrova, M. Oestrich, *Appl. Phys. Lett.* **82**, 4516 (2003).
[18] M. Linnarsson, E. Janzen, B. Monemar, M. Kleverman, A. Thilderkvist, *Phys. Rev. B* **55**, 6938 (1997).
[19] P. Sharma, A. Gupta, K. V. Rao, J. F. Owens, R. Sharma, R. Ahuja, J. M. Osorio Guillen, B. Johansson, G. A. Gehring, *Nature Mater.* **2**, 673 (2003).
[20] K. Ueda, H. Tabata, T. Kawai, *Appl. Phys. Lett.* **79**, 988 (2001).
[21] S. Ramachandran, A. Tiwari, J. Narayan, *Appl. Phys. Lett.* **84**, 5255 (2004).
[22] H. T. Lin, T. S. Chin, J. C. Shih, S. H. Lin, T. M. Hong, R. T. Huang, F. R. Chen, J. J. Kai, *Appl. Phys. Lett.* **85**, 621 (2004).
[23] H. Saeki, H. Tabata, T. Kawai, *Solid State Commun.* **120**, 439 (2001).
[24] E. C. Stoner, *Proc. Roy. Soc. A* **165**, 372 (1938).
[25] K. Sato, L. Bergqvist, J. Kudrnovsky, P. H. Dederichs, O. Eriksson, I. Turek, B. Sanyal, G. Bouzerar, H. Katayama-Yoshida, V. A. Dinh, T. Fukushima, H. Kizaki, R. Zeller, *Rev. Modern Phys.* **82**, 1633 (2010).

[26] T. Jungwirth, W.A. Atkinson, B. Lee, A. H. Macdonald, *Phys. Rev. B* **59**, 9818 (1999).
[27] C. Liu, F. Yun, H. Morkoc, *J. Mater. Sci. Mater. El.* **16**, 555 (2005).
[28] P. Weiss, *J. de Physique* **6**, 661 (1907).
[29] W. Heisenberg, *Z. Phys.* **49**, 619 (1928).
[30] C. Zener, *Phys. Rev.* **81**, 440 (1951).
[31] C. Zener, *Phys. Rev* **82**, 403 (1951).
[32] M. A. Ruderman, C. Kittel, *Physical Review* **96**, 99–102 (1954).
[33] T. Kasuya, *Prog. Theor. Phys* **16**, 45 (1956).
[34] K. Yosida, *Phys. Rev.* **106**, 893 (1957).
[35] S. D. Sarma, E. H. Hwang, A. Kaminski, *Phys. Rev. B* **67**, 155201 (2003).
[36] S. O. Kucheyev, J. S. Williams, C. Jagadish, J. Zou, C. Evans, A. J. Nelson, A. V. Hamza, *Phys. Rev. B* **67**, 094115 (2003).
[37] T. Sakagami, M. Yamashita, T. Sekiguchi, S. Miyashita, K. Obara, T. Shishido, *J. Cryst. Growth* **229**, 98 (2001).
[38] K. Sato, H. Katayama-Yoshida, *Jpn. J. Appl. Phys.* **39**, L555 (2000).
[39] K. Sato, H. Katayama-Yoshida, *Semicond. Sci. Technol.* **17**, 367 (2002).
[40] Q. Wang, P. Jena, *Appl. Phys. Lett.* **84**, 4170 (2004).
[41] T. Fukumura, Z. Jin, M. Kawasaki, T. Shono, T. Hasegawa, S. Koshihara, H. Koinuma, *Appl. Phys. Lett.* **78**, 958 (2001).
[42] S. W. Yoon, S. B. Cho, S. C. We, S. Yoon, B. J. Suh, H. K. Song, Y. J. Shin, *J. Appl. Phys.* **93**, 7879 (2003).
[43] N. A. Spaldin, *Phys. Rev. B* **69**, 125201 (2004).
[44] Q. Wang, Q. Sun, B. K. Rao, P. Jena, *Phys. Rev. B* **69**, 233310 (2004).
[45] Q. Wang, Q. Sun, P. Jena, Y. Kawazoe, *Phys. Rev. B* **70**, 052408 (2004).
[46] C. H. Chien, S. H. Chiou, G. Y. Guo, Y. D. Yao, *J. Magn. Magn. Mater.* **282**, 275 (2004).
[47] E. C. Lee, K. J. Chang, *Phys. Rev. B* **69**, 085205 (2004).
[48] E. J. Kan, L. F. Yuan, J. Yang, *J. Appl. Phys.* **102**, 033915 (2007).
[49] W. Yan, Z. Sun, Q. Liu, Z. Li, Z. Pan, J. Wang, S. Wei, D. Wang, Y. Zhou, X. Zhang, *Appl. Phys. Lett.* **91**, 062113 (2007).
[50] D. Karmakar, S. K. Mandal, R. M. Kadam, P. L. Paulose, A. K. Rajarajan, T. K. Nath, A. K. Das, I. Dasgupta, G. P. Das, *Phys. Rev. B* **75**, 144404 (2007).
[51] D. Ius, B. Sanyal, O. Eriksson, *J. Appl. Phys.* **101**, 09H101 (2007).
[52] P. Gopal, N. A. Spaldin, *Phys. Rev. B* 74, 094418 (2006).
[53] M. H. F. Sluiter, Y. Kawazoe, P. Sharma, A. Inoue, A. R. Raju, C. Rout, U.V. Waghmare, *Phys. Rev. Lett.* **94**, 187204 (2005).
[54] Y. H. Lin, M. Ying, M. Li, X. Wang, C. W. Nan, *Appl. Phys. Lett.* **90**, 222110 (2007).
[55] A. Chakrabarty, C. H. Patterson, *Phys. Rev. B* **84**, 054441 (2011).
[56] T. Zhang, L. X. Song, Z. Z. Chen, E. W. Shi, L. X. Chao, H. W. Zhang, *Appl. Phys. Lett.* **89**, 172502 (2006).
[57] L. Hu, L. Zhu, H. He, L. Zhang, Z. Ye, *J. Mater. Chem. C* **00**, 1–15 (2014).

[58] Q. Xu, H. Schmidt, L. Hartmann, H. Hochmuth, M. Lorenz, A. Setzer, P. Esquinazi, C. Meinecke, M. Grundmann, *Appl. Phys. Lett.* **91**, 092503 (2007).
[59] C. D. Pemmaraju, T. Archer, R. Hanafin, S. Sanvito, *J. Magn. Magn. Mater.* **316**, 185 (2007).
[60] A. L. Rosa, R. Ahuja, *J. Phys.: Condens. Matter* **19**, 386232 (2007).
[61] S. J. Pearton, C. R. Abernathy, M. E. Overberg, G. T. Thaler, D. P. Norton, N. Theodoropoulou, A. F. Hebard, Y. D. Park, F. Ren, J. Kim, L. A. Boatner, *J. Appl. Phys.* **93**, 1 (2003).
[62] W. Prellier, A. Fouchet, B. Mercey, *J. Phys.: Condens. Matter.* **15**, R1583 (2003).
[63] S. A. Chambers, *Surf. Sci. Rep.* **61**, 345 (2006).
[64] J. C. A. Huang, H. S. Hsu, *Appl. Phys. Lett.* **87**, 132503 (2005).
[65] K. P. Bhatti, S. Kundu, S. Chaudhary, S. C. Kashyap, D. K. Pandya, *J. Phys. D: Appl. Phys.* **39**, 4909 (2006).
[66] C. Kittel, *Introduction to Solid State Physics*, (John Wiley & Sons, New York, 1996).
[67] K. Ando, H. Saito, Z. Jin, T. Fukumura, M. Kawasaki, Y. Matsumoto, H. Koinuma, *J. Appl. Phys.* **89**, 7284 (2001).
[68] Z. Jin, T. Fukumura, M. Kawasaki, K. Ando, H. Saito, T. Sekiguchi, Y. Z. Yoo, M. Murakami, Y. Matsumoto, T. Hasegawa, H. Koinuma, *Appl. Phys. Lett.* **78**, 3824 (2001).
[69] T. Fukumura, Z. Jin, A. Ohtomo, H. Koinuma, M. Kawasaki, *Appl. Phys. Lett.* **75**, 3366 (1999).
[70] S. W. Jung, S. J. An, G. C. Yi, C. U. Jung, S. I. Lee, S. Cho, *Appl. Phys. Lett.* **80**, 4561 (2002).
[71] D. P. Norton, S. J. Pearton, A. F. Hebard, N. Theodoropoulou, L. A. Boatner, R. G. Wilson, *Appl. Phys. Lett.* **82**, 239 (2003).
[72] Y. M. Kim, M. Yoon, I. W. Park, Y. J. Park, J. H. Lyou, *Solid State Commun.* **129**, 175 (2004).
[73] S. Ramachandran, A. Tiwari, J. Narayan, J. T. Prater, *Appl. Phys. Lett.* **87**, 172502 (2005).
[74] T. Wakano, N. Fujimura, Y. Morinaga, N. Abe, A. Ashida, T. Ito, *Physica E* **10**, 260 (2001).
[75] H. Schmidt, M. Diaconu, H. Hochmuth, M. Lorenz, A. Setzer, P. Esquinazi, A. Poppl, D. Spemann, K. W. Nielsen, R. Gross, G. Wagner, M. Grundmann, *Superlattices Microstruct.* **39**, 334 (2006).
[76] S. Venkataraj, N. Ohashi, I. Sakaguchi, Y. Adachi, T. Ohgaki, H. Ryoken, H. Haneda, *J. Appl. Phys.* **102**, 014905 (2007).
[77] J. H. Park, M. G. Kim, H. M. Jang, S. Ryu, Y. M. Kim, *Appl. Phys. Lett.* **84**, 1338 (2004).
[78] D. J. Keavney, D. B. Buchholz, Q. Ma, R. P. H. Chang, *Appl. Phys. Lett.* **91**, 012501 (2007).
[79] C. Song, K. W. Geng, F. Zeng, X. B. Wang, Y. X. Shen, F. Pan, Y. N. Xie, T. Liu, H. T. Zhou, Z. Fan, *Phys. Rev. B* **73**, 024405 (2006).

[80] S. K. Mandal, A. K. Das, T. K. Nath, D. Karmakar, *Appl. Phys. Lett.* **89**, 144105 (2006).
[81] Y. B. Zhang, Q. Liu, T. Sritharan, C. L. Gan, S. Li, *Appl. Phys. Lett.* **89**, 042510 (2006).
[82] S. G. Yang, A. B. Pakhomov, S. T. Hung, C. Y. Wong, *IEEE Trans. Magn.* **38**, 2877 (2002).
[83] X. H. Xu, H. J. Blythe, M. Ziese, A. J. Behan, J. R. Neal, A. Mokhtari, R. M. Ibrahim, A. M. Fox, G. A. Gehring, *New J. Phys.* **8**, 135 (2006).
[84] X. Liu, F. Lin, L. Sun, W. Cheng, X. Ma, W. Shi, *Appl. Phys. Lett.* **88**, 062508 (2006).
[85] D. B. Buchholz, R. P. H. Chang, J. H. Song, J. B. Ketterson, *Appl. Phys. Lett.* **87**, 082504 (2005).
[86] D. L. Hou, X. J. Ye, H. J. Meng, H. J. Zhou, X. L. Li, C.M. Zhen, G.D. Tang, *Appl. Phys. Lett.* **90**, 142502 (2007).
[87] M. A. Herman, H. Sitter, *Molecular Beam Epitaxy: Fundamentals and Current Status*, 2nd edn. (Springer, Berlin, 1996).
[88] G. B. Stringfellow, *Organometallic Vapor-Phase Epitaxy: Theory and Practice*, 2nd edn. (Academic, London, 1999).
[89] D. B. Chrisey, G. K. Hubler, *Pulsed Laser Deposition of Thin Films.* Wiley, New York (1994).
[90] C. Song, F. Zeng, K. W. Geng, X. J. Liu, F. Pan, B. He, W. S. Yan, *Phys. Rev. B* **76**, 045215 (2007).
[91] S. J. Lee, H. S. Lee, D. Y. Kim, T. W. Kim, *J. Cryst. Growth* **276**, 121 (2005).
[92] S. J. Han, J. W. Song, C. H. Yang, S. H. Park, J. H. Park, Y. H. Jeong, K. W. Rhie, *Appl. Phys. Lett.* **81**, 4212 (2002).
[93] D. P. Norton, M. E. Overberg, S. J. Pearton, K. Pruessner, J. D. Budai, L. A. Boatner, M. F. Chisholm, J. S. Lee, Z. G. Khim, Y. D. Park, R. G. Wilson, *Appl. Phys. Lett.* **83**, 5488 (2003).
[94] J. H. Kim, H. Kim, D. Kim, Y. E. Ihm, W. K. Choo, *J. Appl. Phys.* **92**, 6066 (2002).
[95] J. P. Fan, X. Li. Li, Z. Y. Quan, X. H. Xu, *Appl. Phys. Lett.* **102**, 102407 (2013).
[96] X. L. Wang, C. Y. Luan, Q. Shao, A. Pruna, C. W. Leung, R. Lortz, J. A. Zapien, A. Ruotolo, *Appl. Phys. Lett.* **102**, 102112 (2013).
[97] W. Liu, X. Tang, Z. Tang, *J. Appl. Phys.* **114**, 123911 (2013).
[98] U. Ilyas, T. L. Tan, P. Lee, R. V. Ramanujan, F. Li, S. Zhang, R. Chen, H. D. Sun, R. S. Rawat, *J. Magn. Magn. Mater.* **344**, 171 (2013).
[99] T. Dietl, *Nat. Mater.* **2**, 646 (2003).
[100] J. M. D. Coey, Curr. Opin. *Solid State Mater. Sci.* **10**, 83 (2006).
[101] A. J. Behan, A. Mokhtari, H. J. Blythe, D. Score, X. H. Xu, J. R. Neal, A. M. Fox, G. A. Gehring, *Phys. Rev. Lett.* **100**, 047206 (2008).
[102] D. Rubi, J. Fontcuberta, *Phys. Rev. B* **75**, 155322 (2007).

[103] S. Lee, Y. C. Cho, S. J. Kim, C. R. Cho, S.Y. Jeong, S. J. Kim, J. P. Kim, Y. N. Choi, J. M. Sur, *Appl. Phys. Lett.* **94**, 212507 (2009).
[104] S. Kuroda, N. Nishizawa, K. Ki Takita, M. Mitome, Y. Bando, K. Osuch, T. Dietl, *Nature Mater.* **6**, 440 (2007).
[105] S. Yin, M. X. Xu, L. Yang, J. F. Liu, H. Rösner, H. Hahn, H. Gleiter, D. Schild, S. Doyle, T. Liu, T. D. Hu, E. T. Muromachi, J. Z. Jiang1, *Phys. Rev. B* **73**, 224408 (2006).
[106] T. C. Kaspar, T. Droubay, S. M. Heald, P. Nachimuthu1, C. M. Wang, V. Shutthanandan, C. A. Johnson, D. R. Gamelin, S. A. Chambers, *New J. Phys.* **10**, 055010 (2008).
[107] A. Ney, K. Ollefs, S. Ye, T. Kammermeier, V. Ney, T. C. Kaspar, S. A. Chambers, F. Wilhelm, A. Rogalev, *Phys. Rev. Lett.* **100**, 157201 (2008).
[108] A. Ney, T. Kammermeier, K. Ollefs, S. Ye, V. Ney, T. C. Kaspar, S. A. Chambers, F. Wilhelm, A. Rogalev, *Phys. Rev. B* **81**, 054420 (2010).
[109] A. Ney, M. Opel, T. C. Kaspar, V. Ney, S. Ye, K. Ollefs, T. Kammermeier, S. Bauer, K. W. Nielsen, S. T. B. Goennenwein, M. H. Engelhard, S. Zhou, K. Potzger, J. Simon, W. Mader, S. M. Heald, J. C. Cezar, F. Wilhelm, A. Rogalev, R. Gross, S. A. Chambers, *New J. Phys.* **12**, 013020 (2010).
[110] Q. Y. Xu, S. Q. Zhou, D. Mark'o, K. Potzger, J. Fassbender, M. Vinnichenko, M. Helm, H. Hochmuth, M. Lorenz, M. Grundmann, H. Schmidt, *J. Phys. D: Appl. Phys.* **42**, 085001 (2009).
[111] S. M. Heald, A. Mokhtari, A. J. Behan, H. J. Blythe, J. R. Neal, A. M. Fox, G. A. Gehring, *Phys. Rev. B* **79**, 075202 (2009).
[112] A. Janotti, C. G. Van de Walle, *Rep. Prog. Phys.* **72**, 126501 (2009).
[113] C. D. Pemmaraju, R. Hanafin, T. Archer, H. B. Braun, S. Sanvito, *Phys. Rev. B* **78**, 054428 (2008).
[114] G. Ciatto, A. Di Trolio, E. Fonda, P. Alippi, A. M. Testa, A. A. Bonapasta, *Phys. Rev. Lett.* **107**, 127206 (2011).
[115] E. Z. Liu, J. Z. Jiang, *J. Appl. Phys.* **107**, 023909 (2010).
[116] M. H. N. Assadi, Y. B. Zhang, S. Li, *Appl. Phys. Lett.* **95**, 072503 (2009).
[117] A. Dinia, G. Schmerber, C. Mény, V. P. Bohnes, E. Beaurepaire, *J. Appl. Phys.* **97**, 123908 (2005).
[118] Q. Wang, Q. Sun, P. Jena, *J. Phys.: Condens. Matter* **22**, 076002 (2010).
[119] C. H. Patterson, *Phys. Rev. B* **74**, 144432 (2006).
[120] W. Yan, Q. Jiang, Z. Sun, T. Yao, F. Hu, S. Wei, *J. Appl. Phys.* **108**, 013901 (2010).
[121] G. Ciatto, A. Di Trolio, E. Fonda, L. Amidani, F. Boscherini, M. Thomasset, P. Alippi, A. A. Bonapasta, *Appl. Phys. Lett.* **101**, 252101 (2012).
[122] A. Ney, A. Kovacs, V. Ney, S. Ye, K. Ollefs, T. Kammermeier, F. Wilhelm, A. Rogalev, R. E. D. Borkowski, *New J. Phys.* **13**, 103001 (2011).
[123] Q. Li, Y. Wang, L. Fan, J. Liu, W. Konga, B. Yea, *Scripta Mater.* **69**, 694 (2013).
[124] M. Sawicki, E. Guziewicz, M. I. Łukasiewicz, O. Proselkov, I. A. Kowalik, W. Lisowski, P. Dluzewski, A. Wittlin, M. Jaworski, A. Wolska,

W. Paszkowicz, R. Jakiela, B. S. Witkowski, L. Wachnicki, M. T. Klepka, F. J. Luque, D. Arvanitis, J. W. Sobczak, M. Krawczyk, A. Jablonski, W. Stefanowicz, D. Sztenkiel, M. Godlewski, T. Dietl, *Phys. Rev. B.* **88**, 085204 (2013).

[125] P. Sati, C. Deparis, C. Morhain, S. Schafer, A. Stepanov, *Phys. Rev. Lett.* **98**, 37204 (2007).

[126] A. Ney, V. Ney, F. Wilhelm, A. Rogalev, K. Usadel, *Phys. Rev. B* **85**, 245202 (2012).

[127] T. Shi, S. Zhu, Z. Sun, S. Wei, W. Liu, *Appl. Phys. Lett.* **90** 102108 (2007).

[128] T. C. Kaspar, T. Droubay, S. M. Heald, M. H. Engelhard, P. Nachimuthu, S. A. Chambers, *Phys. Rev. B* **77**, 201303(R) (2008).

[129] T. Dietl, *Nat. Mater.* **9**, 965 (2010).

[130] M. Sawicki, T. Devillers, S. Gałęski, C. Simserides, S. Dobkowska, B. Faina, A. Grois, A. Navarro-Quezada, K. N. Trohidou, J. A. Majewski, T. Dietl, A. Bonanni, *Phys. Rev. B* **85**, 205204 (2012).

[131] J. M. D. Coey, P. Stamenov, R. D. Gunning, M. Venkatesan, K. Paul, *New J. Phys.* **12**, 053025 (2010).

[132] A. N. Quezada, N. G. Szwacki, W. Stefanowicz, T. Li, A. Grois, T. Devillers, M. Rovezzi, R. Jakieła, B. Faina, J. A. Majewski, M. Sawicki, T. Dietl, A. Bonanni, *Phys. Rev. B* **84**, 155321 (2011).

[133] A. V. Panov, *Appl. Phys. Lett.* **100**, 052406 (2012),

[134] A. Bonanni and T. Dietl, *Chem. Soc. Rev.* **39**, 528 (2010).

[135] T. S. Herng, M. F. Wong, D. Qi, J. Yi, A. Kumar, A. Huang, F. C. Kartawidjaja, S. Smadici, P. Abbamonte, C. S. Hanke, S. Shannigrahi, J. M. Xue, J. Wang, Y. P. Feng, A. Rusydi, K. Zeng, J. Ding, *Adv. Mater.* **23**, 1635 (2011).

[136] A. Kumar, T. S. Herng, K. Zeng and Ding, *J. Appl. Mater. Interfaces* **4**, 5276 (2012).

[137] H. Liu, F. Zeng, S. Gao, G. Wang, C. Song, F. Pan, *Phys. Chem. Chem. Phys.* **15**, 13153 (2013).

[138] Z. A. Khan, S. Ghosh, *Appl. Phys. Lett.* **99**, 042504 (2011).

[139] M. Venkatesan, C. B. Fitzgerald, J. G. Lunney, J. M. D. Coey, *Phys. Rev. Lett.* **93**, 177206 (2004).

[140] T. Herng, D. C. Qi, T. Berlijn, J. Yi, K. Yang, Y. Dai, Y. Feng, I. Santoso, C. S. Hanke, X. Gao, A. Wee, W. Ku, J. Ding, A. Rusydi, *Phys. Rev. Lett.* **105**, 207201 (2010).

[141] Q. Ma, D. Buchholz, R. P. H. Chang, *Phys. Rev. B* **78**, 214429 (2008).

[142] J. Huang, L. Zhu, L. Hu, S. Liu, J. Zhang, H. Zhang, X. Yang, L. Sun, D. Li, Z. Ye, *Nanoscale* **4**, 1627 (2012).

[143] C. Sudakar, J. S. Thakur, G. Lawes, R. Naik, and V. M. Naik, *Phys. Rev. B* **75**, 054423 (2007).

[144] Y. J. Zhao, C. Persson, S. Lany, A. Zunger, *Appl. Phys. Lett.* **85**, 5860 (2004).

[145] D. Huang, Y. J. Zhao, D. H. Chen, Y. Z. Shao, *Appl. Phy. Lett.* **92**, 182509 (2008).

[146] R. Hanafin, S. Sanvito, *J. Magn. Magn. Mater.* **322**, 1209 (2010).
[147] M. H. N. Assadi, Y. B. Zhang, S. Li, *J. Phys.: Condens. Matter* **22**, 486003 (2010).
[148] J. B. Yi, C. C. Lim, G. Z. Xing, H. M. Fan, L. H. Van, S. L. Huang, K. S. Yang, X. L. Huang, X. B. Qin, B. Y. Wang, T. Wu, L. Wang, H. T. Zhang, X. Y. Gao, T. Liu, A. T. S. Wee, Y. P. Feng, J. Ding, *Phys. Rev. Lett.* **104**, 137201 (2010).
[149] T. F. Shi, Z. G. Xiao, Z. J. Yin, X. H. Li, Y. Q. Wang, H. T. He, J. N. Wang, W. S. Yan, S. Q. Wei, *Appl. Phys. Lett.* **96**, 211905 (2010).
[150] V. K. Sharma, G. D. Varma, *J. Phys.: Condens. Matter* **21**, 296001 (2009).
[151] P. Thakur, V. Bisogni, J. C. Cezar, N. B. Brookes, G. Ghiringhelli, S. Gautam, K. H. Chae, M. Subramanian, R. Jayavel, K. Asokan, *J. Appl. Phys.* **107**, 103915 (2010).
[152] J. M. Chen, P. Nachimuthu, R. S. Liu, S. T. Lees, K. E. Gibbons, I. Gameson, M. O. Jones, P. P. Edwards, *Phys. Rev. B* **60**, 6888 (1999).
[153] K. Asokan, J. C. Jan, K. V. R. Rao, J. W. Chiou, H. M. Tsail, S. Mookerjee, W. F. Pong, M. H. Tsai, R. Kumar, S. Husain, J. P. Srivastava, *J. Phys.: Condens. Matter* **16**, 3791 (2004).
[154] A. Tiwari, M. Snure, D. Kumar, J. T. Abiade, *Appl. Phys. Lett.* **92**, 062509 (2008).
[155] K. Samanta, P. Bhattacharya, R. S. Katiyar, *J. Appl. Phys.* **105**, 113929 (2009).
[156] Q. Ma, J. T. Prater, C. Sudakar, R. A. Rosenberg, J. Narayan, *J. Phys.: Condens. Matter* **24**, 306002 (2012).
[157] J. M. D. Coey, M. Venkatesan, C. B. Fitzgerald, *Nat. Mater.* **4**, 173 (2005).
[158] S. Zhuo, X. C. Liu, Z. Xiong, J. H. Yanga, E. W. Shi, *Solid State Commun.* **152**, 257 (2012).
[159] L. Zhang, B. Lu, Z. Ye, J. Lu, J. Huang, *Solid State Commun.* **170**, 53 (2013).
[160] B. Ghosh, M. Sardar, S. Banerjee, *J. Phys. D: Appl. Phys.* **46**, 135001 (2013).
[161] C. Scott, *Nature Mater.* **9**, 956 (2010).
[162] Q. Wang, Q. Sun, G. Chen, Y. Kawazoe, P. Jena, *Phys. Rev. B* **77**, 205411 (2008).
[163] H. S. Hsu, C. P. Lin, H. Chou, J. C. A. Huang, *Appl. Phys. Lett.* **93**, 142507 (2008).
[164] C. L. Chien, C. R. Westgate, *Hall Effect and Its Applications*, (Plenum, New York, 1980).
[165] E. H. Hall, *Am. J. Math.* **2**, 287 (1879).
[166] E. H. Hall, *Philos. Mag.* **12**, 157 (1881).
[167] C. M. Hurd, *The Hall Effect in Metals and Alloys* (Plenum, New York, 1972).
[168] J. S. Higgins, S. R. Shinde, S. B. Ogale, T. Venkatesan, R. L. Greene, *Phys. Rev. B* **69**, 073201 (2004).

[169] N. Nagaosa, J. Sinova, S. Onoda, A. H. MacDonald, N. P. Ong, *Rev. Mod. Phys.* **82**, 1539 (2010).
[170] Q. Xu, L. Hartmann, H. Schmidt, H. Hochmuth, M. Lorenz, R. S. Grund, C. Sturm, D. Spemann, M. Grundmann, *Phys. Rev. B* **73**, 205342 (2006).
[171] Q. Xu, L. Hartmann, H. Schmidt, H. Hochmuth, M. Lorenz, R. Schmidt-Grund, C. Sturm, D. Spemann, M. Grundmann, *J. Appl. Phys.* **101**, 063918 (2007)
[172] T. Fukumura, H. Toyosaki, Y. Yamada, *Semicond. Sci. Technol.* **20**, S103 (2005)
[173] M. Khalid, P. Esquinazi, *Phys. Rev. B* **85**, 134424 (2012).
[174] F. Pan, C. Song, X. J. Liu, Y. C. Yang, F. Zeng, *Mater. Sci. Eng. R* **62**, 1 (2008).
[175] T. Dietl, T. Andrearczyk, A. Lipińska, M. Kiecana, M. Tay, Y. Wu, *Phys. Rev. B* **76**, 155312 (2007).
[176] A. Bonanni, A. Navarro-Quezada, T. Li, M. Wegscheider, Z. Matej, V. Holy, R. T. Lechner, G. Bauer, M. Rovezzi, F. D. Acapito, M. Kiecana, M. Sawicki, T. Dietl, *Phys. Rev. Lett.* **101**, 135502 (2008).
[177] C. Song, F. Zeng, K. W. Geng, X. B. Wang, Y. X. Shen, F. Pan, *J. Magn. Magn. Mater.* **309**, 25 (2007).
[178] P. Thakur, K. H. Chae, J.-Y. Kim, M. Subramanian, R. Jayavel, K. Asokan, *Appl. Phys. Lett.* **91**, 162503 (2007).
[179] A. L. Ankudinov, B. Ravel, J. J. Rehr, S. D. Conradson, *Phys. Rev. B* **58**, 7565 (1998).
[180] M. Kobayashi, Y. Ishida, J. l. Hwang, T. Mizokawa, A. Fujimori, K. Mamiya, J. Okamoto, Y. Takeda, T. Okane, Y. Saitoh, Y. Muramatsu, A. Tanaka, H. Saeki, H. Tabata, T. Kawai, *Phys. Rev. B* **72**, 201201(R) (2005).
[181] K. Ando, *Science* **312**, 1883 (2006).
[182] K. Ando, A. Chiba, H. Tanoue, *J. Appl. Phys.* **83**, 6545 (1998).
[183] J. R. Neal, A. J. Behan, R. M. Ibrahim, H. J. Blythe, M. Ziese, A. M. Fox, G. A. Gehring, *Phys. Rev. Lett.* **96**, 197608 (2006).
[184] B. Raquet, M. Goiran, N. Negre, J. Leotin, B. Aronzon, V. Rylkov, E. Meilikhov, *Phys. Rev. B* **62**, 17144 (2000).
[185] S. R. Shinde, S. B. Ogale, J. S. Higgins, H. Zheng, A. J. Millis, V. N. Kulkarni, R. Ramesh, R. L. Greene, T. Venkatesan, *Phys. Rev. Lett.* **92** 166601 (2004).
[186] H. S. Hsu, J. C. A. Huang, S. F. Chen, C. P. Liu, *Appl. Phys. Lett.* **90**, 102506 (2007).
[187] J. R. Macdonald, *Impedance Spectroscopy*, (Wiley, New York, 1987).
[188] Y. Ohno, D. K. Young, B. Beschoten, F. Matsukura, H. Ohno, D. D. Awschalom, *Nature* **402**, 790 (1999).
[189] M. Rovezzi, F. D. Acapito, A. Navarro-Quezada, B. Faina, T. Li, A. Bonanni, F. Filippone, A. A. Bonapasta, T. Dietl, *Phys. Rev. B* **79**, 195209 (2009).

[190] T. Dietl, D. Awschalom, M. Kaminska, H. Ohno, *Spintronics, Semiconductors and Semimetals*, vol. 82, (Elsevier, Amsterdam, 2008), p. 371.
[191] S. J. Pearton, D. P. Norton, M. P. Ivill, A. F. Hebard, J. M. Zavada, W. M. Chen, I. A. Buyanova, *IEEE, Trans. Elec. Devi.* **54**, 1040 (2007).
[192] T. Zhao, S. R. Shinde, S. B. Ogale, H. Zheng, T. Venkatesan, R. Ramesh, S. D. Sarma, *Phys. Rev. Lett.* **94**, 126601 (2005).
[193] X. L. Wang, Q. Shao, C. W. Leung, R. Lortz, A. Ruotolo, *Appl. Phys. Lett.* **104**, 062409 (2014).
[194] L. T. Chang, C. Y. Wang, J. Tang, T. Nie, W. Jiang, C. P. Chu, S. Arafin, L. He, M. Afsal, L. J. Chen, K. L. Wang, *Nano Lett.* **14**, 1823 (2014).
[195] M. Younas, C. Xu, M. Arshad, L. P. Ho, S. Zhou, F. Azad, M. J. Akhtar, S. Su, W. Azeem, F. C. C. Ling, *ACS Omega* **2**, 8810–8817 (2017).
[196] S. S. Li, Y. K. Su, *Jpn. J. Appl. Phys.* **58**, SBBI01 (2019).
[197] J. Stohr, H. C. Siegmann, *Magnetism from Fundamentals to Nanoscale Dynamics*, (Springer, New York, 2006).
[198] G. A. Prinz, *Science* **282**, 1660 (1998).
[199] I. Zutic, J. Fabian, S. D. Sarma, *Rev. Mod. Phys.* **76**, 323 (2004).
[200] J. Wan, M. Cahay, S. Bandyopadhyay, *J. Appl. Phys.* **102**, 034301 (2007).
[201] R. M. Stroud, A. T. Hanbicki, Y. D. Park, G. Kioseoglou, A. G. Petukhov, B. T. Jonker, G. Itskos, A. Petrou, *Phys. Rev. Lett.* **89**, 166602 (2002).
[202] S. J. Pearton, D. P. Norton, K. Ip, Y. W. Heo, T. Steiner, *Prog. Mater. Sci.* **50**, 293 (2005).
[203] C. Song, X. J. Liu, F. Zeng, F. Pan, *Appl. Phys. Lett.* **91**, 042106 (2007).
[204] S. Datta, B. Das, *Appl. Phys. Lett.* **56**, 665 (1990).
[205] H. Ohno, D. Chiba, F. Matsukura, T. Omiya, E. Abe, T. Dietl, Y. Ohno, K. Ohtani, *Nature* **408**, 944 (2000).
[206] D. Chiba, F. Matsukura, H. Ohno, *J. Phys. D: Appl. Phys.* **39**, R215 (2006).
[207] Y. Bychkov, E. L. Rashba, *J. Phys. C* **17**, 6093 (1984).
[208] V. I. Litvinov, *Phys. Rev. B* **68**, 155314 (2003).
[209] K. Sato, H. Katayama-Yoshida, *Mat. Res. Soc. Symp. Proc.* **666**, F4.6 (2001).
[210] S. J. Pearton, C. R. Abernathy, D. P. Norton, A. F. Hebard, Y. D. Park, L. A. Boatner, J. D. Budai, *Mater. Sci. Eng.* **R40**, 137 (2003).
[211] K. C. Hall, M. E. Flatte, *Appl. Phys. Lett.* **88**, 162503 (2006).
[212] B. Huang, D. J. Monsma, I. Appelbaum, *Appl. Phys. Lett.* **91**, 072501 (2007).
[213] Feiderling, M. Keim, G. Reuscher, W. Ossau, G. Schmidt, A. Waag, L. W. Molenkamp, *Nature* **402**, 787 (1999).
[214] T. Diet, H. Ohno, *Rev. Mod. Phys.* **86**, 187 (2014).
[215] S. Mankovsky, S. Polesya, S. Bornemann, J. Minar, F. Hoffmann, C. H. Back, H. Ebert, *Phys. Rev. B* **84**, 201201(2011).
[216] A. Stroppa, G. Kresse, *Phys. Rev. B* **79**, 201201 (2009).
[217] I. D. Marco, P. Thunstrom, M. I. Katsnelson, J. Sadowski, K. Karlsson, S. Lebegue, J. Kanski, O. Eriksson, *Nat. Commun.* **4**, 2645 (2013).

Index

Printed in the United States
By Bookmasters